HIDDEN MARKOV MODELS FOR BIOINFORMATICS

Computational Biology

VOLUME 2

HIDDEN MARKOV MODELS FOR BIOINFORMATICS

by

TIMO KOSKI
Department of Mathematics,
University of Linköping, Sweden

KLUWER ACADEMIC PUBLISHERS
DORDRECHT / BOSTON / LONDON

A C.I.P. Catalogue record for this book is available from the Library of Congress.

ISBN 1-4020-0135-5

Published by Kluwer Academic Publishers,
P.O. Box 17, 3300 AA Dordrecht, The Netherlands.

Sold and distributed in North, Central and South America
by Kluwer Academic Publishers,
101 Philip Drive, Norwell, MA 02061, U.S.A.

In all other countries, sold and distributed
by Kluwer Academic Publishers,
P.O. Box 322, 3300 AH Dordrecht, The Netherlands.

Printed on acid-free paper

Printed in the Netherlands.

Contents

Foreword

This text is based on a set of notes produced for courses given for graduate students in mathematics, computer science and biochemistry during the academic year 1998–1999 at the University of Turku in Turku and at the Royal Institute of Technology (KTH) in Stockholm. The course in Turku was organized by Professor Mats Gyllenberg's group[1] and was also included within the postgraduate program **ComBi**[2], a Graduate School in Computational Biology, Bioinformatics, and Biometry, directed by Professor Esko Ukkonen at the University of Helsinki. The purpose of the courses was to give a thorough and systematic introduction to probabilistic modelling in bioinformatics for advanced undergraduate and graduate students who had a fairly limited background in probability theory, but were otherwise well trained in mathematics and were already familiar with at least some of the techniques of algorithmic sequence analysis. Portions of the material have also been lectured at shorter graduate courses and seminars both in Finland and in Sweden. The initial set of notes circulated also for a time outside those two countries via the World Wide Web.

The intermediate course in probability theory and techniques of discrete mathematics held by the author at the University College of Södertörn (Huddinge, Sweden) during the academic year 1997–1998 has also influenced the presentation. The opportunity to give this course is hereby gratefully acknowledged.

The text was further elaborated and updated during the Bioinformatics term at the Stochastic Centre[3] of Chalmers University of Technology in Gothenburg during the Spring of 2000. The author is indebted to professors

[1] `http://www.utu.fi/ml/sovmat/bio/`

[2] `http://www.cs.helsinki.fi/combi/`

[3] `http://www.md.chalmers.se/Centres/SC/bioinfotheme.html`

Mats Rudemo and Olle Nerman for the opportunity to participate in the activity. Further updating has been done subsequently in response to more recent developments.

The author is grateful to a number of students who have followed the courses and seminars for pointing out several mistakes and errors as well as for making suggestions for improved presentation and selection of topics.

Some of the research underlying this text has been supported by the Academy of Finland, the Swedish Natural Science Research Council (NFR) as well as by the Knut and Alice Wallenberg foundation (KAW) in Stockholm.

In an abstract manner of speaking we are here dealing with probabilistic operations on strings. Probability is hereby interpreted in a Bayesian sense, which means that a probability assigned to a string is dependent on background knowledge and a model family, whose parameters can also be assigned probabilities.

The model families and the methods of probabilistic learning are in this text presented in an order of increasing complexity and flexibility in three steps starting from the multinomial processes to the Markov chains to the hidden Markov models, including a number of sophisticated modifications (WMM, IMM, frame-dependent models, mixture transition models, and hidden semi-Markov models) which seem necessary in successful biological sequence analysis. For each of the model families some of its known applications in biological sequence analysis are pointed out.

The first set of written notes were distributed weekly in the class. Hence the references were assigned independently to each handout, the parts of which have now have been expanded to chapters. This structure has not been revamped. Referencing is thus performed like this: For more extensive and broader treatments of bioinformatics and computational biology we refer to (Baldi and Brunak 1998, von Heijne 1987, Pevzner 2000, Waterman 1995).

- P. Baldi and S. Brunak (1998): *Bioinformatics. The Machine Learning Approach.* A Bradford Book. The MIT Press. Cambridge Massachusetts; London.

- D. Gusfield (1997): *Algorithms on Strings, Trees and Sequences. Computer Science and Computational Biology.* Cambridge University Press. Cambridge.

- G. von Heijne (1987): *Sequence Analysis in Molecular Biology. Treasure Trove or Trivial Pursuit.* Academic Press, San Diego, New York, et al.

- P.A. Pevzner (2000): *Computational Molecular Biology.* The MIT Press. Cambridge Massachusetts; London.

- M.S. Waterman (1995): *Introduction to Computational Biology. Maps, sequences and genomes.* Chapman and Hall, London, Glasgow, Weinheim, New York, Melbourne, Madras.

Chapter 1

Prerequisites in probability calculus

1.1 Background

A number of formulae and results in probability calculus are quickly recapitulated to establish the conventions, notations, and terminology used in this text. This constitutes a very brief summary of probability calculus to the extent required for these notes. In a restricted sense this summary makes this set of notes self-contained. Of course, it does no harm to have more extensive knowledge about the issues touched upon here found, e.g., in the references (Blom et al. 1994, Gray and Davisson 1986, Gut 1995 and Helms 1997).

1.2 Formulae and Definitions

1.2.1 Alphabet, Sequence

$\mathcal{X}$ and $\mathcal{Y}$ are two discrete *alphabets* (also *state spaces*), whose generic elements are called *symbols* and denoted by x_i and y_j, respectively.

$$\mathcal{X} = \{x_1, \cdots, x_L\} \qquad \mathcal{Y} = \{y_1, \cdots, y_J\}.$$

$|\mathcal{X}| :=$ the number of elements in $\mathcal{X}$ = cardinality of $\mathcal{X} = L \leq \infty$, $|\mathcal{Y}| = J \leq \infty$. Unless otherwise stated the alphabets considered here are finite.

Example 1.2.1 In analysis and modelling of DNA sequences (discussed in Chapters 9 and 11) the most immediate instance of an alphabet is

$$\mathcal{X} = \{A, T, C, G\}.$$

These are the four bases present in subunits of DNA called *nucleotides*: adenine (A), thymine (T), cytosine (C) and guanine (G). The particular order of these bases arranged along the sugarphosphate backbone is called the *DNA sequence.* ■

More formally we define a sequence as follows.

Definition 1.2.1 A *sequence* or a *string* $\mathbf{x}$ of length m is an ordered list of m symbols from an alphabet $\mathcal{X}$ written contiguously from left to right

$$\mathbf{x} = x_{l_1} x_{l_2} \ldots x_{l_i} \ldots x_{l_m}; \qquad x_{l_i} \in \mathcal{X}, \qquad i = 1, \ldots, m.$$

We say that x_{l_i} is the symbol in position i of the sequence which we are reading from left to right. The length of the symbol is denoted by $|\mathbf{x}| = m$. The *empty sequence* $\emptyset$ has length 0 and contains no symbols. ■

A DNA sequence of length 15 is thus, for example,

$$\mathbf{x} = \text{ AAAACGTTTAAAAAA}.$$

DNA's genetic code can be represented as a single alphabetic sequence composed of these four symbols. It is by means of this code that the gene controls the formation of other substances in the cell. Progress in biotechnology as well as in computer hardware and software have made it possible to determine the nucleotide sequences for a large and increasing number of organisms. Such data are now available in computer readable form, so it is possible to look for and analyze patterns within sequences using probabilistic modelling and computer algorithms.

As stated in the introduction, the text will be concerned with some of the numerous *model families* assigning a probability $P(\mathbf{x})$ to a string (or by extension to a set of strings) that have been applied in bioinformatics. The model families and the methods of probabilistic learning are in this text presented in an order of increasing complexity and flexibility from the multinomial processes via the Markov chains to hidden Markov models. For each of the model families some of the respective applications in biological sequence analysis are presented. In addition certain modifications of these standard models, introduced for biological sequence modelling, are discussed.

1.2.2 Random Variables and their Distributions

X is a (**discrete**) random variable that assumes values in $\mathcal{X}$ and Y is a (**discrete**) random variable that assumes values in $\mathcal{Y}$.

Remark 1.2.1 These are measurable maps $X(\omega)$, $\omega \in \Omega$, from a basic probability space $(\Omega, \mathcal{F}, P)$ (= outcomes, a sigma field of of subsets of Ω and probability measure P on $\mathcal{F}$). For the most part in the current text this foundational piece of mathematics is suppressed. A very readable survey of these ideas is given by Gray and Davisson (1986). ■

Events of the form $\{\omega \in \Omega \mid X(\omega) = x_i\}$ are thus more simply written as

$$\{X = x_i\} : X \text{ assumes the value } x_i,$$

$$\{Y = y_j\} : Y \text{ assumes the value } y_j.$$

Then the probability of the event $\{X = x_i\}$ is

$$f_X(x_i) := P(X = x_i). \tag{2.1}$$

The probability of the event $\{Y = y_j\}$ is

$$f_Y(y_j) := P(Y = y_j). \tag{2.2}$$

Furthermore,

$$f_X := (f_X(x_1), \cdots, f_X(x_L))$$

designates a discrete probability distribution on $\mathcal{X}$ and

$$f_Y := (f_X(y_1), \cdots, f_Y(y_J))$$

designates a discrete probability distribution on $\mathcal{Y}$. This means that

$$f_X(x_i) \geq 0, \tag{2.3}$$

$$\sum_{i=1}^{L} f_X(x_i) = 1, \tag{2.4}$$

and similarly for f_Y. If $A \subseteq X$ then

$$P_X(A) := \sum_{x_i \in A} f_X(x_i) \tag{2.5}$$

is the probability of the event that X assumes a value in A, a subset of $\mathcal{X}$. From (2.5) one easily finds the complement rule

$$P(A^c) = 1 - P(A), \tag{2.6}$$

where A^c is the complement of A, i.e., those outcomes which do not lie in A. Also

$$P(A \cup B) = P(A) + P(B) - P(A \cap B), \tag{2.7}$$

is immediate. This constitutes a powerful calculus, since it applies ad verbatim also to compound alphabets with elements that are themselves sequences from a finite alphabet or from a Cartesian product of alphabets.

The notation f_X encompasses *tabular* probability distributions, which are simply tables of numbers with the properties (2.3) and (2.4), and *algorithmic* distributions, which are algorithms for computing probabilities, e.g., like the forward–backward procedure in chapter 14. A *parametric* distribution is a special kind of algorithmic distribution: it consists of a few numerical parameters and a relatively simple algorithm, usually a formula for computing probabilities given some specific values of these parameters. The examples immediately following belong to the parametric type.

Example 1.2.2 (Bernoulli random variables) Consider the binary alphabet $\mathcal{X} = \{0, 1\}$. Let p be a number between zero and one. If X is a random variable assuming values in $\{0, 1\}$ and if

$$f_X(1) = P(X = 1) = p$$

then we call X a Bernoulli random variable with a *probability of success p*. We write $X \in \text{Be}(p)$. We refer to p as a *parameter* of the distribution f_X. The parameter p will be called the *chance of success* in Chapter 3. ■

Example 1.2.3 (Geometric random variables) Consider an infinite sequence of Bernoulli random variables $\{X_n\}_{n=1}^{\infty}$ with a common probability of success p. This models an infinite sequence of zeros and ones. Let

$$X = \text{ the first time } (= k) \text{ at which success occurs.}$$

Then $\mathcal{X} = \{1, 2, 3, \ldots, \infty\}$. It can be proved (Helms 1997) that X is a well defined random variable with respect to a suitable probability space. We say that X has *a geometric probability distribution* if

$$f_X(k) = p \cdot (1 - p)^{k-1}; \qquad k = 1, 2, \ldots, . \tag{2.8}$$

This is written as $X \in \text{Ge}(p)$. ■

Example 1.2.4 (Uniform random variables) Let $\mathcal{X} = \{x_1, x_2, \ldots, x_L\}$ be a finite alphabet and

$$f_X(x_k) = \frac{1}{L}; \qquad k = 1, 2, \ldots, L. \tag{2.9}$$

This is called the *uniform probability distribution.* If X is a random variable assuming values in this alphabet and $P(X = k) = 1/L$, then we say that

'X is selected at random'

and we write $X \in \mathrm{U}(1, L)$. ∎

Example 1.2.5 (Poisson distributed random variable) Let

$$f_X(k) = e^{-\lambda} \cdot \frac{\lambda^k}{k!}, \qquad \lambda > 0, \qquad k = 0, 1, 2, \ldots.$$

This is a probability distribution on the non-negative integers called the *Poisson distribution.* A random variable X with this distribution is said to be Poisson distributed with parameter λ and we write $X \in \mathrm{Po}(\lambda)$. ∎

1.2.3 Joint Probability Distributions

A two dimensional *joint (simultaneous) probability distribution* is a probability defined on the alphabet $\mathcal{X} \times \mathcal{Y}$

$$f_{X,Y}(x_i, y_j) := P(X = x_i, Y = y_j). \tag{2.10}$$

Hence $0 \leq f_{X,Y}(x_i, y_j)$ and $\sum_{i=1}^{L} \sum_{j=1}^{L} f_{X,Y}(x_i, y_j) = 1$.
Marginal distribution for X:

$$f_X(x_i) = \sum_{j=1}^{J} f_{X,Y}(x_i, y_j). \tag{2.11}$$

Marginal distribution for Y:

$$f_Y(y_j) = \sum_{i=1}^{L} f_{X,Y}(x_i, y_j). \tag{2.12}$$

These notions can be extended to define the joint (simultaneous) probability distribution and the marginal distributions of n random variables.

1.2.4 Conditional Probability Distributions

The conditional probability for $X = x_i$ given $Y = y_j$ is

$$f_{X|Y}(x_i \mid y_j) := \frac{f_{X,Y}(x_i, y_j)}{f_Y(y_j)}. \tag{2.13}$$

The conditional probability for $Y = y_j$ given $X = x_i$ is

$$f_{Y|X}(y_j \mid x_i) := \frac{f_{X,Y}(x_i, y_j)}{f_X(x_i)}. \tag{2.14}$$

Here we assume $f_Y(y_j) > 0$ and $f_X(x_i) > 0$. If $f_X(x_i) = 0$, we can make the definition of $f_{Y|X}(y_j \mid x_i)$ arbitrarily through $f_X(x_i) \cdot f_{Y|X}(y_j \mid x_i) = f_{X,Y}(x_i, y_j)$. In other words

$$f_{Y|X}(y_j \mid x_i) = \frac{\text{prob. for the event } \{X = x_i, Y = y_j\}}{\text{prob. for the event } \{X = x_i\}}. \tag{2.15}$$

Hence

$$\sum_{i=1}^{L} f_{X|Y}(x_i \mid y_j) = 1.$$

1.2.5 A Chain Rule

Let Z be a (discrete) random variable that assumes values in $\mathcal{Z} = \{z_k\}_{k=1}^{K}$. In view of (2.15) we have also, if $f_Z(z_k) > 0$,

$$f_{X,Y|Z}(x_i, y_j \mid z_k) = \frac{f_{X,Y,Z}(x_i, y_j, z_k)}{f_Z(z_k)}.$$

Then we obtain as an identity

$$f_{X,Y|Z}(x_i, y_j \mid z_k) = \frac{f_{X,Y,Z}(x_i, y_j, z_k)}{f_{Y,Z}(y_j, z_k)} \cdot \frac{f_{Y,Z}(y_j, z_k)}{f_Z(z_k)}$$

and again by definition of conditional probability the right hand side is equal to

$$f_{X|Y,Z}(x_i \mid y_j, z_k) \cdot f_{Y|Z}(y_j \mid z_k).$$

In other words,

$$f_{X,Y|Z}(x_i, y_j \mid z_k) = f_{X|Y,Z}(x_i \mid y_j, z_k) \cdot f_{Y|Z}(y_j \mid z_k). \tag{2.16}$$

This *chain rule* (with obvious extensions to n variables) will be invoked in numerous situations in the sequel, especially in Chapter 13.

1.2.6 Independence

X and Y are *independent* random variables if and only if

$$f_{X,Y}(x_i, y_j) = f_X(x_i) \cdot f_Y(y_j) \tag{2.17}$$

for *all pairs* (x_i, y_j) in $\mathcal{X} \times \mathcal{Y}$. In other words all events $\{X = x_i\}$ and $\{Y = y_j\}$ are to be independent. We say that $X_1, X_2, \ldots, X_n$ are *independent* random variables if and only if the joint distribution

$$f_{X_1,X_2,\ldots,X_m}(x_{i_1}, x_{i_2} \ldots, x_{i_m}) = P(X_1 = x_{i_1}, X_2 = x_{i_2}, \ldots, X_m = x_{i_m}) \tag{2.18}$$

is equal to

$$f_{X_1,X_2,\ldots,X_m}(x_{i_1}, x_{i_2} \ldots, x_{i_m}) = f_{X_1}(x_{i_1}) \cdot f_{X_2}(x_{i_2}) \cdots f_{X_m}(x_{i_m}) \tag{2.19}$$

for every m, $2 \leq m \leq n$, and every $x_{i_1}, x_{i_2} \ldots, x_{i_m} \in \mathcal{X}^m$. We are here assuming for simplicity that $X_1, X_2, \ldots, X_n$ take values in the same alphabet. One way of assigning a probability to a sequence $\mathbf{x} = x_{i_1} x_{i_2} \ldots x_{i_n}$ is thus by

$$P(\mathbf{x}) = f_{X_1}(x_{i_1}) \cdot f_{X_2}(x_{i_2}) \cdots f_{X_n}(x_{i_n}) = \prod_{l=1}^{n} f_{X_l}(x_{i_l}).$$

1.2.7 Conditional Independence

The random variables X and Y are called *conditionally independent* given Z if

$$f_{X,Y|Z}(x_i, y_j|z_k) = f_{X|Z}(x_i|z_k) \cdot f_{Y|Z}(y_j|z_k) \tag{2.20}$$

for all triples $(z_k, x_i, y_j) \in \mathcal{Z} \times \mathcal{X} \times \mathcal{Y}$ (cf. (2.16)). We write (2.20) symbolically as

$$X \perp Y|Z.$$

We may also say that $X \longrightarrow Z \longrightarrow Y$ forms a *Markov Chain.* Conditional independence is a core element of the definition of a hidden Markov model.

1.2.8 Probability Models with Independence

Example 1.2.6 (Independent Bernoulli random variables) Take n independent Be (p) random variables $\{X_i\}_{i=1}^{n}$ (Example 1.2.2). Then

$$\begin{aligned} f_{X_1,X_2,\ldots,X_n}(x_{i_1}, x_{i_2} \ldots, x_{i_n}) &= P(X_1 = x_{i_1}, X_2 = x_{i_2}, \ldots, X_n = x_{i_n}) \\ &= p^{\sum_{j=1}^{n} x_{i_j}} \cdot (1-p)^{n-\sum_{j=1}^{n} x_{i_j}} = p^k \cdot (1-p)^{n-k} \end{aligned}$$

if $\sum_{j=1}^{n} x_{i_j} = k$. ■

Example 1.2.7 (Binomial random variables) Let A be some event defined in terms of some basic experiment. Introduce the random variable

$$X = \textit{the number of times the event } A \textit{ occurs in } n \textit{ repetitions} \text{ of some basic experiment.}$$

Then the alphabet of X is $\mathcal{X} = \{0, 1, \cdots, n\}$. If the repetitions are modeled as independent then X has the *binomial distribution*:

$$f_X(k) = P(X = k) = \binom{n}{k} p^k (1-p)^{n-k}. \qquad (2.21)$$

Recall that $\binom{n}{k} = \frac{n!}{(n-k)!k!}$ (the *binomial coefficients*, Chapter 5 in (Graham et al. 1989)), $0! := 1$. We say that X is a binomial random variable with parameters p and n and write

$$X \in \text{Bin}(n, p).$$

Clearly the distribution of X is such that

$$X = X_1 + X_2 + \ldots + X_n,$$

as an equality in distribution, where $X_1, X_2, \ldots, X_n$ are independent Be (p) variables as in the preceding example. ■

If np is small, i.e., the event A occurs 'rarely', the binomial distribution can be approximated by a Poisson distribution Po(λ) with $\lambda = np$ (*the law of small numbers*). The approximation $\text{Bin}(n, p) \approx \text{Po}(np)$, or, more appropriately, its generalizations have turned out to be useful for a number of problems in computational biology (Waterman 1995). Poisson approximation is the topic of the Exercises 10–12 at the end of this chapter.

1.2.9 Multinomial Probability Distribution

Let $X_1, X_2, \ldots, X_n$ be independent random variables assuming values in

$$\mathcal{X} = \{x_1, \cdots, x_L\}$$

with the common distribution

$$p_l = P(X_i = x_l), \qquad l = 1, 2, \ldots, L.$$

Thus

$$P(X_1 = x_{i_1}, X_2 = x_{i_2}, \ldots, X_n = x_{i_n}) = p_{i_1} \cdot p_{i_2} \cdots p_{i_n}.$$

For $l = 1, 2, \ldots, L$

$$n_l = \text{ the number of times the symbol } x_l \text{ is found in } \mathbf{x} = x_{i_1} x_{i_2} \ldots x_{i_n}.$$

Thus $n_1 + n_2 + \ldots + n_L = n$. Then from the preceding,

$$P(\mathbf{x}) = P(X_1 = x_{i_1}, X_2 = x_{i_2}, \ldots, X_n = x_{i_n}) = p_1^{n_1} \cdot p_2^{n_2} \cdots p_L^{n_L}.$$

This model for assigning a probability to a sequence will for convenience of reference be called a *multinomial process* or a *Markov chain of zero order*. If we introduce the random variables $Y_1, Y_2, \ldots, Y_L$ as functions of the random variables $X_1, \ldots, X_n$ by

$$Y_l = \text{ the number of times } X_i = x_l, \qquad i = 1, 2, \ldots, n,$$

then the joint distribution of $Y_1, Y_2, \ldots, Y_L$ is

$$P(Y_1 = n_1, Y_2 = n_2, \ldots, Y_L = n_L) = C(n_1, n_2, \ldots, n_L) \cdot p_1^{n_1} \cdots p_L^{n_L}. \quad (2.22)$$

By independence (symmetry) this is equal to the sum of probabilities of all those outcomes of $X_1, \ldots, X_n$ that have exactly $n_1, n_2, \ldots, n_L$ as their frequency counts. Therefore a combinatorial argument (Briggs 1993, p. 97) shows that

$$C(n_1, n_2, \ldots, n_L) = \frac{n!}{n_1! n_2! \ldots n_L!}, \quad (2.23)$$

which is called the *multinomial coefficient*. The probability in (2.22) is called the *multinomial distribution*. Note that the binomial distribution is a special case. The *Whittle distribution* introduced in Chapter 8 is an extension of the multinomial distribution to Markov chains.

In the sequel we shall frequently regard the probability assignment

$$P(\mathbf{x}) = P(X_1 = x_{i_1}, X_2 = x_{i_2}, \ldots, X_n = x_{i_n}) = p_1^{n_1} \cdot p_2^{n_2} \cdots p_L^{n_L},$$

as a statement of conditional independence (in an extended sense)

$$P(\mathbf{x} \mid \underline{p}) = P(X_1 = x_{i_1}, X_2 = x_{i_2}, \ldots, X_n = x_{i_n} \mid \underline{p}) = p_1^{n_1} \cdot p_2^{n_2} \cdots p_L^{n_L},$$

where $\underline{p} = (p_1, p_2, \cdots, p_L)$ is seen as an outcome of a continuous random variable. Since $\underline{p}$ is an unknown parameter, it is supposed constant and is therefore in some literature called a random element $\mathbf{p}$ so that we may write or $X_i \perp X_j | \underline{\mathbf{p}}$.

1.2.10 A Weight Matrix Model for a Family of Sequences

A weight matrix $\mathcal{M}$ is a simple model often used in biological modelling as a representation for a *family of sequences.* The sequences are supposed to have equal length (=n) and to have no *gaps* (no positions are blank). A weight matrix $\mathcal{M}$ has as entries the tabular probabilities $p_i(x_j)$ for that a string should have the symbol x_j at position i:

$$\mathcal{M}: \quad \begin{array}{ccc} p_1(x_1) & \dots & p_n(x_1) \\ p_1(x_2) & \dots & p_n(x_2) \\ \vdots & \vdots & \vdots \\ p_1(x_L) & \dots & p_n(x_L) . \end{array} \tag{2.24}$$

Thus $\sum_{j=1}^{L} p_i(x_j) = 1$ for every column i. The assignment of probabilities for a string $\mathbf{x} = x_{l_1} x_{l_2} \dots x_{l_n}$ conditioned on $\mathcal{M}$ is now given by

$$P(\mathbf{x}|\mathcal{M}) = \prod_{j=1}^{L} \prod_{i=1}^{n} p_i(x_j)^{I_{i,x_j}(\mathbf{x})}, \tag{2.25}$$

where the indicator $I_{i,x_j}(\mathbf{x})$, a function of $\mathbf{x}$, is 0 if $x_j \neq x_{l_i}$, i.e., if the symbol x_j does not appear in position i in the string $\mathbf{x}$ and is 1 otherwise. Thus the symbols in the different positions are conditionally independent given $\mathcal{M}$.

A sequence of strings $\mathbf{x}^1, \dots, \mathbf{x}^t$, which are also taken as mutually independent conditioned on $\mathcal{M}$, which is a model, is by multiplication of the expressions in (2.25) assigned the probability

$$P\left(\mathbf{x}^1, \dots, \mathbf{x}^t|\mathcal{M}\right) = \prod_{s=1}^{t} P(\mathbf{x}^s|\mathcal{M}) = \prod_{j=1}^{L} \prod_{i=1}^{n} p_i(x_j)^{n_i(x_j)}, \tag{2.26}$$

where $n_i(x_j)$ is the number of times the symbol x_j appears on position i in $\mathbf{x}^1, \dots, \mathbf{x}^t$.

The weight matrix model is in bioinformatics often known as a *profile* (Gusfield 1997, ch. 14). In practice the probabilities $p_i(x_j)$ are *estimated* or *learned* using a set of sequences known to belong to a family of sequences, e.g., a family of proteins ($L = 20$). One important application is in *database searching* (Gusfield 1997, ch. 15). The notion of a profile (for a family of biological sequences) will be encountered and extended in several different ways in the later chapters.

The elementary computational operations in (2.25) – (2.26) will be frequently used in later chapters when dealing with more refined probability models for biological sequences and sequence families.

1.2.11 Simplifying Notations

For ease of writing we are sometimes going to drop the symbols in the argument and to write for (2.13)

$$f_{X|Y} = \frac{f_{X,Y}}{f_Y}, \tag{2.27}$$

for (2.14)

$$f_{Y|X} = \frac{f_{X,Y}}{f_X}, \tag{2.28}$$

and for (2.17)

$$f_{X,Y} = f_Y \cdot f_X. \tag{2.29}$$

Eq. (2.29) implies thus

$$X \text{ and } Y \text{ are independent } \Longleftrightarrow \begin{cases} f_{X|Y} = f_X, \\ \\ f_{Y|X} = f_Y. \end{cases}$$

Conditional independence (2.20) thus becomes

$$f_{Z,X|Y} = f_{Z|Y} \cdot f_{X|Y}. \tag{2.30}$$

Furthermore, this is equivalent to

$$f_{Z|X,Y} = f_{Z|Y}. \tag{2.31}$$

Independence of $X_1, X_2, \ldots, X_n$ gives

$$\prod_{l=1}^{n} f_{X_l}.$$

1.3 Learning and Bayes' Rule

1.3.1 Bayes' Rule

Since

$$f_{X,Y} = f_{X|Y} \cdot f_Y = f_{Y|X} \cdot f_X$$

we have in a formal way

$$f_{X|Y} = \frac{f_{Y|X} \cdot f_X}{f_Y}.$$

But the marginal distribution f_Y is from (2.12) and (2.28) written as

$$f_Y(y_j) = \sum_{i=1}^{L} f_{Y|X}(y_j \mid x_i) f_X(x_i). \tag{3.1}$$

Hence we have obtained

Bayes' Rule

$$f_{X|Y}(x_i \mid y_j) = \frac{f_{Y|X}(y_j \mid x_i) \cdot f_X(x_i)}{\sum_{i=1}^{L} f_{Y|X}(y_j \mid x_i) f_X(x_i)}. \tag{3.2}$$

Bayes' rule gives a fundamental operation for *up-date of probability distributions* in response to observed information. The rule shows how knowledge about the occurrence of the event $Y = y_j$ is to be used to transform the probabilities f_X on $\mathcal{X}$ to a new probability distribution $f_{X|Y}$. In this context we note the possibility of interpreting probability as a degree of rational belief, which makes (3.2) a rule for inductive reasoning.

Terminology for Bayes' Rule

f_X : A **Prior Distribution** on $\mathcal{X}$;

$f_{X|Y}$: A **Posterior Distribution** on $\mathcal{X}$.

If X and Y are independent then the prior distribution and posterior distribution are identical and there is no *learning* about X from Y and vice versa. Bayes' rule can be seen as just a formal identity derived from certain premises and definitions. One of the stages of probabilistic modelling and learning is to find explicit forms of the distributions, appropriate in some specific application, to be used in Bayes' rule.

1.3.2 A Missing Information Principle and Inference

Let us suppose that we have a pair of random variables (X, Y). These are thought to occur simultaneously and to assume values in $\mathcal{X} \times \mathcal{Y}$. Let us now suppose that we have observed the event $\{Y = y_j\}$ but are missing information about the corresponding outcome of X (so that X is *hidden* from us).

The *missing information principle* says for our purposes that the posterior $f_{X|Y}(x \mid y_j)$ contains all the probabilistic information about X. This missing information principle is a cornerstone of probabilistic modelling of biological sequences as will become evident in the sequel.

In many situations we want to summarize the posterior $f_{X|Y}(x \mid y_j)$ by a single symbol ('*estimate*') in the alphabet $\mathcal{X}$. This operation of probabilistic *inference* is called *(point) estimation.* The following two methods of estimation would seem natural:

Maximum A Posteriori (**MAP**) Estimate, $\widehat{X}_{\text{MAP}}$, of X given $Y = y_j$:

$$\widehat{X}_{\text{MAP}} = \arg\max_{x \in \mathcal{X}} f_{X|Y}(x \mid y_j). \tag{3.3}$$

Maximum Likelihood (**ML**) Estimate, $\widehat{X}_{\text{ML}}$, of X given $Y = y_j$:

$$\widehat{X}_{\text{ML}} = \arg\max_{x \in \mathcal{X}} f_{Y|X}(y_j \mid x). \tag{3.4}$$

1.4 Some Distributions for DNA Analysis

1.4.1 Fragment Accuracy

One interesting application of the probability calculus presented so far is the method for evaluating the accuracy of shotgun sequencing given in (Churchill and Waterman 1992). In shotgun sequencing a large DNA molecule is broken into a collection of *fragments* which are sampled as randomly as possible. A short introduction to the basics of shotgun sequencing is given by (Myers 1999).

The fragments are cloned and sequenced individually. The sampled fragments are then used for assembly by determining their relative orientations and overlaps and aligned (the technical meaning of this phrase is made more precise in Exercise 9 and in Chapter 5) to form a column-by-column matrix.

The algorithmic principles of fragment assembly are stated in (Peltola et al. 1984).

Suppose that a collection of fragment sequences $\mathbf{x}^1, \ldots, \mathbf{x}^m$ has been aligned by some procedure. The symbol $\emptyset$ is used as a place holder for nonaligned positions beyond the ends of the fragment. The depth of coverage of position i is defined as the number of fragments contributing sequence information at position i

$$d = \sum_{l=1}^{t} I\left(x_i^l \neq \emptyset\right),$$

where $I\left(x_i^l \neq \emptyset\right) = 1$ if $x_i^l \neq \emptyset$ and zero otherwise. In addition

$$\overline{p} = \sum_{x_j \neq x_l, x_j \in \mathcal{X}} P\left(x_i^l = x_j \mid x_l\right),$$

where $P\left(x_i^l = x_j \mid x_i\right)$ is the conditional probability that the shotgun sequenced fragment $\mathbf{x}^l$ is equal to x_j in position i, given that the true sequence symbol is x_l. It is assumed that $\overline{p}$ is the same for all x_l and i.

For a position covered at depth d, we compute the probability that at most one half of the bases are correct. This is the binomial probability (see Example 1.2.7) given by

$$P\left(\text{error}|d\right) = \sum_{k=0}^{\lfloor d/2 \rfloor} \binom{d}{k} \overline{p}^{d-k}(1-\overline{p})^k,$$

where the probability of success is taken as $1-\overline{p}$ and $\lfloor d/2 \rfloor$ is largest integer smaller than or equal to $d/2$. In shotgun sequencing the distribution of d can according to (Churchill and Waterman, 1992) be taken as $\text{Po}\left(\lambda\right)$. Positions not covered by any fragments are ignored and thus d is at least 1. Then we obtain

$$P\left(\text{error}\right) = \frac{1}{1-e^{-\lambda}} \sum_{d=1}^{\infty} \frac{\lambda^d}{d!} e^{-\lambda} P\left(\text{error}|d\right) \tag{4.5}$$

as a practical measure of the accuracy of shotgun sequencing.

1.4.2 The Distribution of the Number of Fragments

The distribution of fragment lengths from restriction digests in DNA is important in many algorithms used in bioinformatics. The fragment lengths are revealing for DNA *polymorphisms* (roughly speaking, variations of marker positions in DNA sequences of different individuals in a population (Gusella 1986)) that have been used for constructing gene maps. In (Bishop et al. 1983) the distribution of the *number of fragments* is presented. Let N designate the number of fragments of length greater than x base pairs. Fragments shorter than this will not be detected. Let the number of cuts of the DNA occurring within the probe used (Gusella 1986) be a random variable with the distribution $\text{Po}(P/m)$. Here P is the length of the probe and m is the mean length of the fragments. We condition on there being $N = n - 1$ cuts within the probe. Then the probability of seeing i fragments of length greater than x is given (by a well known result independently due to many different authors) as

$$P(i \text{ gaps } | n) = \binom{n}{i} \sum_{j=0}^{n-i} (-1)^j \binom{n-i}{j} \left(1 - (i+j)\frac{x}{P}\right)_+^{n-1}, \qquad (4.6)$$

where $a_+ = \max(a, 0)$. A short proof of this is found, e.g., in (Holst 1980 p. 624). Therefore, by the preceding rules of probability calculus we get the probability of i fragments of length greater than x as

$$\begin{aligned} P(N = i) &= \sum_{n=1}^{\infty} \frac{(P/m)^{n-1} e^{-P/m}}{(n-1)!} P(i \text{ gaps } | n) \\ &= \sum_{j=0}^{\infty} \frac{(-1)^j}{i! j! m^{i+j}} \left[P - (i+j)(x-m)\right] \left(1 - (i+j)x\right)_+^{i+j-1} e^{-(i+j)/m}. \end{aligned} \qquad (4.7)$$

1.5 Expectation

If $\mathcal{X}$ is a discrete alphabet with a finite set of real numbers as symbols ($x_i \in \mathcal{X} \subset R$) we define the expectation, $E(X)$, of the random variable X as

$$E(X) := \sum_{i=1}^{L} x_i f_X(x_i). \qquad (5.8)$$

In the case $L = \infty$ we require absolute convergence of the series. If $\mathcal{X}$ is an arbitrary alphabet and $g(\cdot)$ is a map (function) of $\mathcal{X}$ to R, $g : \mathcal{X} \longmapsto R$,

$$E(g(X)) := \sum_{i=1}^{L} g(x_i) f_X(x_i). \qquad (5.9)$$

A speciality of information theory, the topic of Chapter 2, is the use of the following particular case of (5.9) with $g(\cdot) = -\log_2(f_X(\cdot))$,

$$H(X) := E[-\log_2(f_X(X))] := -\sum_{i=1}^{L} \log_2(f_X(x_i)).f_X(x_i) \tag{5.10}$$

The quantity $H(X)$ is the *entropy* of X or of f_X, and

$$H(Y) := E[-\log_2(f_Y(Y))] := -\sum_{j=1}^{J} \log_2(f_Y(y_j))f_Y(y_j). \tag{5.11}$$

Note even the same principle in (the simultaneous entropy)

$$\begin{aligned} H(X,Y) &:= E\left[-\log_2\left(f_{X,Y}(X,Y)\right)\right] \\ &= -\sum_{i=1}^{L}\sum_{j=1}^{J} f_{X,Y}(x_i,y_j)\log_2\left(f_{X,Y}(x_i,y_j)\right) \end{aligned} \tag{5.12}$$

and in (*mutual information*)

$$\begin{aligned} I(X;Y) &:= E\left(\log_2\left(\frac{f_{X,Y}(X,Y)}{f_X(X)f_Y(Y)}\right)\right) \\ &= \sum_{i=1}^{L}\sum_{j=1}^{J} f_{X,Y}(x_i,y_j)\log_2\left(\frac{f_{X,Y}(x_i,y_j)}{f_X(x_i)f_Y(y_j)}\right). \end{aligned} \tag{5.13}$$

The meaning of these expressions will be discussed in the next chapter.

1.6 Jensen's Inequality

Let $\phi(\cdot)$ is a *convex* function on a convex subset of the real line, $\phi : \mathcal{X} \longmapsto R$. Convexity means by definition that

$$\phi\left(\lambda_1 x_1 + \lambda_2 x_2 + \ldots + \lambda_n x_n\right) \leq \lambda_1 \phi\left(x_1\right) + \lambda_2 \phi\left(x_2\right) + \ldots + \lambda_n \phi\left(x_n\right),$$

where $\lambda_1 \geq 0$, $\lambda_2 \geq 0, \ldots, \lambda_n \geq 0$ and $\lambda_1 + \lambda_2 + \ldots + \lambda_n = 1$. The basic analytic properties of convex functions and criteria of convexity are found in (Khuri 1993 pp. 78 – 87). But using the language of expectation and the formula (5.9) this is

Jensen's Inequality

$$\phi\left(E\left(X\right)\right) \leq E\left(\phi(X)\right), \tag{6.14}$$

if X is a discrete random variable assuming values in $\mathcal{X}$.

Example 1.6.1 $\phi(x) = -\log x$ is a convex function. Then if X is a random variable with positive real values we have

$$\log\left(E\left(X\right)\right) \geq E\left(\log(X)\right).$$

■

1.7 Conditional Expectation

The *conditional expectation* of X (X is real valued), given $Y = y_j$, is

$$E\left[X|Y = y_j\right] = \sum_{i=1}^{L} x_i f_{X|Y}(x_i|y_j). \tag{7.15}$$

From the preceding definitions we obtain the useful formula

$$E(X) = \sum_{j=1}^{J} E\left[X|Y = y_j\right] \cdot f_Y(y_j).$$

The *conditional expectation* of $g(X)$, $g : \mathcal{X} \longmapsto R$, given $Y = y_j$, is

$$E\left[g(X)|Y = y_j\right] = \sum_{i=1}^{L} g(x_i) f_{X|Y}(x_i|y_j). \tag{7.16}$$

Since this is another expectation for convex ϕ we have Jensen's inequality

$$\phi\left(E[X|Y = y_j]\right) \leq E\left[\phi(X)|Y = y_j\right]. \tag{7.17}$$

An example from information theory is given with $g(\cdot) = -\log_2 f_{X|Y}(\cdot|y_j)$ as

$$H(X|Y = y_j) := -\sum_{i=1}^{L} \log_2\left(f_{X|Y}(x_i|y_j)\right) \cdot f_{X|Y}(x_i|y_j),$$

which is called the *conditional entropy* of X given $Y = y_j$.

1.8 Law of Large Numbers

Let $\mathcal{X}$ be a real and finite alphabet and let the *variance*, $\mathrm{Var}(X)$, of X be

$$\begin{aligned}\mathrm{Var}(X) &:= E\left([X - E(X)]^2\right)\\ &= \sum_{i=1}^{L}(x_i - E(X))^2 f_X(x_i)\end{aligned} \tag{8.18}$$

(by (5.9)). Hence $\mathrm{Var}(X) = E(X^2) - (E(X))^2$ (($E(X)$ is given in (5.8)). Chebyshev's inequality is another partial intuitive interpretation of variance.

Proposition 1.8.1 (Chebyshev's inequality) *For $k > 0$,*

$$P(\mid X - E(X) \mid > k) \leq \frac{\mathrm{Var}(X)}{k^2}. \tag{8.19}$$

Proof of (8.19):

$$\begin{aligned}\mathrm{Var}(X) &= \sum_{i=1}^{L}(x_i - E(X))^2 f_X(x_i)\\ &= \sum_{\{x_i \mid |x_i - E(X)| > k\}} (x_i - E(X))^2 f_X(x_i)\\ &\quad + \sum_{\{x_i \mid |x_i - E(X)| \leq k\}} (x_i - E(X))^2 f_X(x_i)\\ &\geq k^2 \sum_{\{x_i \mid |x_i - E(X)| > k\}} f_X(x_i)\\ &= k^2 P(\mid X - E(X) \mid > k),\end{aligned} \tag{8.20}$$

where we have used the property that $\sum_{\{x_i \mid |x_i - E(X)| \leq k\}}(x_i - E(X))^2 f_X(x_i) \geq 0$ and the formula (2.5). ∎

Let X_k be independent identically distributed copies of X, in the sense that $f_{X_k} = f_X$, for $k = 1, 2, \cdots$, and let $\mathcal{X}$ be a real alphabet. Set $\mu = E(X)$, $\mathrm{Var}(X) = \sigma^2$. Let

$$\overline{x}_N := \frac{1}{N}\sum_{k=1}^{N} X_k. \tag{8.21}$$

Then $E(\overline{x}_N) = \mu$ and

$$\text{Var}(\overline{x}_N) = \frac{1}{N^2}\sum_{k=1}^{N}\text{Var}(X_k) = \frac{1}{N^2}N\sigma^2 = \frac{1}{N}\sigma^2.$$

Hence Chebyshev's inequality yields

$$\begin{aligned} P\left(|\overline{x}_N - \mu| > k\right) &\leq \frac{\text{Var}(\overline{x}_N)}{k^2} \\ &= \frac{1}{N}\frac{\sigma^2}{k^2}. \end{aligned} \tag{8.22}$$

As $N \longrightarrow \infty$ we have thus obtained the following proposition.

Proposition 1.8.2 Law of Large Numbers
If X_k are independent identically distributed copies of X, for $k = 1, 2, ..$, and if $\mathcal{X}$ is a real and finite alphabet and $\mu = E(X)$, $\text{Var}(X) = \sigma^2$. Then

$$P\left(|\overline{x}_N - \mu| > k\right) \longrightarrow 0. \tag{8.23}$$

for any $k > 0$. ∎

In words this means that the arithmetic mean of a sequence of independent, equally distributed random variables converges to their common expectation $= \mu$. The Law of Large Numbers (8.23) can also be stated as follows: for every $\epsilon > 0$ there exists a $N_\epsilon = N(\epsilon, k, \sigma^2)$ such that $N > N_\epsilon$ implies

$$P\left(|\overline{x}_N - \mu| > k\right) < \epsilon.$$

The law of large numbers for binomial random variables means here that

$$P\left(\left|\frac{X}{n} - p\right| > k\right) \longrightarrow 0. \tag{8.24}$$

as n increases to $+\infty$ for any $k > 0$.

1.9 Exercises

1. **Basic properties of conditional independence**
   Show that

   (a)
   $$X \perp Y|Z \Longleftrightarrow (X,Z) \perp (Y,Z)\,|Z.$$

   (b) If $U = u(X)$ is a function of X and $X \perp Y|Z$ then

   (i)
   $$U \perp Y|Z.$$

   (ii)
   $$X \perp Y|\,(Z,U)\,.$$

   (c) If $X \perp Y|Z$ and $X \perp W|\,(Y,Z)$, then
   $$X \perp (W,Y)\,|Z.$$

   (Dawid 1980).

2. **A Markov chain**
   A sequence of random variables, $\{X_n\}_{n=0}^{\infty}$ assumes values in a finite state space $S = \{1,2,\ldots,J\}$ with J states. We set
   $$Y_t = (X_{t+1}, X_{t+2}\ldots)$$
   as the future of the chain and
   $$Z_t = (\ldots, X_{t-2}, X_{t-1})$$
   for the past of the chain. Then the Markov property can be stated as
   $$Y_t \perp Z_t|X_t.$$
   In particular
   $$(X_t, X_{t+1}) \perp Z_{t-1}|X_{t-1}.$$
   Show, using this and the results above, that
   $$X_t \perp (Z_{t-1}, Y_{t+1})\,|\,(X_{t-1}, X_{t+1})\,.$$

This is the *nearest neighbor property* of a Markov chain: given the states at all times other than t, the conditional distribution of X_t is, in fact, determined by the states at times $t-1$ and $t+1$ only.
Hint: Convince yourself of the following

$$X_t \perp Z_{t-1} | (X_{t-1}, X_{t+1}),$$

also

$$(Z_{t-1}, X_t) \perp Y_{t+1} | X_{t+1},$$

so that (why ?)

$$X_t \perp Y_{t+1} | (X_{t-1}, X_{t+1}, Z_{t-1})$$

and then use the properties of conditional independence established in the preceding exercise.(Dawid 1980)

3. **Fisher's Device**
Let Y_j be independent $\in \text{Po}(\Psi_j)$, $\Psi_j > 0$, $j = 1, 2, \ldots, k$. Let

$$N = Y_1 + Y_2 + \ldots + Y_k.$$

Show that

$$P(Y_1 = n_1, \ldots, Y_k = n_k | N = n) = \frac{n!}{n_1! \cdots n_k!} \theta_1^{n_1} \cdots \theta_k^{n_k}$$

where

$$\theta_j = \frac{\Psi_j}{\sum_{j=1}^{k} \Psi_j}.$$

4. **Trinomial Probabilities**
The trinomial distribution is

$$P(n_1, n_2, n_3) = \frac{n!}{n_1! n_2! n_3!} \theta_1^{n_1} \theta_2^{n_2} \theta_3^{n_3}.$$

Show that

$$P(n_2, n_3 | n_1) = \binom{n_2 + n_3}{n_2} \pi_2^{n_2} (1 - \pi_2)^{n_3},$$

where

$$\pi_2 = \frac{\theta_2}{\theta_2 + \theta_3}.$$

Hint: Note that the claimed distribution is a binomial distribution.

5. Consider the alphabet $\mathcal{X} = \{1, 2, 3, \ldots\}$ consisting of the positive integers, i.e., $x_j = j, j = 1, 2, 3, \ldots$. Explain why the function $f_X(j)$ defined for each $j \geq 1$ as

$$f_X(j) = \frac{1}{j(j+1)} = \frac{1}{j} - \frac{1}{j+1}$$

can be regarded as a probability distribution for some random variable X? If X is a random variable with this distribution, does X have an expectation $E[X]$?

6. Let $X \in \mathrm{U}(0, 100)$. Given that $X = x$, an integer $Y \in \mathrm{U}(0, x)$. Compute the expectation $E[Y]$ and the variance $\mathrm{Var}[Y]$.

7. **Mullen's ratio theorem**

   Let X and Y be be independent random variables assuming values in a finite discrete alphabet of real numbers. The values of X are always positive. Show that

$$E\left(\frac{Y}{X}\right) \geq \frac{E(Y)}{E(X)}$$

   (K. Mullen, *The American Statistician*, Vol. 21, 1967, pp. 30–31.)

8. Two strings $\mathbf{x}$ and $\mathbf{y}$ with $|\mathbf{x}| = |\mathbf{y}| = n$ of symbols from an alphabet $\mathcal{X} = \{x_1, \ldots, x_J\}$ are considered. Hence, e.g.,

$$\mathbf{x} = (x_{l_1} x_{l_2} \ldots x_{l_i} \ldots x_{l_m}); x_{l_i} \in \mathcal{X}, i = 1, \ldots, m.$$

   We define $S(\mathbf{x}, \mathbf{y})$ be the number of *matches*, or of positions, where x_{l_i} and y_{l_i} are equal, $1 \leq i \leq n$.

   Let $P(S(\mathbf{x}, \mathbf{y}) = k)$ be the probability that the number of matches of $\mathbf{x}$ and $\mathbf{y}$ is equal to k, assuming that the two strings are chosen by some random rule. Assume in fact that $\mathbf{x}$ is selected by independent sampling with replacement from $\mathcal{X}$ with the probabilities

$$p_j^{(1)} = \frac{n_j^{(1)}}{n}, \qquad j = 1, \ldots, J$$

   and $\mathbf{y}$ is selected (independently of $\mathbf{x}$) by independent sampling with replacement from $\mathcal{X}$ with the probabilities

$$p_j^{(2)} = \frac{n_j^{(2)}}{n}, \qquad j = 1, \ldots, J.$$

Prove that

$$P(S(\mathbf{x},\mathbf{y}) = k) = \binom{n}{k} p^k (1-p)^{n-k},$$

where

$$p = \sum_{j=1}^{J} \frac{n_j^{(1)} \cdot n_j^{(2)}}{n^2}.$$

9. **Null Model for Pairwise Alignments**

Definition 1.9.1 A *(global) alignment* of two sequences $\mathbf{x}$ and $\mathbf{y}$ is obtained by first inserting chosen spaces ('-'), either into or at ends of $\mathbf{x}$ and $\mathbf{y}$, and then placing the two resulting sequences one above the other so that every symbol or space in every in either sequence is opposite to a unique symbol or a unique space in the other sequence. We denote the aligned sequences by $\mathbf{x}^*$ and $\mathbf{y}^*$. ■

Example 1.9.1 Let $\mathbf{x} = ATAAGC$, $\mathbf{y} = AAAAACG$. To obtain an alignment in the sense defined, one of the many possible ones, we may write

$$\begin{matrix} \mathbf{x}^* \\ \mathbf{y}^* \end{matrix} = \begin{matrix} -ATAAGC- \\ AAAAA-CG \end{matrix} \tag{9.1}$$

■

This means in the first place that two aligned sequences have the same length. Pairwise alignment is the topic in Chapter 5.

Two strings $\mathbf{x}$ and $\mathbf{y}$ with $|\mathbf{x}| = n$ and $|\mathbf{y}| = m$, $n \leq m$, of symbols from an alphabet $\mathcal{X} = \{x_1, \ldots, x_J\}$ are to be aligned. Let us suppose that the sequences are selected as in the preceding exercise with the probabilities for $\mathbf{x}$ being

$$p_j^{(1)} = \frac{n_j^{(1)}}{n}, \qquad j = 1, \ldots, J$$

and $\mathbf{y}$ is selected (independently of $\mathbf{x}$) and by independent sampling with replacement from $\mathcal{X}$ with the probabilities

$$p_j^{(2)} = \frac{n_j^{(2)}}{m}, \qquad j = 1, \ldots, J.$$

Let us suppose that g_1 and g_2 spaces '$-$' are inserted to $\mathbf{x}$ and $\mathbf{y}$, respectively, to get the pairwise alignment.

Let us define $P(S(\mathbf{x}^*, \mathbf{y}^*) = k)$ as the probability that the number of matches, e.g., $\begin{smallmatrix} A \\ A \end{smallmatrix}$ above, in $\mathbf{x}^*$ and $\mathbf{y}^*$ is equal to k. Show that

$$P(S(\mathbf{x}^*, \mathbf{y}^*) = k) = \binom{m - g_1}{k} p^k (1-p)^{m-g_1-k},$$

where

$$p = \frac{\sum_{j=1}^{J} n_j^{(1)} \cdot n_j^{(2)}}{n \cdot m}.$$

10. **Variation distance**

Let X and Y be two random variables with the non-negative integers as alphabet. Define $d(X,Y)$ as

$$d(X,Y) := \frac{1}{2} \sum_{m=0}^{\infty} |P(X=m) - P(Y=m)|. \tag{9.2}$$

Show that this is a *distance function* or a *metric* between the random variables X and Y. A metric is a function such that

1. $d(X,Y) \geq 0$, with equality $d(X,Y) = 0$ iff $X = Y$ $(P(X \neq Y) = 0)$.
2. $d(X,Y) = d(Y,X)$. ('symmetry')
3. $d(X,Z) \leq d(X,Y) + d(Y,Z)$. ('triangle inequality')

11. **Contraction**

Let X, Y and Z be three random variables with the non-negative integers as alphabet. Z is independent of (X,Y). Show that the metric in (9.2) has the property

$$d(X+Z, Y+Z) \leq d(X,Y). \tag{9.3}$$

Hint: Use the identity

$$P(X + Z \in A) = \sum_{m=0}^{\infty} P(X \in A - m, Z = m),$$

where A is any subset (event) of the set of non-negative integers, and $A - m$ is the event A with m excluded.

12. **Poisson Approximation**

(a) Let $X \in \mathrm{Be}(p)$, $0 < p < 1$ and $Y \in \mathrm{Po}(p)$. Show using the metric in (9.2) that

$$d(X, Y) \leq \frac{1}{2}p^2. \tag{9.4}$$

Hint: Compute first that $d(X, Y) = p \cdot (1 - e^{-p})$.

(b) Let $X_1, \ldots, X_n$ be independent and identically distributed $U_i \in \mathrm{Be}(p)$, so that $X = X_1 + \ldots + X_n \in \mathrm{Bin}(n, p)$. Let $Y \in \mathrm{Po}(np)$. Show using the metric in (9.2) that

$$d(X, Y) \leq \frac{1}{2}n \cdot p^2.$$

Hint: Use an inductive extension of (9.3).

The argument underlying this exercise is taken from (Serfling 1978). Poisson approximation can be generalized to deal with sums of dependent Bernoulli distributed variables. The bound on $d(X, Y)$ in (9.4) can be improved, see (Serfling 1978). Applications of Poisson approximation to technical problems in bioinformatics are found in (Waterman 1995, Chapter 12.3).

13. **Fragment Accuracy**
Can the sum in (4.5)

$$P(\text{error}) = \frac{1}{1 - e^{-\lambda}} \sum_{d=1}^{\infty} \frac{\lambda^d}{d!} e^{-\lambda} P(\text{error}|d)$$

be given a closed form expression?

14. **Number of Fragments**

(a) Prove the result in (4.7).

(b) Show that the mean and variance of N, the number of fragments of length greater than x, are given by

$$E[N] = h(1), \qquad \mathrm{Var}[N] = h(2) + h(1) - h(1)^2,$$

where

$$h(r) = \frac{[P - r(x - m)]}{m^r} [P - rx]^{r-1} e^{-r/m}.$$

In fact $h(r)$ is the *descending factorial moment* (Blom et al. 1994, p. 24)

$$h(r) = E\left[N(N-1)\cdots(N-r+1)\right].$$

(Bishop et al. 1983).

1.10 References and Further Reading:

- More extensive accounts of probability calculus at an intermediate level are found in

G. Blom, L. Holst and D. Sandell (1994): *Problems and Snapshots from the World of Probability.* Springer Verlag.

R.M. Gray and L.D. Davisson (1986): *Random Processes. A Mathematical Approach for Engineers.* Prentice-Hall Information and System Science Series. Prentice Hall, Englewood Cliffs, New Jersey. Also available at `http://ee.stanford.edu/~gray/sp.html`.

A. Gut (1995): *An Intermediate Course in Probability.* Springer Verlag.

L.L. Helms (1997): *Introduction to Probability Theory With Contemporary Applications.* W.H. Freeman and Company. New York.

- Treatments of discrete mathematics useful in this course are

N.L. Briggs (1993): *Discrete Mathematics. Revised Edition*, Oxford Science Publications, Clarendon Press, Oxford.

R.L. Graham, D.E. Knuth and O. Patashnik (1989): *Concrete Mathematics.* Addison Wesley Publishing Company, Reading Mass., Menlo Park California, etc. .

- A reference for calculus is

A.I. Khuri (1993): *Advanced Calculus with Applications in Statistics.* John Wiley and Sons, New York.

- Bioinformatics and computational biology

D. Gusfield (1997): *Algorithms on Strings, Trees and Sequences. Computer Science and Computational Biology.* Cambridge University Press. Cambridge.

M.S. Waterman (1995): *Introduction to Computational Biology. Maps, sequences and genomes.* Chapman and Hall, London, Glasgow, Weinheim, New York, Melbourne, Madras.

- Conditional independence

A.P. Dawid (1980): Conditional independence for statistical operations. *Annals of Statistics*, 8, pp. 598–617.

- Poisson approximation

  R.J. Serfling (1978): Some elementary results on Poisson approximation in a sequence of Bernoulli trials. *SIAM Review*, 20, pp. 567–579.

- Fragment assembly and fragment accuracy

  G.A. Churchill and M.S. Waterman (1992): The Accuracy of DNA Sequences: Estimating Sequence Quality. *Genomics*, 14, pp. 89–98.

  G. Myers (1999): Whole-Genome DNA Sequencing. *Computing in Science and Engineering*, 1, pp. 33–43.

  H. Peltola, H. Söderlund and E. Ukkonen (1988): SEQAID: a DNA sequence assembling program based on a mathematical model. *Nucleic Acids Research*, 12, pp. 307–321.

- The distribution of the number of fragments

  T. Bishop, J.A. Williamson and M.H. Skolnik (1983): A Model for Restriction Fragment Length Distributions. *American Journal of Human Genetics*, 35, pp. 795–815.

  J.F. Gusella (1986): DNA polymorphism and human disease. *Annual Reviews of Biochemistry*, 55, pp. 831–854.

  L. Holst (1980): On the Lengths of the Pieces of a Stick Broken at Random. *Journal of Applied Probability*, 17, pp. 623–634.

Chapter 2

Information and the Kullback Distance

2.1 Introduction

The advent and maintenance of life is related to the generation, processing, and propagation of *information* at the molecular level. One may claim that the concept of information will have to be added to the list of basic concepts such as *energy, matter,* etc.. Even though information is encoded in terms of specific spatial and/or temporal patterns or structures of matter, it cannot be identified with its physical carrier. To deal with information Shannon's theory provides tools which are based on measuring information in terms of bits (one *binary information unit* is referred to as one bit), or, in terms of the minimal amount of complexity of structures needed to encode a given piece of information. Therefore it is no surprise that there have been numerous attempts at applying the notion of information measure to biology (Yockey 1992). In this chapter we will concentrate on the more mathematical aspects that yield understanding of the nature and significance of the Kullback Distance (Kullback 1959). The Kullback distance turns out to play an important role in theoretical analysis of learning sequence models from data.

A number of formulae and results in information theory are recapitulated in this chapter in order to establish the conventions, notations, and terminology used in this text. For more extensive accounts of mutual information and Shannon's theory of information we refer to (Abramson 1963, Ash 1967,

Blahut 1987, Cover and Thomas 1991) as well as to (MacKay 1995, Åslund 1961).

2.2 Mutual Information

Let us recall the Bayesian 'update' of probability as given in

$$f_{X|Y}(x_i \mid y_j) = \frac{f_{Y|X}(y_j \mid x_i) \cdot f_X(x_i)}{f_Y(y_j)}. \tag{2.1}$$

Let us set

$$I_{i,j} := \log \frac{f_{X|Y}(x_i \mid y_j)}{f_X(x_i)}.$$

This measures clearly the ratio of posterior beliefs to prior beliefs on a logarithmic scale and can be interpreted as the amount of uncertainty about the event $X = x_i$ removed by observing the event $Y = y_j$. By uncertainty of an event we thus mean the negative logarithm of a probability of an event. (A given event can in general be assigned several different probabilities in terms of different models). Here we assume for simplicity of presentation that all required probability ratios are well defined. In view of the basic definition (2.13) in Chapter 1 we have also

$$I_{i,j} := \log \frac{f_{X,Y}(x_i, y_j)}{f_Y(y_j) \cdot f_X(x_i)}.$$

The expectation of this quantity turns out to have several important properties, which will be summarized below. Thus we introduce

The mutual information:

$$I(X;Y) := \sum_{i=1}^{L} \sum_{j=1}^{J} f_{X,Y}(x_i, y_j) I_{i,j}. \tag{2.2}$$

If the mutual information between X and Y is expressed using logarithms to the base 2,

$$I(X;Y) = \sum_{i=1}^{L} \sum_{j=1}^{J} f_{X,Y}(x_i, y_j) \log_2 \left[\frac{f_{X,Y}(x_i, y_j)}{f_X(x_i) f_Y(y_j)} \right] \tag{2.3}$$

is sometimes said to have the dimension [bits/symbol]. One application of $I(X;Y)$ to bioinformatics is found in (Altschul 1991).

2.3 Properties of Mutual Information

An important special case of $I(X;Y)$ is $I(X;X)$.

2.3.1 Entropy

Taking $X = Y$ in the definition (2.2) and interpreting $f_{X|X}(x_i, x_j) = 1$ if $i = j$ and zero otherwise we obtain

$$I(X;X) = \sum_{i=1}^{L}\sum_{j=1}^{J} f_{X|X}(x_i|x_j) f_X(x_j) I_{i,j} = -\sum_{i=1}^{L} f_X(x_i)\log(f_X(x_i)).$$

Hence we define the *entropy* of X as

$$H(X) := -\sum_{i=1}^{L} f_X(x_i)\log(f_X(x_i)). \tag{3.4}$$

Example 2.3.1 (Binary entropy function) For the special case $\mathcal{X} = \{x_1, x_2\}$, with $p := f_X(x_1)$,

$$h(p) := -p\log_2(p) - (1-p)\log_2(1-p) \tag{3.5}$$

is the *(binary) entropy function.* This is also the entropy of a Bernoulli random variable $X \in \mathrm{Be}(p)$ with $\mathcal{X} = \{0,1\}$). The maximum value of the binary entropy function is $h\left(\frac{1}{2}\right) = 1$ and defines the unit of binary information, i.e., *one bit.* ■

Example 2.3.2 (Entropy of random DNA) Let X assume values in $\mathcal{X} = \{A, T, C, G\}$ and let $X \in \mathrm{U}(1,4)$. In our terminology X is a nucleotide chosen at random. Then

$$H(X) = -\left(\frac{1}{4}\log_2\frac{1}{4} + \frac{1}{4}\log_2\frac{1}{4} + \frac{1}{4}\log_2\frac{1}{4} + \frac{1}{4}\log_2\frac{1}{4}\right) = \log_2 4 = 2. \tag{3.6}$$

We can, in this case, think of the entropy $H(X) = 2$ [bits] as the number of yes/no questions needed to identify an outcome of X. ■

The *equivocation* between X and Y is defined as

$$H(X|Y) := -\sum_{i=1}^{L}\sum_{j=1}^{J} f_{X,Y}(x_i, y_j)\log(f_{X|Y}(x_i|y_j)), \tag{3.7}$$

or

$$H(X|Y) = \sum_{j=1}^{J} H(X|Y = y_j)\cdot f_Y(y_j).$$

The *prevarication* between X and Y is defined as

$$H(Y|X) := -\sum_{i=1}^{L}\sum_{j=1}^{J} f_{X,Y}(x_i, y_j)\log(f_{Y|X}(y_j|x_i)). \tag{3.8}$$

Properties:

1. $H(X) \geq 0$.
2. $H(X|Y) \leq H(X)$ with equality if and only if X and Y are independent.
3. $H(X) \leq \log_2(|\mathcal{X}|) = \log_2 L$ with equality if and only if X is uniformly distributed, i.e., $f_X(x_i) = \frac{1}{L}$ for all i.
4. $H(X)$ is *concave* in f_X. (A function H is concave if $-H$ is convex.)

2.3.2 Some Further Formulas

More on Mutual Information

Properties of $I(X;Y)$:

1. $I(X;Y) \geq 0$ with equality if and only if X and Y are independent.
2. $$\begin{aligned} I(X;Y) &= H(X) - H(X|Y) \\ &= H(Y) - H(Y|X) \\ &= H(X) + H(Y) - H(X,Y). \end{aligned} \tag{3.9}$$

   In this expression

$$H(X,Y) = -\sum_{i=1}^{L}\sum_{j=1}^{J} f_{X,Y}(x_i, y_j)\log f_{X,Y}(x_i, y_j). \tag{3.10}$$

3.

$$I((X_1,Y_1);\cdots;(X_n,Y_n)) = \sum_{i=1}^{n} I(X_i;Y_i) \tag{3.11}$$

if the *pairs* (X_i, Y_i) are independent of each other.

Chain Rule for Entropy

$$H(X,Y) = H(X) + H(Y|X) = H(Y) + H(X|Y), \tag{3.12}$$

where $H(X,Y)$ is given in (3.10).

2.4 Shannon's Source Coding Theorems

2.4.1 AEP

The *Asymptotic Equipartition Property* (**AEP**) or '*Almost all events are almost equally surprising*' (Cover and Thomas 1991, p. 56) is a fundamental result due to Shannon. Its extensions are also known as *Shannon–McMillan–Breiman theorems.* The extensions of **AEP** to ergodic Markov chains and to ergodic hidden Markov models are required in the asymptotic learning theory in Chapters 8 and 17, respectively.

Proposition 2.4.1 *If $X_1, X_2, \cdots,$ are independent and identically distributed with the distribution f_X, then*

$$-\frac{1}{n}\log f_{X_1,X_2,\cdots,X_n}(X_1,X_2,\cdots,X_n) \longrightarrow H(X). \tag{4.1}$$

in probability.

Proof: This is a consequence of the Law of Large Numbers, Proposition 1.8.2 in Chapter 1 and that X_i, $i = 1, 2, \ldots,$ are independent and have the distribution f_X. ■

The **typical set** $A_\epsilon^n(X) \subseteq \mathcal{X}^n := \mathcal{X} \times \cdots \times \mathcal{X}$ is the set of sequences $x_{i_1}x_{i_2}\cdots x_{i_n}$ such that

$$e^{-n(H(X)+\epsilon)} \le f_{X_1,X_2,\cdots,X_n}(x_{i_1},x_{i_2},\cdots,x_{i_n}) \le e^{-n(H(X)-\epsilon)}. \tag{4.2}$$

The source coding theorem in the next subsection is based on the following properties of the typical set:

1. If $(x_{i_1}x_{i_2}\cdots x_{i_n}) \in A^n_\epsilon(X)$, then
$$P(X_1 = x_{i_1}, X_2 = x_{i_2}, \cdots, X_n = x_{i_n}) \propto e^{-n(H(X)\pm\epsilon)}.$$
2. $P(A^n_\epsilon(X)) > 1-\epsilon$ for n sufficiently large.
3. $\mid A^n_\epsilon(X) \mid \leq e^{n(H(X)+\epsilon)}$. ■

2.4.2 The Source Coding Theorem

We now apply the **AEP** to the problem of coding strings of length n $\mathbf{x} = x_{i_1}\ldots x_{i_n}$ from the alphabet $\mathcal{X}$ into codewords (strings) of length r with symbols from an alphabet of size K. We assume that the number of codewords K^r is much smaller than the number $(= L^n)$ of strings of length n from $\mathcal{X}$. Hence this has the effect of compacting data. But all strings $\mathbf{x} = x_{i_1}x_{i_2}\ldots x_{i_n}$ cannot be thus coded and those not coded are assigned to a default codeword. The key to success is to give codewords only to typical sequences.

During decompacting of the data, if the default codeword is observed, a decoding failure is announced. We will show that the probability of decoding failure can be made arbitrarily small. We compute here in natural logarithms.

Proposition 2.4.2 *Let $x_{i_1}\ldots x_{i_n}$ be modeled as outcomes of $X_1,\cdots,X_n$, which are are independent, identically distributed with the distribution f_X. Let the string $\mathbf{x} = x_{i_1}\ldots x_{i_n}$ be encoded into a codeword of length r of symbols from an alphabet of size K. Then if*
$$r\log K > nH(X), \tag{4.3}$$
for any $\epsilon > 0$ the probability of decoding failure can be made less than ϵ for n sufficiently large.

Proof: Consider the set $A^n_\delta(X)$ of typical strings $x_{i_1}x_{i_2}\cdots x_{i_n}$ such that
$$e^{-n(H(X)+\delta)} \leq f_{X_1,X_2,\cdots,X_n}(x_{i_1},x_{i_2},\cdots,x_{i_n}) \leq e^{-n(H(X)-\delta)}.$$
The number of typical strings is bounded above by
$$|A^n_\delta(X)| \leq e^{n(H(X)+\delta)}.$$
By assumption (4.3) we can choose a $\delta > 0$ so that
$$r\log K > n\left(H(X)+\delta\right).$$

Therefore, because the number of codewords is K^r, we see that the number of codewords is at least as large as $e^{n(H(X)+\delta)}$. Hence we can assign a codeword to each string $\mathbf{x}$ in $A_\delta^n(X)$. Therefore the set of strings causing a decoding failure is contained in the complement of $A_\delta^n(X)$. By the **AEP** we know that for any $\delta > 0$, $P(A_\delta^n(X)) > 1 - \delta$ for n sufficiently large, and therefore the probability of the complement (equal to the probability of decoding failure) can be smaller than ϵ. ■

Hence the compression ratio r/n (codesymbols per symbol) is bounded below by the threshold

$$\frac{H(X)}{\log K}.$$

If logarithms to the base 2 are used and $K = 2$ this shows that $H(X)$ has the dimension [bits/symbol]. Clearly for (4.3) to be possible it should be that there is the strict inequality $\log L > H(X)$. There is a converse theorem (Blahut 1987 p. 77) showing that compression ratios below this threshold lead to probability of decoding failure arbitrarily close to one.

2.4.3 Lossless Compression Codes and Entropy

The important operational meaning of entropy $H(X)$ has been expressed in principle in the first coding theorem above, which verifies the possibility of data compression. The coding operation discussed in the theorem used fixed length codestrings and was devoid of any algorithmic structure. Therefore decoding envisaged there would be in practice very time consuming since it would mean looking up in a possible large codebook.

Codes based on effective algorithmic considerations (Storer 1988) have also coding theorems providing lower bounds on the respective compression ratios. The practical codes have variable length codestrings. These codes are lossless, i.e., they have no decoding failure but there is a chance that they may sometimes produce longer codestrings. Below we list a number of inequalities for $H(X)$ bounding the mean length L of the codestrings

$$L := \sum_{i=1}^{L} f_X(x_i) \cdot l_i$$

for various codes for compaction of data. Here l_i is the number of bits used to code the ith symbol in an alphabet using a technique of source coding.

For example in the *Shannon– Fano code*[1] we take

$$l_i = \left\lceil \log_2 \left(\frac{1}{f_X(x_i)} \right) \right\rceil \tag{4.4}$$

and it can be proved that

$$L < H(X) + 1,$$

which again shows the operational importance of $H(X)$.

A code is called *uniquely decodable* if any string composed of a finite number of symbols from $\mathcal{X}$ gets a unique codestring. A necessary and sufficient condition on the lengths of codestrings l_i in a uniquely decodable code is the *Kraft inequality*:

$$\sum_{i=1}^{L} 2^{-l_i} \leq 1. \tag{4.5}$$

This is also known as *Szilard's inequality*, see (Åslund 1961). The expected length of a uniquely decodable code is bounded as

$$L = \sum_{i=1}^{L} f_X(x_i) \cdot l_i \geq H(X). \tag{4.6}$$

2.5 Kullback Distance

2.5.1 Definition and Examples

As shown above, the entropy $H(X)$ is a measure of the uncertainty in bits (binary information units) of the random variable X. It is also in view of (4.6) a lower bound for the number of bits (binary digits) needed on the average to describe the random variable.

Now we introduce the *relative entropy* which by definition can be seen as measuring the extra codelength (or redundancy) of a Shannon–Fano type of code (4.4) based on the distribution $g(x)$ when the 'true' distribution is $f(x)$. The relative entropy is then defined as follows (Kullback 1959).

Definition 2.5.1 Let $\mathcal{X} = \{x_1, \cdots, x_L\}$ be an alphabet and let

$$\mathbf{f} := (f(x_1), \cdots, f(x_L))$$

[1] $\lceil x \rceil :=$ ceiling of $x \stackrel{\text{def}}{=}$ least integer $\geq x$.

and

$$\mathbf{g} := (g(x_1), \cdots, g(x_L))$$

be two probability distributions defined on $\mathcal{X}$. Then their *relative entropy*, or the *Kullback distance* between $\mathbf{f}$ and $\mathbf{g}$, is defined by

$$D(\mathbf{f} \mid \mathbf{g}) = \sum_{i=1}^{L} f(x_i) \log \frac{f(x_i)}{g(x_i)}. \tag{5.1}$$

■

Here we use the conventions $0 \cdot \log \frac{0}{g(x_i)} = 0$ and $f(x_i) \log \frac{f(x_i)}{0} = \infty$. The logarithm is the natural logarithm unless otherwise stated. Note that this is not a distance in the sense of being a metric, since, e.g., the symmetry property of a metric obviously need not hold.

Example 2.5.1 (Information Content) If X is a random variable with the distribution $\mathbf{f} = (f(x_1), \cdots, f(x_L))$, any probability distribution on an alphabet of L symbols, and if $\mathbf{g}$ is the uniform distribution on this alphabet, then

$$D(\mathbf{f} \mid \mathbf{g}) = \log L - H(X).$$

This quantity is sometimes known as the *information content (of* $\mathbf{f}$*)*, see, e.g., (Schneider et al. 1986). ■

Example 2.5.2 (Two Bernoulli distributions) Let $\mathcal{X} = \{0, 1\}$ and $0 \leq p \leq 1$ and $0 \leq g \leq 1$. Let $\mathbf{f} = (1-p, p)$ and $\mathbf{g} = (1-g, g)$ be the two Bernoulli distributions Be(p) and Be(g), respectively. Then

$$D(\mathbf{f} \mid \mathbf{g}) = (1-p) \cdot \log \frac{1-p}{1-g} + p \cdot \log \frac{p}{g}. \tag{5.2}$$

We can also rewrite this as

$$D(\mathbf{f} \mid \mathbf{g}) = -(1-p) \cdot \log(1-g) - p \cdot \log g - h(p),$$

where $h(p)$ is the entropy function (3.5) in natural logarithm. ■

2.5.2 Calibration

Let us set in (5.2) of the preceding example

$$D\left(\mathrm{Be}(p) \mid \mathrm{Be}(g)\right) = -(1-p)\cdot\log(1-g) - p\cdot\log g - h(p), \tag{5.3}$$

which is easily plotted as a function of g for any fixed p or vice versa. These plots suggest a number of the general properties of the relative entropy like that it is nonnegative, becomes zero only if $p = g$ and that the relative entropy is quadratic in small neighborhoods of p.

The cases in which $D(\mathbf{f} \mid \mathbf{g}) = 0$ and $D(\mathbf{f} \mid \mathbf{g}) = \infty$ are easily interpreted in an intuitive sense. McCulloch (1989) suggests a way to understand the other values of $D(\mathbf{f} \mid \mathbf{g})$ by introducing the following idea of *calibration.* Let $D(\mathbf{f} \mid \mathbf{g}) = k$ be the value of the Kullback distance between any two probability distributions on the same alphabet. Then we look for *calibration of* k or a number $g(k)$ such that the equation

$$D\left(\mathrm{Be}\left(\frac{1}{2}\right) \mid \mathrm{Be}\left(g(k)\right)\right) = k \tag{5.4}$$

is satisfied. This calibration tells us that the Kullback distance between $\mathbf{f}$ and $\mathbf{g}$ is the same as between a Bernoulli distribution $\mathrm{Be}\left(\frac{1}{2}\right)$, a random or 'fair' choice of value in $\{0,1\}$, and a Bernoulli distribution $\mathrm{Be}\left(g(k)\right)$. Therefore, the closer to $\frac{1}{2}$ the calibration is, the more similar $\mathbf{f}$ and $\mathbf{g}$ are. The calibration can be found explicitly, since

$$D\left(\mathrm{Be}\left(\frac{1}{2}\right) \mid \mathrm{Be}\left(g\right)\right) = -\frac{1}{2}\log\left(4g(1-g)\right), \tag{5.5}$$

which gives an algebraic equation of second degree from which the calibration can be solved as

$$g(k) = \frac{1}{2}\left(1 + \sqrt{1 - e^{-2k}}\right). \tag{5.6}$$

2.5.3 Properties

Next we give a general proof of the important property that the relative entropy is nonnegative.

Proposition 2.5.3 *For any probability distributions* $\mathbf{f}$ *and* $\mathbf{g}$ *on the same alphabet*

$$D(\mathbf{f} \mid \mathbf{g}) \geq 0. \tag{5.7}$$

Proof: Let X be a random variable that has the distribution $\mathbf{f}$. We write $p(x)$ and $g(x)$ for the generic values of the probability of $x \in \mathcal{X}$ in the two distributions. Then we have

$$D(\mathbf{f} \mid \mathbf{g}) = E\left[\log \frac{p(X)}{g(X)}\right]$$

and this is equal to

$$D(\mathbf{f} \mid \mathbf{g}) = -E\left[\log \frac{g(X)}{p(X)}\right].$$

Since $\phi(x) = -\log x$ is a convex function we have that

$$-E\left[\log \frac{g(X)}{p(X)}\right] \geq -\log E\left[\frac{g(X)}{p(X)}\right],$$

where we have used Jensen's inequality. But

$$E\left[\frac{g(X)}{p(X)}\right] = \sum_{i=1}^{L} f(x_i) \frac{g(x_i)}{f(x_i)} = 1$$

and since $\log 1 = 0$, we have proved our assertion. ■

The following way of rewriting (5.7) is known as

Gibbs' inequality :

$$\sum_{i=1}^{L} p_i \log \frac{1}{p_i} \leq \sum_{i=1}^{L} p_i \log \frac{1}{g_i} \tag{5.8}$$

for any $p_i \geq 0$ and $g_i \geq 0$ with $\sum_{i=1}^{L} p_i = \sum_{i=1}^{L} g_i = 1$. ■

There is a sharper lower bound for the Kullback distance. For this we prove a generalization of Gibbs' inequality that will be useful for several purposes in the sequel.

Lemma 2.5.4 Log sum inequality

If $a_1, \ldots, a_n$ and $b_1, \ldots, b_n$ are non-negative numbers, then

$$\sum_{i=1}^{n} a_i \log \frac{a_i}{b_i} \geq \left(\sum_{i=1}^{n} a_i\right) \log \frac{\sum_{i=1}^{n} a_i}{\sum_{i=1}^{n} b_i}. \tag{5.9}$$

Proof: We assume that the numbers are positive, as can be done without loss of generality. Let us note that $\phi(t) = t \log t$ is a strictly convex function. This means that

$$\sum_{i=1}^{n} \lambda_i \phi(t_i) \geq \phi\left(\sum_{i=1}^{n} \lambda_i t_i\right),$$

where $\sum_{i=1}^{n} \lambda_i = 1$, $\lambda_i \geq 0$. Then we set $\lambda_i = b_j / \sum_{i=1}^{n} b_i$ and $t_i = a_i/b_i$ and obtain

$$\sum_{i=1}^{n} \frac{a_i}{\sum_{i=1}^{n} b_i} \log \frac{a_i}{b_i} \geq \sum_{i=1}^{n} \frac{a_i}{\sum_{i=1}^{n} b_i} \log \sum_{i=1}^{n} \frac{a_i}{\sum_{i=1}^{n} b_i}$$

as was to be proved. ■

Then the desired lower bound is as follows.

Proposition 2.5.5 Variation Distance Bound
For any probability distributions **f** *and* **g** *on the same alphabet*

$$D(\mathbf{f} \mid \mathbf{g}) \geq \frac{1}{4} \left(\sum_{i=1}^{L} \mid f(x_i) - g(x_i) \mid \right)^2 . \tag{5.10}$$

Proof: The main steps of the proof are given below as a set of exercises. ■

In fact a sharper constant equal to $1/(2 \cdot \log 2)$ in the right hand side of (5.10) has been established, see (Cover and Thomas 1991). Recalling the variation distance $d(X, Y)$ defined in (9.2) in Chapter 1 we may write here

$$D(\mathbf{f} \mid \mathbf{g}) \geq d(X, Y)^2 , \tag{5.11}$$

with the identification of **f** and **g** as the distributions of X and Y, respectively.

2.6 The Score and the Fisher Information

Let us consider an indexed family of discrete probability distributions, or a *model family,*

$$\{\mathbf{f}_\theta = (f(x_1; \theta), \cdots, f(x_L; \theta)) \mid \theta \in \Lambda\},$$

where the indexing is in θ, a parameter with values in a subset Λ of the real line. As examples of this we may note $\mathrm{Be}(p)$ $(p = \theta)$ or $\mathrm{Po}(\lambda)$ $(\lambda = \theta)$. Let X be a random variable with a distribution $\mathbf{f}_\theta$ in the family. We assume that $\mathbf{f}_\theta$ is differentiable in θ.

Definition 2.6.1 *The* score V *is the random variable*

$$V(X;\theta) := \frac{\partial}{\partial\theta}\log f(X;\theta). \tag{6.1}$$

■

A large part of the material in the sequel is in one way or the other connected to some or other score. By this definition it is clear that

$$EV(X;\theta) = \sum_{i=1}^{L} f(x_i;\theta)\cdot\frac{\partial}{\partial\theta}\log f(x_i;\theta) = 0.$$

Therefore $EV\left(X;\theta\right)^2 = \mathrm{Var}V\left(X;\theta\right)$. This variance is a known measure of information called the *Fisher information.*

Definition 2.6.2 *The Fisher information is* $J(\theta)$ *defined by*

$$J(\theta) = EV\left(X;\theta\right)^2 = \sum_{i=1}^{L} f(x_i;\theta)\left(\frac{\partial}{\partial\theta}\log f(x_i;\theta)\right)^2. \tag{6.2}$$

■

If $X_1,\ldots,X_n$ are independent random variables with the distribution $\mathbf{f}_\theta$, then

$$J_n(\theta) = EV\left(X_1,\ldots,X_n;\theta\right)^2 = n\cdot J\left(\theta\right). \tag{6.3}$$

Another important and interesting property of the Fisher information is that

$$\lim_{\theta\to\theta_o}\frac{D\left(\mathbf{f}_{\theta_o}\mid\mathbf{f}_\theta\right)}{\left(\theta-\theta_o\right)^2} = \frac{1}{\log 4}J\left(\theta_o\right). \tag{6.4}$$

Hence

$$D\left(\mathbf{f}_{\theta_o}\mid\mathbf{f}_\theta\right)\approx\left(\theta-\theta_o\right)^2\cdot J\left(\theta_o\right),$$

which is a more quantitative expression of the observation mentioned above that the Kullback distance is a quadratic function in the vicinity of the reference measure.

2.7 Exercises on Mutual Information and Codelengths

1. **A principle of maximum entropy**
Let $\mathcal{X} = \{1, 2, 3, 4, 5, 6\}$ and $\theta_j \geq 0$, $\sum_{j=1}^{6} \theta_j = 1$ be any probability distribution on $\mathcal{X}$. Show that the entropy

$$-\sum_{j=1}^{6} \theta_j \log \theta_j$$

is maximized by the uniform probability distribution.
Hint: Lagrange multipliers.

2. Let X be a random variable that assumes values in the finite alphabet $\mathcal{X} = \{x_1, \ldots, x_k\}$. Let $\mathbf{x} = (x_{j_1} \ldots x_{j_n})$ be a finite sequence modeled as independent samples of X. Let

$$n_j = \text{ the number of times the symbol } x_j \text{ is present in } \mathbf{x}$$

with $n_1 + n_2 + \cdots n_k = n$ and let

$$\mathbf{f}(\mathbf{x}) = \left(\frac{n_1}{n}, \ldots, \frac{n_k}{n}\right)$$

denote the corresponding empirical distribution on $\mathcal{X}$. There are situations which constrain $\mathbf{f}(\mathbf{x})$ to lie in some fixed class of empirical distributions denoted by $\mathcal{E}$. Let $\mathcal{X}_n$ be the set of all sequences $\mathbf{x}$ so that the corresponding empirical distribution lies in $\mathcal{E}$ or formally

$$\mathcal{X}_n = \{\mathbf{x} = (x_{j_1}, \ldots, x_{j_n})\,|x_{j_l} \in \mathcal{X}, j = 1, \ldots, n; \mathbf{f}(\mathbf{x}) \in \mathcal{E}\}.$$

Let $\theta_j \geq 0$, $\sum_{j=1}^{k} \theta_j = 1$ and

$$L(\mathbf{x}) = -\sum_{j=1}^{k} n_j \log \theta_j.$$

(a) Show that

$$\sum_{\mathbf{x}\in\mathcal{X}_n} L(\mathbf{x}) = \sum_{j=1}^{k} \sum_{\mathbf{f}\in\mathcal{E}} \frac{n!}{n_1! \cdots n_k!} n_j \log \theta_j.$$

(b) Set

$$\beta_j = \sum_{f \in \mathcal{E}} \frac{n!}{n_1! \cdots n_k!} n_j.$$

Then show that

$$\sum_{\mathbf{x} \in \mathcal{X}_n} L(\mathbf{x})$$

is minimized, when the logarithm is to base 2, as function of $\theta_1, \ldots, \theta_k$, if

$$\widehat{\theta}_j = \frac{\beta_j}{\sum_{j=1}^{k} \beta_j}, j = 1, \ldots, k.$$

(*Interpretation*: The minimal value of $\sum_{\mathbf{x} \in \mathcal{X}_n} L(\mathbf{x})$ is the minimum codelength required to represent the sequences in $\mathcal{X}_n$ with a binary code.)
Hint: Lagrange multipliers are not necessary.

(c) Specialize the preceding to $\mathcal{X} = \{0, 1\}$ and

$$\mathcal{X}_n = \text{ all sequences of binary digits of length } n.$$

Then there is only one parameter $\theta = P(X = 1|\theta)$ to choose. Show that the value of this parameter minimizing $\sum_{\mathbf{x} \in \mathcal{X}_n} L(\mathbf{x})$ is

$$\widehat{\theta} = 1/2.$$

3. Prove that

(a) $I(X;Y) = I(Y;X)$.

(b) $I(X;Y) \geq 0$.

(c) $H(X) \geq H(X|Y)$.

(d) $H(X) \leq \log L$, if X is any discrete random variable assuming values in an alphabet with L symbols. (Logarithm is to the base 2).

(e) $I((X_1, Y_1); \cdots; (X_n, Y_n)) = \sum_{i=1}^{n} I(X_i; Y_i)$, if the *pairs* (X_i, Y_i) are independent of each other.

(f) Prove $H(X) \geq H(X|Y)$ using Jensen's inequality (Åslund 1961 p. 13)

(g) Prove the *Data Processing Inequality*: If $X \longrightarrow Y \longrightarrow Z$ forms a *Markov Chain* (in the sense of Exercise 2 in chapter 1) then

$$I(X;Y) \geq I(X;Z). \tag{7.1}$$

4. X is a random variable that has the uniform distribution on the alphabet $\{-2,-1,1,2\}$. The random variable Y is defined by $Y = X^2$. Compute $I(X,Y)$ using logarithms to the base 2.

5. Let X be a random variable having the probability distribution

$$p_X(x) = \frac{1}{2^x}, \qquad x = 1, 2, \ldots,$$

Compute the entropy $H(X)$.

6. Select an integer $X \in \mathrm{U}(1,\ldots,n)$. An integer $Y|X = x \in \mathrm{U}(1,\ldots,x)$. Compute the joint entropy $H(X,Y)$ without using the joint distribution of the random variable (X,Y).

7. Binary digits 0 and 1 are transmitted over a communications channel with noise. Let X denote the transmitted bit and Y denote the received bit. $P(Y = 0|X = 1]$ is equal to q. The conditional probability $P(Y = 1|X = 0]$ is also equal to q, $0 \leq q \leq 1$. The probability distribution of X is Be(p), $0 \leq p \leq 1$. Show that

$$I(X;Y) = h(p \cdot q + (1-p) \cdot (1-q)) - h(q).$$

8. Show that

$$d_E(X,Y) = H(X|Y) + H(Y|X) \tag{7.2}$$

is a *metric* (see Exercise 10 in Chapter 1) between the random variables X and Y (Horibe 1973).

9. Use *Fano's Inequality*

$$h(P_e) + P_e \log_2(|\mathcal{X}| - 1) \geq H(X|Y), \tag{7.3}$$

where $P_e := P(g(Y) \neq X)$ and $g(\cdot)$ is any function of Y. Then prove that the entropy metric $d_E(X,Y)$, Exercise 8, is continuous in the sense that

$$P(X_n \neq Y_n) \rightarrow 0$$

if and only if

$$d_E(X_n, Y_n) \rightarrow 0.$$

10. Verify the three properties of the typical set $A_\epsilon^n(X)$.

11. Let $X \in \mathrm{Bin}(n,p)$ and consider the entropy of X as a function of p,

$$f(p) = H(X).$$

Show that $f(p) = f(1-p)$ and that $f(p)$ is a concave function, i.e., $-f(p)$ is convex. Find $p^* = \arg\max f(p)$.

12. Let $X \in \mathrm{Ge}(p)$, see example 2.2.3 in Chapter 1. Show that

$$H(X) = \frac{1}{p} h(p).$$

2.8 Kullback Distance and Fisher Information

1. Prove that if $X_1, \ldots, X_n$ are independent random variables with the distribution f_θ, then

$$J_n(\theta) = EV(X_1, \ldots, X_n; \theta)^2 = n \cdot J(\theta). \tag{8.1}$$

2. Prove that if $X_1, \ldots, X_n$ are independent random variables with the distribution f_θ, then

$$-\frac{1}{n} \sum_{i=1}^{n} \frac{\partial^2}{\partial \theta^2} \log f(X_i; \theta) \to J(\theta)$$

in probability.

3. Prove that

$$\lim_{\theta \to \theta_o} \frac{D(\mathbf{f}_{\theta_o} \mid \mathbf{f}_\theta)}{(\theta - \theta_o)^2} = \frac{1}{\log 4} J(\theta_o). \tag{8.2}$$

4. Consider the Bernoulli distribution $\mathrm{Be}(\theta)$ with parameter θ, $0 \le \theta \le 1$. Show that the Fisher information is

$$J(\theta) = \frac{1}{\theta \cdot (1-\theta)},$$

whenever this is defined.

5. If $p(\mathbf{n}|\underline{\theta})$ is the multinomial distribution (see Chapter 1), i.e.,

$$p(\mathbf{n}|\underline{\theta}) = \frac{n!}{n_1! \cdot n_2! \cdot \ldots \cdot n_L!} \prod_{j=1}^{L} \theta_j^{n_j},$$

show by using Stirling's approximation formula that

$$\ln p(\mathbf{n}|\underline{\theta}) \approx C - nD(\widehat{p}|\underline{\theta}),$$

where $\widehat{p}$ is the empirical distribution

$$\widehat{p} = \left(\frac{n_1}{n}, \ldots, \frac{n_L}{n}\right),$$

and

$$\underline{\theta} = (\theta_1, \ldots, \theta_L).$$

6. Discuss the **AEP** for the multinomial distribution using the preceding exercise.

7. Prove (5.10), i.e.,

$$D(\mathbf{f} \mid \mathbf{g}) \geq \frac{1}{4}\left(\sum_{i=1}^{L} \mid f(x_i) - g(x_i) \mid\right)^2. \tag{8.3}$$

It suffices to verify the steps $(a)-(c)$ below.

(a) Show first that

$$D(\mathbf{f} \mid \mathbf{g}) \geq -2\log \sum_{i=1}^{L} f(x_i)^{1/2} \cdot g(x_i)^{1/2}.$$

Hint: First consider the identities

$$D(\mathbf{f} \mid \mathbf{g}) = 2\sum_{i=1}^{L} f(x_i)^{1/2} \cdot f(x_i)^{1/2} \log \frac{f(x_i)^{1/2}}{g(x_i)^{1/2}}$$

$$= 2\sum_{i=1}^{L} f(x_i)^{1/2} \left[\sum_{i=1}^{L} \frac{f(x_i)^{1/2}}{\sum_{i=1}^{L} f(x_i)^{1/2}} \cdot f(x_i)^{1/2} \log \frac{f(x_i)^{1/2}}{g(x_i)^{1/2}}\right].$$

Next use here the log sum inequality (5.9) with

$$a_i = \frac{f(x_i)^{1/2} f(x_i)^{1/2}}{\sum_{i=1}^{L} f(x_i)^{1/2}}$$

and

$$b_i = \frac{f(x_i)^{1/2} g(x_i)^{1/2}}{\sum_{i=1}^{L} f(x_i)^{1/2}}.$$

(b) Show next that

$$-2\log \sum_{i=1}^{L} f(x_i)^{1/2} \cdot g(x_i)^{1/2} \geq \sum_{i=1}^{L} \left(f(x_i)^{1/2} - g(x_i)^{1/2} \right)^2.$$

Hint: Use the standard inequality $\log x \leq x - 1$ valid for all $x > 0$ in the expression in the left hand side.

(c) Then finally prove

$$\sum_{i=1}^{L} \left(f(x_i)^{1/2} - g(x_i)^{1/2} \right)^2 \geq \frac{1}{4} \left(\sum_{i=1}^{L} \mid f(x_i) - g(x_i) \mid \right)^2.$$

Hint: First note that

$$\left(\sum_{i=1}^{L} \mid f(x_i) - g(x_i) \mid \right)^2$$

$$= \left(\sum_{i=1}^{L} \mid f(x_i)^{1/2} - g(x_i)^{1/2} \mid \cdot \left(f(x_i)^{1/2} + g(x_i)^{1/2} \right) \right)^2$$

By the well known Cauchy–Schwartz inequality (cf. (7.3) in Chapter 16) applied on the second expression in the right hand side

$$\left(\sum_{i=1}^{L} \mid f(x_i) - g(x_i) \mid \right)^2$$

$$\leq \sum_{i=1}^{L} \mid f(x_i)^{1/2} - g(x_i)^{1/2} \mid^2 \cdot \sum_{i=1}^{L} \left(f(x_i)^{1/2} + g(x_i)^{1/2} \right)^2.$$

Then by expanding the square we obtain

$$\sum_{i=1}^{L} \left(f(x_i)^{1/2} + g(x_i)^{1/2}\right)^2 \leq 4.$$

Hence we have obtained

$$\left(\sum_{i=1}^{L} \mid f(x_i) - g(x_i) \mid\right)^2 \leq 4 \cdot \sum_{i=1}^{L} (f(x_i)^{1/2} - g(x_i)^{1/2})^2,$$

as was to be proved.

8. Let

$$\mathbf{f}^n(\mathbf{x}) = \prod_{i=1}^{n} \mathbf{f}_X(x_i), \mathbf{g}^n(\mathbf{x}) = \prod_{i=1}^{n} \mathbf{g}_X(x_i).$$

Show that

$$D(\mathbf{f}^n|\mathbf{g}^n) = nD(\mathbf{f}|\mathbf{g}). \tag{8.4}$$

9. **Large deviations for binomial random variables**
In an Exercise in Chapter 1 it was shown that the number of matches in a sequence *alignment* of unrelated sequences can be modeled by binomial random variables. It is therefore of interest to compute the significance of an observation with an unusual number k of successes under this null model. We shall next present an accurate approximation (bound) for this purpose. This is done in a series of steps (a)–(d) below.

(a) Prove Markov's inequality

$$P(|X| \geq t) \leq \frac{E|X|}{t}$$

for $t > 0$ and any random variable X assuming values in a discrete alphabet. (*Hint*: Check the proof of Chebyshev's inequality.)

(b) Let $S_n \in \text{Bin}(n.p)$. Show that for any positive β

$$P\left(e^{\beta \cdot S_n} \geq e^{a\beta n}\right) \leq \left[e^{-a \cdot \beta}\left(1 - p + p \cdot e^{\beta}\right)\right]^n$$

Hint: Recall that

$$S_n = X_1 + X_2 + \ldots + X_n,$$

where $X_1, X_2, \ldots, X_n$ are independent $\text{Be}(p)$ variables.

(c) Prove that if $1 > a > p > 0$ and if

$$\left[e^{-a\cdot\beta}\left(1-p+p\cdot e^{\beta}\right)\right]$$

is minimized by an optimal choice of β, then its value is $e^{-D(\mathrm{Be}(a)|\mathrm{Be}(p))}$ using the notation in (5.3).

(d) Show using the preceding steps that if $1 > a > p > 0$, then

$$P\left(S_n \geq an\right) \leq e^{-nD(\mathrm{Be}(a)|Be(p))}.$$

This is the *large deviation bound on binomial probabilities* due to Harald Cramér. Here we have a method of assessing the significance of observing, e.g., $a = k/n$ matches in an alignment under the assumption that the sequences are unrelated. (Arratia and Gordon 1989).

2.9 References and Further Reading:

N. Abramson (1963): *Information Theory and Coding.* McGraw-Hill Book Company. New York.

S.F. Altschul (1991): Amino Acid Substitution Matrices from an Information Theoretic Perspective. *Journal of Molecular Biology*, 219, pp. 555–565.

R. Arratia and L. Gordon (1989): Tutorial on Large Deviations for the Binomial Distribution. *Bulletin of Mathematical Biology*, pp. 125–141.

R. Ash (1967): *Information Theory.* Interscience Publishers, New York.

R.E. Blahut (1987): *Principles and Practice of Information Theory.* Addison-Wesley Publishing Company, Reading Massachusetts.

T.M. Cover and J.A. Thomas (1991): *Elements of Information Theory.* John Wiley and Sons, New York.

Y. Horibe (1973): A Note on Entropy Metrics. *Information and Control*, Vol. 22, pp. 403–404.

S. Kullback (1959): *Information Theory and Statistics.* John Wiley and Sons, New York.

D.J.C. MacKay (1995): *A Short Course in Information Theory – Outline.* `http://wol.ra.phy.cam.ac.uk/mackay/itprnn/1997/`

R. McCulloch (1989): Local Model Influence. *Journal of American Statistical Association*, 84, pp. 473–478.

T.D. Schneider, G.D. Stormo, L. Gold and A. Ehrenfeucht (1986): Information Content of Binding Sites on Nucleotide Sequences. *Journal of Molecular Biology*, 188, pp. 415–431.

M. Storer (1989): *The Data Compression Book. Methods and Theory.* Computer Science Press, Rockville, Maryland.

H.P. Yockey (1992): *Information Theory and Molecular Biology.* Cambridge University Press, Cambridge.

N. Åslund (1961): Informationsteoriens Fundamentalsatser I–II. *Nordisk Matematisk Tidskrift*, Vol. 9., pp. 5–25, p. 97–108, respectively.

Chapter 3

Probabilistic Models and Learning

3.1 Introduction

By 'learning from data' one often means the process of inferring a general law or principle from the observations of particular instances. The general law is a piece of knowledge about the mechanism of nature that generates the data. The learning can be done by use of 'MODELS', which serve as the language in which the constraints predicated on the data can be described.

The kind of data we are interested in here are those encountered in a probabilistic learning approach to bioinformatics, i.e., biological sequences (such as DNA, RNA, amino acids). There are different models which can be useful for their respective purposes when dealing with this data cf. (Doolittle 1996). The aim of the methods to be introduced is to extract useful knowledge on the basis of a single sequence $\mathbf{x}$ or several sequences using probabilistic models. We occasionally call $\mathbf{x}$, for convenience of reference, a *training sequence*.

It is desirable that a probabilistic model should assign a probability distribution with support containing a whole set of sequences, not just the training sequence. Otherwise the model is *overfitted* or tailored to training data and lacks capacity for *generalization*. In psychological literature generalization means the ability to infer correct structure from examples. Probabilistic modelling will make it possible to address questions like 'which sequences in a data base fit a model?', 'how well do the sequences fit?', and 'which in a family of models best describes a set of sequences?'.

Such activity has been known as *statistical model fitting* but is nowadays often called *machine learning*. The text (Bishop 1995) shows how to mix terminology and techniques of both. An early and very clearly written treatment of machine learning is found in the reprint (Nilsson 1990). Machine learning is desiring to automate the whole process of finding good model families and models. The ability of machines to perform the complete process can be put into doubt. Machine learning methods are regarded as useful for areas like molecular biology, where there is a large amount of data but little established theory which is computationally feasible. Next we discuss the concepts of probability calculus with a view for machine learning. In this vein we shall eventually present comparison of model families, stochastic complexity, and Bayesian codelengths. This suggests links between information theory, probability, and machine learning. These ideas of compact storage of items, prediction, and learning have been recognized in microbial systematics (Priest and Austin 1993) for a very long time.

3.2 Bayesian probability

3.2.1 Chance and Probability

The Bayesian probability of an event is a person's *degree of belief* in that event. In most other elementary textbooks the concept of probability is introduced as an objective physical property of the reality (the probability that a coin will land on the head), whereas a Bayesian probability is a property of the person (often called the 'Statistician') who assigns the probability (her/his degree of belief that the coin will land heads).

To keep these two concepts of probability distinct we refer to the the physical property as the *chance* of that event and we refer to a degree of belief in an event as a *Bayesian, or personal, probability*. Alternatively, when the context is clear, we can refer to a Bayesian probability simply as probability. The mathematical rules will turn out to be the same as those already encountered.

One important difference between chance and personal probability is that to measure the latter we do not need repeated trials. For example, imagine, following (Heckerman 1997), the repeated tosses of a sugar cube onto a wet surface. Every time the cube is tossed its dimensions will be changed slightly. Hence it will be a difficult task to obtain the chances of the outcomes of this kind of cube tossing. In personal probability we simply restrict our attention to the next toss of the cube and assign a probability.

Bayesian probability is, mathematically speaking, a function of two arguments. These are the event A (such as the next toss of the cube) about which you are uncertain, and your knowledge H, e.g., the outcomes of the tosses so far, the physical properties of the cube, and so on, when you make the uncertainty statement. We write this as

$$P(A \mid H); \text{ read: the probability of } A \text{ given } H.$$

The interpretation of $P(A \mid H)$ is that it is your subjective belief in the truth of A if you know that H is true. It is instructive to think of $P(A \mid H)$ as a measurement, such as a measurement of length or of temperature. But $P(A \mid H)$ measures belief. Clearly we have, in the preceding elaboration of probability calculus and information theory, suppressed the knowledge variable H in our manipulations with the probability calculus.

One main criticism of Bayesian probability is that these measurements seem arbitrary. Why should one's degrees of belief satisfy the rules of probability? On what scale should probabilities be measured? How is one to assign probabilities other than zero and one?

There are several different answers to these questions that will not be discussed in detail here, instead we refer to (Cox 1961, Halldén 1973, Jeffreys 1998 (first edition 1939)), and, last but not least, to E. Jaynes' legendary monograph on the World Wide Web (Jaynes).

These studies have in common the result that there are certain natural properties of a continuous measure of explicitly conditioned belief that will logically yield the axioms of Bayesian probability. The set of properties assumed vary from one researcher in the foundations of probability to the next, but the axioms of probability obtained are the same.

There are also very critical appraisals of Bayesian probability, one fairly recent printed clash of opinion is found in (Hearst 1997). (Hartigan 1983) and (Neapolitan 1990) give clear presentations of the most important amongst the many schools of thought on the meaning of probability. Hamming (1991, Chap. 8) contributes with original views on these problems, too. Bayesian probability exercises also a strong influence in artificial intelligence (Bender 1996, Cheeseman 1988, Neapolitan 1990).

3.2.2 Coherence

The function $P(A \mid H)$ obeys the three axioms:

Convexity

$$0 \leq P(A \mid H) \leq 1$$

and

$$P(A \mid H) = 1,$$

if H is known by you logically to imply A.

Addition

$$P(A \cup B \mid H) = P(A \mid H) + P(B \mid H) - P(A \cap B \mid H).$$

Multiplication

$$P(A \cap B \mid H) = P(A \mid H) \cdot P(B \mid A \cap H).$$

In the preceding two chapters we have, of course, tacitly applied these rules, since if $A \subseteq \mathcal{X}$ and X is a random variable with the distribution f_X, then

$$P_X(A) := \sum_{x_i \in A} f_X(x_i) \tag{2.1}$$

satisfies the axioms above. These axioms are also satisfied by the conditional probabilities defined earlier. The axiom of multiplication has previously appeared as the chain rule (2.16) in Chapter 1. The system of axioms can be elaborated by requiring the events to form a sigma field and by strengthening the axiom of addition to countable additivity. These matters are not discussed here. What is important is the idea of *coherence* apparent in these axioms (Lindley 1987). By coherence we mean that the beliefs may be personal, but they must fit together. For example, the multiplication rule states that we may only assess two of the involved probabilities independently, a third one is given by the rule.

3.3 Models with Conditional Independence

There is one type of learning which we will be extensively concerned with: this is inferring, analysing, and using a family of models indexed by parameters. Formally we want to find $P(A \mid H)$ when A is a sequence and H is a knowledge base that consists of a sequence (training data) and a model M which is determined by a finite set of parameters. The first families of models to be studied are binary and multinomial processes of conditional independence.

3.3.1 Modelling and Learning for Tosses of a Thumb tack

The Model Family

The modelling of thumb tack (drawing pin) tosses in this subsection is taken from (Heckerman 1997). The mathematics involved is found in greater detail in (von Mises and Geiringer 1964) but, in fact, goes back to P.S. Laplace.

We consider a sequence of flips of a thumb tack. If we throw a thumb tack in the air, it will come to rest either on its point (0) or on its head (1). Suppose we flip the thumb tack n times (fixing n in advance), making sure that the physical properties of the thumb tack and the conditions under which it is flipped remain stable over time. We let $\mathbf{x}$ denote the sequence of outcomes of the flips, $\mathbf{x} = x_{i_1} x_{i_2} \ldots x_{i_n}$, $x_{i_l} \in \{0, 1\}$. Let now Θ be a random variable (quantity), whose values are numbers, denoted by θ, between zero and one, $0 \leq \theta \leq 1$. These values θ correspond to the possible values of the *chance of obtaining heads* in tossing a thumb tack.

MODEL FAMILY:

CONDITIONED ON $\Theta = \theta$, THE SYMBOLS IN $\mathbf{x}$ ARE INDEPENDENT.

Or more completely: the symbols in $\mathbf{x}$ are outcomes of independent Bernoulli $\text{Be}(\theta)$ random variables. Hence a model in the family is given by the probability assignment

$$
\begin{aligned}
P(\mathbf{x} \mid \Theta = \theta) &= \prod_{l=1}^{n} \theta^{x_{i_l}} \cdot (1-\theta)^{1-x_{i_l}} \\
&= \theta^{\sum_{l=1}^{n} x_{i_l}} \cdot (1-\theta)^{n-\sum_{l=1}^{n} x_{i_l}} = \theta^{k} \cdot (1-\theta)^{n-k},
\end{aligned}
$$

if $\sum_{l=1}^{n} x_{i_l} = k$. One *problem of inference* or model fitting is to find the model (within a pre-established family) that is best in some sense given some observed data. In the thumb tack example we understand this as follows. We have observed n outcomes of flips of a thumb tack $\mathbf{x}$ and wish to determine which of the models in the family that best describes this set of flips.

The Posterior Density

To progress with this we express our uncertainty about Θ using a probability density function $f_{\Theta}(\theta)$, which is called the *prior*. Formally this means

$$f_{\Theta}(\theta) \geq 0, \qquad 0 \leq \theta \leq 1$$

and $f_{\Theta}(\theta) = 0$ elsewhere, and

$$\int_0^1 f_{\Theta}(\theta)\, d\theta = 1.$$

Also $P(a < \Theta \leq b) = \int_a^b f_{\Theta}(\theta)\, d\theta$. The prior can be included as a part of the description of the model family.

By an extension of Bayes' rule to continuous random variables we obtain the *posterior*

$$f_{\Theta|\mathbf{x}}(\theta \mid \mathbf{x}) = \frac{P(\mathbf{x} \mid \Theta = \theta) \cdot f_{\Theta}(\theta)}{\int_0^1 P(\mathbf{x} \mid \Theta = \theta) \cdot f_{\Theta}(\theta)\, d\theta}, 0 \leq \theta \leq 1 \tag{3.1}$$

and zero elsewhere. Owing to the standardization $f_{\Theta|\mathbf{x}}(\theta \mid \mathbf{x})$ is another probability density for Θ.

The posterior $f_{\Theta|\mathbf{x}}(\theta \mid \mathbf{x})$ expresses our updated belief in the statement that θ is the true chance of obtaining heads given that we have observed $\mathbf{x}$.

One way to proceed further from here is to use an explicit form for $f_{\Theta}(\theta)$. There could be several choices, but some are at least analytically more advantageous. Let us consider the *uniform prior* given by

$$f_{\Theta}(\theta) = \begin{cases} 1, & 0 \leq \theta \leq 1, \\ 0 & \text{elsewhere.} \end{cases}$$

The uniform prior is often interpreted as a representation of complete ignorance (Jeffreys 1998).

By the *Beta integral* found in (8.10) in Appendix 3.8.3 we can calculate

$$\int_0^1 P(\mathbf{x} \mid \Theta = \theta) \cdot f_\Theta(\theta)\, d\theta = \int_0^1 \theta^k \cdot (1-\theta)^{n-k}\, d\theta = \frac{k!(n-k)!}{(n+1)!}.$$

Then we have

$$f_{\Theta|\mathbf{x}}(\theta \mid \mathbf{x}) = \begin{cases} \dfrac{(n+1)!}{k!(n-k)!} \cdot \theta^k (1-\theta)^{n-k}, & 0 \le \theta \le 1 \\ 0 & \text{elsewhere.} \end{cases} \tag{3.2}$$

This is a *Beta density*, see Appendix 3.8.3.

The Maximum Likelihood Estimate

To understand better the alluded properties of $f_{\Theta|\mathbf{x}}(\theta \mid \mathbf{x})$ we introduce the *maximum likelihood estimate* $\widehat{\theta}_{\mathrm{ML}}$ *of* θ defined by

$$\widehat{\theta}_{\mathrm{ML}} = \arg\max_{0 \le \theta \le 1} P(\mathbf{x} \mid \Theta = \theta) = \arg\max_{0 \le \theta \le 1} \theta^k \cdot (1-\theta)^{n-k}.$$

The rationale for this is that we *try to find the model within the family that gives the training sequence* $\mathbf{x}$ *the highest possible probability.* The probability $P(\mathbf{x} \mid \Theta = \theta)$ regarded as a function of θ is known as the *likelihood function*

$$L(\theta) = P(\mathbf{x} \mid \Theta = \theta).$$

The likelihood function $L(\theta)$ thus compares the plausibilities of different models for given $\mathbf{x}$. A straightforward maximization of the likelihood function gives

$$\widehat{\theta}_{\mathrm{ML}} = \frac{k}{n}. \tag{3.3}$$

By a Taylor expansion of $\log P(\mathbf{x} \mid \Theta = \theta)$ around $\widehat{\theta}_{\mathrm{ML}}$ and by using the definition of the Fisher information $J(\theta)$ in chapter 2 (and the exercises there), we can show that $f_{\Theta|\mathbf{x}}(\theta \mid \mathbf{x})$ is approximately

$$\approx e^{-\frac{1}{2} nJ\left(\widehat{\theta}_{\mathrm{ML}}\right) \cdot \left(\theta - \widehat{\theta}_{\mathrm{ML}}\right)^2} = e^{-\frac{1}{2} \frac{n}{\widehat{\theta}_{\mathrm{ML}} \cdot (1-\widehat{\theta}_{\mathrm{ML}})} \cdot \left(\theta - \widehat{\theta}_{\mathrm{ML}}\right)^2}. \tag{3.4}$$

A concise proof of this result in a general case under certain regularity conditions is given in (Hartigan 1983, chapter 11). We can empirically, say for

$\mathbf{x}$ drawn from a pseudo-random number generator, plot the posterior density (3.2) as a function of θ and observe the property (3.4), when the length of a typical (in the sense of chapter 2) string $\mathbf{x}$ increases. In fact this holds independently of the prior density. This behaviour is clearly present in a typical simulation.

The Personal Probability for the Outcome of the Next Toss

We may be primarily interested in evaluating probabilities $P(A \mid H)$ without the infusion of the auxiliary notion of a parameter (Roberts 1965). In the thumb tack model we may be concerned with

$$P(X_{n+1} = \text{head}|\mathbf{x}),$$

if X_{n+1} is a random variable modelling the next toss, given n flips of the thumb tack as recorded in $\mathbf{x}$. The evaluation of this probability is left as an exercise of this chapter. The computation of $P(X_{n+1} = \text{head}|\mathbf{x})$ has also been discussed from the pedagogical point of view as a potential item in a high school curriculum of statistics in (Lindley 1970). Lindley sees here a natural transition from rules of probability to the rules of inference that should be easier at a high school level than the frequentist approach.

3.3.2 Learning of the Multinomial Process

The Model Family

Let $X_1, X_2, \ldots, X_n$ be independent random variables assuming values in

$$\mathcal{X} = \{x_1, \cdots, x_L\}$$

with the common distribution

$$\theta_l = P(X_i = x_l), \qquad l = 1, 2, \ldots, L.$$

Hence $\theta_1 + \theta_2 + \ldots + \theta_L = 1$. Let $\mathbf{x} = x_{i_1} x_{i_2} \ldots x_{i_n}$ be a string of symbols from $\mathcal{X}$ and let for $l = 1, 2, \ldots, L$

$$n_l = \text{ the number of times the symbol } x_l \text{ is found in } x_{i_1} x_{i_2} \ldots x_{i_n}.$$

We set

$$\underline{\theta} = (\theta_1, \theta_2, \ldots, \theta_L)$$

and consider $\underline{\Theta}$ as a random variable (element) that assumes values in the simplex

$$S_L = \{\underline{\theta} \mid \theta_1 + \theta_2 + \ldots + \theta_L = 1, \theta_l \geq 0, l = 1, \ldots, L\}.$$

With this we introduce the following model family, sometimes known as the *multinomial process*, which is frequently used as a 'null model' in biological sequence analysis, see, e.g., (Goldman 1993).

THE MODEL FAMILY: MULTINOMIAL PROCESS

CONDITIONED ON $\underline{\Theta} = \underline{\theta}$, THE SYMBOLS IN $\mathbf{x}$ ARE INDEPENDENT.

Thus, as shown before,

$$P(\mathbf{x} \mid \underline{\theta}) = \theta_{i_1} \cdot \theta_{i_2} \cdots \theta_{i_n} = \theta_1^{n_1} \cdot \theta_2^{n_2} \cdots \theta_L^{n_L}.$$

Again we find a prior $\phi_{\underline{\Theta}}(\underline{\theta})$ for Θ. Let us consider the *Dirichlet prior* given by

$$\phi_{\underline{\Theta}}(\underline{\theta}) = \begin{cases} \frac{\Gamma(\alpha)}{\Gamma(\prod_{j=1}^{L} \alpha q_j)} \prod_{j=1}^{L} \theta_j^{\alpha q_j - 1}, & \underline{\theta} \in S_L, \\ 0 & \text{elsewhere,} \end{cases}$$

where *the hyperparameters* are $\alpha > 0$, $q_j \geq 0$, $\sum_{j=1}^{L} q_j = 1$, $\Gamma(z)$ is Euler's gamma function as given in Appendix 3.8. The prior $\phi_{\underline{\Theta}}$ is in (8.7), from the Appendix given the symbol

$$\mathrm{Dir}(\alpha q_1, \ldots, \alpha q_L).$$

By extension of Bayes' rule we get the *posterior*

$$\phi_{\underline{\Theta}|\mathbf{x}}(\underline{\theta}|\mathbf{x}; \underline{\alpha}) = \frac{P(\mathbf{x} \mid \underline{\Theta} = \underline{\theta}) \cdot \phi_{\underline{\Theta}}(\underline{\theta})}{\int_{S_L} P(\mathbf{x} \mid \underline{\Theta} = \underline{\theta}) \cdot \phi_{\underline{\Theta}}(\underline{\theta})\, d\underline{\theta}}, \underline{\theta} \in S_L \qquad (3.5)$$

and zero elsewhere. A rationale for choosing this prior is recapitulated in (Gyllenberg and Koski 2001). Using the Dirichlet integral expounded in Appendix 3.8 we get

Proposition 3.3.1 *The posterior density $\phi_{\underline{\Theta}|\mathbf{x}}(\underline{\theta}|\mathbf{x}; \underline{\alpha})$ is a Dirichlet density*

$$\mathrm{Dir}(n_1 + \alpha q_1, \ldots, n_L + \alpha q_L)$$

or

$$\phi_{\underline{\Theta}|\mathbf{x}}\left(\underline{\theta}|\mathbf{x};\underline{\alpha}\right)=\frac{\Gamma\left(n+\alpha\right)}{\prod_{i=1}^{L}\Gamma\left(\alpha q_i+n_i\right)}\prod_{i=1}^{L}\theta_i^{n_i+\alpha q_i-1}. \tag{3.6}$$

■

This property says that the posterior density is in the same family of densities as the prior. Hence the prior is called *closed under sampling* or a *conjugate prior*. A fully Bayesian solution with priors for the hyperparameters is suggested in (Good 1967). A survey of the main approaches to prior selection is found in (Berger 1985, Chapter 3)

Mean Posterior Estimate

One useful property of the Dirichlet density is that we can compute explicitly the expectation of any θ_i with respect to the posterior density. In fact this expectation is by (8.9) in Appendix 3.8 and (3.6)

$$\widehat{\theta}_i=\int_{S_L}\theta_i\phi\left(\theta_1,\ldots,\theta_L|\mathbf{x};\underline{\alpha}\right)d\theta_1\ldots d\theta_L=\frac{n_i+\alpha q_i}{n+\alpha}. \tag{3.7}$$

This is also Exercise 2 below. This result can be seen as a *regularization* adding pseudo-counts αq_i to the vector of observed counts $\underline{n}$ and then normalising so that $\sum_{i=1}^{L}\widehat{\theta}_i=1$. We refer to (Berg and von Hippel 1987) for an analysis of these formulae in terms of the concepts of physics and molecular biology. If we have $n=0$ the estimate is simply q_i.

Maximum Likelihood

The maximum likelihood estimate of $\underline{\theta}$ (a finite table of probabilities) is given by

$$\widehat{\underline{\theta}}_{\mathrm{ML}}=\arg\max_{\underline{\theta}\in S_L}P\left(\mathbf{x}\mid\underline{\theta}\right)=\arg\max_{\underline{\theta}\in S_L}\theta_1^{n_1}\cdot\theta_2^{n_2}\cdots\theta_L^{n_L}.$$

The presence of S_L imposes a constrained problem of maximization. We take the natural logarithm of $P\left(\mathbf{x}\mid\underline{\theta}\right)$, which gives us the *log likelihood function*

$$l\left(\theta_1,\theta_2,\ldots,\theta_L\right)=\log P\left(\mathbf{x}\mid\underline{\theta}\right)$$

We may equivalently seek the maximum of $l\left(\theta_1,\theta_2,\ldots,\theta_L\right)$. Since the constraint $\theta_1+\theta_2+\ldots+\theta_L=1$ must be met, we consider the new auxiliary function in $L-1$ free variables

$$\widetilde{l}\left(\theta_1,\theta_2,\ldots,\theta_{L-1}\right)=l\left(\theta_1,\theta_2,\ldots,1-\left(\theta_1+\theta_2+\ldots+\theta_{L-1}\right)\right).$$

This gives

$$\widetilde{l}(\theta_1, \theta_2, \ldots, \theta_{L-1}) = n_1 \cdot \log \theta_1 + \ldots + n_L \cdot \log(1 - (\theta_1 + \theta_2 + \ldots + \theta_{L-1})).$$

Vi differentiate partially $\widetilde{l}(\theta_1, \theta_2, \ldots, \theta_{L-1})$ with respect to $\theta_1, \theta_2, \ldots, \theta_{L-1}$ and set the partial derivatives equal to zero. This gives us the system of equations

$$\frac{\partial}{\partial \theta_1} \widetilde{l}(\theta_1, \theta_2, \ldots, \theta_{L-1}) = \frac{n_1}{\theta_1} - \frac{n_L}{1 - (\theta_1 + \theta_2 + \ldots + \theta_{L-1})} = 0,$$

$$\vdots$$

$$\frac{\partial}{\partial \theta_{L-1}} \widetilde{l}(\theta_1, \theta_2, \ldots, \theta_{L-1}) = \frac{n_{L-1}}{\theta_{L-1}} - \frac{n_L}{1 - (\theta_1 + \theta_2 + \ldots + \theta_{L-1})} = 0.$$

This leads to the equalities

$$\frac{n_1}{\theta_1} = \frac{n_2}{\theta_2} = \ldots = \frac{n_L}{1 - (\theta_1 + \theta_2 + \ldots + \theta_{L-1})}.$$

Let us denote the common value of these ratios as λ so that

$$\theta_1 = \frac{n_1}{\lambda}, \theta_2 = \frac{n_2}{\lambda}, \ldots, \theta_L = \frac{n_L}{\lambda}.$$

We determine λ from the constraint $\theta_1 + \theta_2 + \ldots + \theta_L = 1$, which gives

$$1 = \theta_1 + \theta_2 + \ldots + \theta_L = \frac{n_1}{\lambda} + \frac{n_2}{\lambda} + \ldots + \frac{n_L}{\lambda}$$

or

$$\lambda = n_1 + n_2 + \ldots + n_L = n.$$

Hence we have obtained the solution to $\nabla \widetilde{l}(\theta_1, \theta_2, \ldots, \theta_{L-1}) = 0$ written in a componentwise form as

$$\widehat{\theta}_i = \frac{n_i}{n}, i = 1, \ldots, L.$$

Strictly taken we have yet to prove that this yields a maximum. For this we could check the matrix of second order partial derivatives of $\widetilde{l}$ (Khuri 1993 p. 283). There is a more instructive way to prove that the estimate found above actually gives the maximum. In fact the proof of the next proposition shows that it is not even necessary to differentiate to prove that we have found the maximum likelihood estimate.

Proposition 3.3.2 *The maximum likelihood estimate* $\widehat{\theta}_{\text{ML}}$ *of* $\underline{\theta}$ *is*

$$\underline{\widehat{\theta}}_{\text{ML}} = \left(\frac{n_1}{n}, \frac{n_2}{n}, \ldots, \frac{n_L}{n}\right).$$

Proof: Clearly the candidate solution $\underline{\widehat{\theta}}_{\text{ML}}$ belongs to S_L and is thus admissible. Since $P(\mathbf{x} \mid \underline{\theta}) = \prod_{i=1}^{L} \theta_i^{n_i}$, the following identity is evident

$$H\left(\underline{\widehat{\theta}}_{\text{ML}}\right) = -\frac{1}{n} \log P\left(\mathbf{x} \mid \underline{\widehat{\theta}}_{\text{ML}}\right), \tag{3.8}$$

where

$$H\left(\underline{\widehat{\theta}}_{\text{ML}}\right) = -\sum_{i=1}^{L} \widehat{\theta}_i \log \widehat{\theta}_i \tag{3.9}$$

is the (empirical) Shannon entropy in natural logarithms.

Next we take an arbitrary $\underline{\theta}$ in S_L. Then we have in view of (3.9) for an arbitrary $\underline{\theta}$ in S_L and if $D\left(\underline{\widehat{\theta}}_{\text{ML}} | \underline{\theta}\right)$ is the Kullback distance another evident identity

$$P(\mathbf{x} \mid \underline{\theta}) = \prod_{i=1}^{L} \theta_i^{n_i} = e^{-n\left(D\left(\underline{\widehat{\theta}}_{\text{ML}} | \underline{\theta}\right) + H\left(\underline{\widehat{\theta}}_{\text{ML}}\right)\right)}. \tag{3.10}$$

Thus from (3.8) and (3.10)

$$\begin{aligned} \frac{P\left(\mathbf{x} \mid \underline{\widehat{\theta}}_{\text{ML}}\right)}{P(\mathbf{x} \mid \underline{\theta})} &= e^{-nH\left(\underline{\widehat{\theta}}_{\text{ML}}\right)} \cdot e^{n\left(D\left(\underline{\widehat{\theta}}_{\text{ML}} | \underline{\theta}\right) + nH\left(\underline{\widehat{\theta}}_{\text{ML}}\right)\right)} \\ &= e^{nD\left(\underline{\widehat{\theta}}_{\text{ML}} | \underline{\theta}\right)} \geq 1, \end{aligned} \tag{3.11}$$

where the last inequality follows because the Kullback distance has been proved to be nonnegative. Equality holds if and only if $\underline{\widehat{\theta}}_{\text{ML}} = \underline{\theta}$. Thus

$$P\left(\mathbf{x} \mid \underline{\widehat{\theta}}_{\text{ML}}\right) \geq P(\mathbf{x} \mid \underline{\theta})$$

for every $\underline{\theta}$ in S_L and the assertion is proved. ■

Here we may note that another way of referring to α in (3.7) is to talk about the *flattening constant* (Bender 1996, pp. 554 - 555). The flattening constant determines a linear interpolation between the maximum likelihood estimate $\frac{n_i}{n}$ and the prior estimate q_i. Hence α has the interpretation as the degree of confidence we distribute between the data and the prior.

Lod Score and Database Search

The profile or weight matrix model for sequence families was first introduced in Section 1.2.10 in Chapter 1. Let us this time consider $\mathcal{M}_0$ and $\mathcal{M}_1$, two weight matrix models for two different *families of sequences*. The sequences under consideration are constrained to have equal length (n) with no gaps. We may, e.g., think of $\mathcal{M}_0$ and $\mathcal{M}_1$ as stemming from substrings of longer sequences. (By a *substring* of $\mathbf{x} = x_{l_1}x_{l_2}\ldots x_{l_n}$ starting at position r and ending at position s we mean the contiguous symbols $x_{l_r}\ldots x_{l_s}$ in $\mathbf{x}$. We may also call $\mathbf{x}$ the *superstring* of $x_{l_r}\ldots x_{l_s}$.) The substring focused on may be seen as a *pattern* which to some degree characterizes the family.

The weight matrices $\mathcal{M}_0$ and $\mathcal{M}_1$ have as entries the probabilities $q_i\left(x_j\right)$ and $p_i\left(x_j\right)$, respectively, that a string in the family has the symbol x_j in position i, $i = 1,\ldots,n$. Since the symbols in different positions are assumed independent, the procedure of maximum likelihood estimation given above can evidently be extended to estimate $\left(p_i\left(x_j\right)\right)_{j=1}^{L}$ and $\left(q_i\left(x_j\right)\right)_{j=1}^{L}$ for $i = 1,\ldots,n$ as the appropriate relative frequencies based on the sequences held in the databases for the respective family, cf. Exercise 15 at the end of this chapter.

Then under the assumptions of the weight matrix model the '*lod score*' (to borrow a phrase from genetics, (Lange 1997)) for a string $\mathbf{x} = x_{l_1}x_{l_2}\ldots x_{l_n}$ is defined as the logarithm of the ratio of likelihoods by

$$\log\frac{P\left(\mathbf{x}|\mathcal{M}_1\right)}{P\left(\mathbf{x}|\mathcal{M}_0\right)} = \sum_{j=1}^{L}\sum_{i=1}^{n} I_{x_j}\left(x_{l_i}\right)\cdot\log\frac{p_i\left(x_j\right)}{q_i\left(x_j\right)}, \tag{3.12}$$

using (2.25) in Chapter 1. The indicator $I_{x_j}\left(x_{l_i}\right)$ is 0 if $x_j \neq x_{l_i}$ i.e. if the symbol x_j does not appear in position i in the string $\mathbf{x}$ and is equal to 1 otherwise. Lod score compares the plausibility of the model $\mathcal{M}_1$ to that of the model $\mathcal{M}_0$ as an explanation of $\mathbf{x}$. Lod score can be used as a criterion of deciding to which of the underlying sequence families $\mathbf{x}$ should be assigned, so that if

$$\log\frac{P\left(\mathbf{x}|\mathcal{M}_1\right)}{P\left(\mathbf{x}|\mathcal{M}_0\right)} > \log c$$

for some threshold $c > 0$, then $\mathcal{M}_1$ is the family chosen for $\mathbf{x}$, and vice versa.

The lod score is regarded as ∞, if there is one $q_i\left(x_j\right) = 0$ with $I_{x_j}\left(x_{l_i}\right) \neq 0$. This removes all plausibility from $\mathcal{M}_0$ as a model for $\mathbf{x}$. This is not reasonable, unless we believe that there will never in the future be x_j in position i under the model $\mathcal{M}_0$. Hence regularization, in the sense discussed above, making every $q_i(x_j)$ and $p_i(x_j)$ positive is a way to compare the models $\mathcal{M}_0$ and $\mathcal{M}_1$ in a less preconceived manner.

(Altschul 1991, Altschul et al. 1994) have studied lod score in *searching of biological databases.* A sequence $\mathbf{x}$, used for searching, is a substring of length n of a much longer superstring, say $\widehat{\mathbf{x}}$. The weight matrix model $\mathcal{M}_1$ is based on the statistics of substrings (patterns) of length n of strings assigned to a family. The columns in the weight matrix $\mathcal{M}_0$ are all equal to a single distribution $(q(x_j))_{J=1}^{L}$, which is obtained as the overall probability that the symbol should appear anywhere in the database holding the strings in the family.

The lod score in (3.12) is computed for all or many substrings of $\widehat{\mathbf{x}}$. The superstring $\widehat{\mathbf{x}}$ is taken as a member of the family $\mathcal{M}_1$ if the lodscore for some substring has a high value, i.e., matches the pattern expressed in $\mathcal{M}_1$ better the random statistics in $\mathcal{M}_0$. The ensuing problem of statistical significance of the maximal lodscore value is solved in (Karlin and Altschul 1990) for large n.

3.3.3 General Summary

A formal Bayesian modelling articulates the information in a sequence with evidence other than that of the sequence. It is thought that there is always such evidence or that there is no such thing as the 'right analysis' if there is none. The evidence is assessed by judgement and is expressed in probability theory terms:

(1) a probability distribution which specifies the probability of any sequence conditioned by certain parameters;

(2) a prior expresses uncertainty about the parameters.

When (1) is combined with the training sequence we obtain the *likelihood function* of the sequence. The likelihood function is combined with (2) via Bayes' rule to produce a *posterior distribution* for the parameters of the model and this is the output of the formal Bayesian analysis. This is also argued by Liu and Lawrence (1999) as the procedure for analysis of biological sequences.

3.4 Comparison of Model Families

3.4.1 Bayes Factor

Suppose now that we wish to compare two different model families proposed for a given sequence $\mathbf{x}$. Under the model family M_i the sequence is related to the parameters $\underline{\theta}_i$ by a distribution $P_i(\mathbf{x} \mid \underline{\theta}_i)$ and the prior densities for the parameters are $\phi_i(\underline{\theta}_i)$ for $i = 1, 2$. Then we can compare the two families by computing the ratio of the posterior probabilities $P(M_i \mid \mathbf{x})$ of the two models.

For comparison of the two model families we do not want to compare individual models within the respective families. To avoid this we do **model averaging** by

$$q_i(\mathbf{x}) = \int P_i(\mathbf{x} \mid \underline{\theta}_i) \cdot \phi_i(\underline{\theta}_i)\, d\underline{\theta}_i, \qquad i = 1, 2. \tag{4.1}$$

The quantity $q_i(\mathbf{x})$ is sometimes referred to as the *evidence* for M_i. Then Bayes' rule gives the ratio (*posterior odds*)

$$\frac{P(M_1 \mid \mathbf{x})}{P(M_2 \mid \mathbf{x})} = \frac{P(M_1)}{P(M_2)} \cdot \frac{q_1(\mathbf{x})}{q_2(\mathbf{x})}, \tag{4.2}$$

where $P(M_i)$ is the prior probability (or belief) in model M_i for $i = 1, 2$. The ratio of the evidences

$$B(\mathbf{x}) = \frac{q_1(\mathbf{x})}{q_2(\mathbf{x})} \tag{4.3}$$

is known as the *Bayes factor* (Jeffreys 1998) and can be seen representing the weight of evidence in the sequence $\mathbf{x}$ in favour of model family M_1 against model family M_2. If the posterior odds in favour of model family M_1 will be greater than the prior odds $P(M_1)/P(M_2)$, then model family M_1 fits ('explains') the sequence better than model family M_2. Examples of model family comparison will be given in the sequel. The posterior odds $\frac{P(M_1|\mathbf{x})}{P(M_2|\mathbf{x})}$ is a criterion for *learning a model family from data.* An extensive review of Bayes factors is (Kass and Raftery 1994).

3.4.2 Inductive Learning, Updates

Let $\mathbf{x}^n = x_{j_1}x_{j_2}\ldots x_{j_n}$ and $P(M \mid \mathbf{x}^n)$ be computed as above or by some other way. Suppose the symbols are obtained sequentially. If we compute

$P(M \mid \mathbf{x}^{n+1})$ after $x_{j_{n+1}}$ has been acquired, we have by Bayes' rule (3.2) and the chain rule (2.16), both from in Chapter 1,

$$\begin{aligned} P\left(M \mid \mathbf{x}^{n+1}\right) &= \frac{P\left(x_{j_{n+1}} \mid \mathbf{x}^{n}, M\right) \cdot P\left(\mathbf{x}^{n} \mid M\right) P(M)}{P\left(\mathbf{x}^{n+1}\right)} \\ &= P\left(x_{j_{n+1}} \mid \mathbf{x}^{n}, M\right) \cdot \frac{P\left(\mathbf{x}^{n} \mid M\right) P(M)}{P\left(\mathbf{x}^{n+1}\right)} \\ &= P\left(x_{j_{n+1}} \mid \mathbf{x}^{n}, M\right) \cdot \frac{P\left(\mathbf{x}^{n} \mid M\right) P(M)}{P\left(x_{j_{n+1}} \mid \mathbf{x}^{n}\right) P\left(\mathbf{x}^{n}\right)} \\ &= \frac{P\left(x_{j_{n+1}} \mid \mathbf{x}^{n}, M\right)}{P\left(x_{j_{n+1}} \mid \mathbf{x}^{n}\right)} \cdot P\left(M \mid \mathbf{x}^{n}\right). \end{aligned} \tag{4.4}$$

Hence in the update $P(M \mid \mathbf{x}^n)$ becomes the new prior. The Bayesian probability has thus a capacity of *inductive learning*, i.e., of rationally changing of the degrees of personal belief to correspond to data or a changed state of knowledge. To paraphrase Cheeseman (1988), the fundamental viewpoint of Bayesian probability is that *knowledge is an update process.*

Iterating in the given manner we obtain

$$P\left(M \mid \mathbf{x}^{n+1}\right) \propto \prod_{l=0}^{n} P\left(x_{j_{l+1}} \mid \mathbf{x}^{l}, M\right)$$

Here $P\left(x_{j_{l+1}} \mid \mathbf{x}^{l}, M\right)$ is the posterior predictive distribution

$$P\left(x_{j_{l+1}} \mid \mathbf{x}^{l}, M\right) = \int P\left(x_{j_{l+1}} \mid \underline{\theta}\right) \cdot \phi\left(\underline{\theta} \mid \mathbf{x}^{l}, M\right) d\underline{\theta}, \tag{4.5}$$

where $\phi\left(\underline{\theta} \mid \mathbf{x}^{l}, M\right)$ is the posterior density for the parameters in the model M. Hence the process of learning on the basis of the string $\mathbf{x}^{n+1}$ can be seen as evaluating sequentially the posterior predictive distributions (see an exercise below). In this sense we do not necessarily insist that the model family is the "true" theory of the real world domain involved. In fact we apply a method of generating predictive distributions for the next symbol to come up.

3.5 Some Asymptotics for Evidence

We reconsider the model family and the generic string $\mathbf{x}$ in subsection 3.3.2 above. The purpose is to evaluate in an asymptotic manner the evidence in

(4.1) now written as

$$q(\mathbf{x}) = \int_{S_L} P(\mathbf{x} \mid \underline{\theta}) \cdot \phi(\underline{\theta})\, d\underline{\theta},$$

where ϕ is any prior probability density which is everywhere positive on S_L. We recall the maximum likelihood estimate

$$\widehat{\underline{\theta}}_{\mathrm{ML}} = \left(\frac{n_1}{n}, \frac{n_2}{n}, \ldots, \frac{n_L}{n}\right).$$

and

$$\widehat{\theta}_i = \frac{n_i}{n}, \qquad i = 1, \ldots, L.$$

We define the Kullback 'sphere' with radius δ as

$$C\left(\widehat{\underline{\theta}}_{\mathrm{ML}}, \delta\right) = \left\{\underline{\theta} \in S_L \middle| D\left(\widehat{\underline{\theta}}_{\mathrm{ML}} \middle| \underline{\theta}\right) < \delta\right\} \tag{5.1}$$

for any $\delta > 0$. In addition the prior probability of any (measurable) $V \subseteq S_L$ is denoted by

$$P_\phi(V) = \int_V \phi(\underline{\theta})\, d\underline{\theta}.$$

Proposition 3.5.1 *If*

$$q(\mathbf{x}) = \int_{S_L} P(\mathbf{x} \mid \underline{\theta}) \cdot \phi(\underline{\theta})\, d\underline{\theta}, \tag{5.2}$$

where ϕ is any prior probability density which is everywhere positive on S_L, then

$$P_\phi\left(C\left(\widehat{\underline{\theta}}_{\mathrm{ML}}, \delta\right)\right) > 0, \tag{5.3}$$

$$-\frac{1}{n}\log q(\mathbf{x}) \leq H\left(\widehat{\underline{\theta}}_{\mathrm{ML}}\right) + \delta - \frac{1}{n}\log P_\phi\left(C\left(\widehat{\underline{\theta}}_{\mathrm{ML}}, \delta\right)\right) \tag{5.4}$$

and

$$\begin{aligned} -\frac{1}{n}\log q(\mathbf{x}) &\geq H\left(\widehat{\underline{\theta}}_{\mathrm{ML}}\right) \\ &- \frac{1}{n}\log\left(e^{-n\delta} + (1 - e^{-n\delta})P_\phi\left(C\left(\widehat{\underline{\theta}}_{\mathrm{ML}}, \delta\right)\right)\right). \end{aligned} \tag{5.5}$$

Proof: From (3.9) and (3.10) we obtain, since $H\left(\underline{\widehat{\theta}}_{\mathrm{ML}}\right)$ does not depend on $\underline{\theta}$, the bound

$$q(\mathbf{x}) \geq e^{-n\left(\delta + H\left(\underline{\widehat{\theta}}_{\mathrm{ML}}\right)\right)} \int_{\underline{\theta} \in C\left(\underline{\widehat{\theta}}_{\mathrm{ML}}, \delta\right)} \phi(\underline{\theta})\, d\underline{\theta}, \tag{5.6}$$

by definition of $C\left(\underline{\widehat{\theta}}_{\mathrm{ML}}, \delta\right)$ and since the integral over the complement of $C\left(\underline{\widehat{\theta}}_{\mathrm{ML}}, \delta\right)$ in S_L is nonnegative. $D\left(\underline{\widehat{\theta}}_{\mathrm{ML}} | \underline{\theta}\right)$ is by construction a real valued continuous function of $\underline{\theta} \in S_L$. Thus the prior density $\phi(\underline{\theta})$ assigns a positive measure to any (Euclidean) neighborhood of $\underline{\widehat{\theta}}_{\mathrm{ML}}$ containing $C\left(\underline{\widehat{\theta}}_{\mathrm{ML}}, \delta\right)$ and hence to $C\left(\underline{\widehat{\theta}}_{\mathrm{ML}}, \delta\right)$ itself. The assertion in (5.4) is thus there. Since $D\left(\underline{\widehat{\theta}}_{\mathrm{ML}} | \underline{\theta}\right) \geq 0$,

$$\begin{aligned} q(\mathbf{x}) &\leq e^{-nH\left(\underline{\widehat{\theta}}_{\mathrm{ML}}\right)} \int_{\underline{\theta} \in C\left(\underline{\widehat{\theta}}_{\mathrm{ML}}, \delta\right)} \phi(\underline{\theta})\, d\underline{\theta} \\ &+ e^{-nH\left(\underline{\widehat{\theta}}_{\mathrm{ML}}\right)} \cdot e^{-n\delta} \int_{\underline{\theta} \in C\left(\underline{\widehat{\theta}}_{\mathrm{ML}}, \delta\right)^c} \phi(\underline{\theta})\, d\underline{\theta}, \end{aligned}$$

where $C(\underline{\widehat{\theta}}_{\mathrm{ML}}, \delta)^c$ is the set theoretic complement of the neighborhood. This yields

$$q(\mathbf{x}) \leq e^{-nH\left(\underline{\widehat{\theta}}_{\mathrm{ML}}\right)} \left(e^{-n\delta} + (1 - e^{-n\delta}) P_\phi\left(C\left(\underline{\widehat{\theta}}_{\mathrm{ML}}, \delta\right)\right)\right),$$

which gives the final claim as the asserted. ∎

From the above it follows, since the Kullback 'sphere' has a constant radius with regard to n, that

$$\left| -\frac{1}{n} \log q(\mathbf{x}) - H\left(\underline{\widehat{\theta}}_{\mathrm{ML}}\right) \right| \to 0, \tag{5.7}$$

as $n \to \infty$. This demonstrates that the influence (per symbol) of the prior beliefs in the evidence vanishes in the average as the length of the string increases.

In addition we may express the result in the proposition as

$$-\frac{1}{n} \log q(\mathbf{x}) \approx H\left(\underline{\widehat{\theta}}_{\mathrm{ML}}\right) - \frac{1}{n} \log P_\phi\left(C\left(\underline{\widehat{\theta}}_{\mathrm{ML}}, \delta\right)\right).$$

or in view of (3.8) and (3.9)

$$-\frac{1}{n}\log q(\mathbf{x}) \approx -\frac{1}{n}\log P\left(\mathbf{x} \mid \widehat{\underline{\theta}}_{\mathrm{ML}}\right) - \frac{1}{n}\log P_{\phi}\left(C\left(\widehat{\underline{\theta}}_{\mathrm{ML}}, \delta\right)\right).$$

The evidence (for a model family) in $\mathbf{x}$ becomes smaller if $\widehat{\underline{\theta}}_{\mathrm{ML}}$ hits a Kullback sphere with a small prior probability. There are much sharper and quite general expansions of $-\frac{1}{n}\log q(\mathbf{x})$ (Clarke and Barron 1990).

3.6 Evidence and Bayesian Codelengths

Next we demonstrate a source coding interpretation of $-\log q(\mathbf{x})$ in (5.2). This is a *Bayesian codelength* as an instance of the Shannon–Fano code lenghts mentioned in Chapter 2. The quantity $-\log q(\mathbf{x})$ is also known as *stochastic complexity* of $\mathbf{x}$ with regard to the model family M (Rissanen 1997).

Let now $l(\mathbf{x})$ denote an arbitrary real valued function of any string $\mathbf{x}$ with n symbols picked from the discrete alphabet $\mathcal{X}$. We assume that $l(\mathbf{x})$ satisfies the Kraft inequality:

$$\sum_{\mathbf{x}} 2^{-l(\mathbf{x})} \leq 1. \tag{6.1}$$

Owing to this form of Kraft's inequality the *logarithms are to the base* 2 in the rest of this chapter.

Lemma 3.6.1 *Let* $\mathbf{P}$ *be any probability distribution and let* $\mathbf{p}(\mathbf{x})$ *be the probability distribution induced on the space of n-strings* $\mathbf{x}$ *by a random variable* X

$$\mathbf{P}(X = \mathbf{x}) = \mathbf{p}(\mathbf{x}).$$

Then for any positive c there holds

$$\mathbf{P}(l(X) \leq l_{\mathbf{p}}(X) - c) \leq 2^{-c} \tag{6.2}$$

where $l(\cdot)$ *satisfies* (6.1) *and*

$$l_{\mathbf{p}}(X) = -\log \mathbf{p}(X)$$

Proof: Let c be a given positive number. Take, supposing this is possible, $\mathbf{x}$ such that

$$l(\mathbf{x}) \leq l_{\mathbf{p}}(\mathbf{x}) - c,$$

which says that $\mathbf{x}$ is such that

$$\mathbf{p}(\mathbf{x}) \leq 2^{-c} \cdot 2^{-l(\mathbf{x})}.$$

Hence by (6.1)

$$\sum_{\mathbf{x}: l(\mathbf{x}) \leq l_{\mathbf{p}}(\mathbf{x}) - c} \mathbf{p}(\mathbf{x}) \leq 2^{-c},$$

as was to be proved. The result is trivially true with a given c if there is no string to sum over in the left hand side. ■

Let us set for any $\epsilon > 0$ and any $l(\cdot)$ that satisfies (6.1)

$$A := \{\mathbf{x} | l(\mathbf{x}) - (-\log q(\mathbf{x})) \leq 2 \cdot \log(\epsilon)\}, \tag{6.3}$$

where

$$q(\mathbf{x}) = \int_{S_L} P(\mathbf{x} \mid \underline{\theta}) \cdot \phi(\underline{\theta}) \, d\underline{\theta}.$$

Furthermore we set

$$P_{\underline{\theta}}(A) = \sum_{\mathbf{x} \in A} P(\mathbf{x} \mid \underline{\theta}). \tag{6.4}$$

In fact the next proposition, which is due to (Dawid 1992), holds for any family of models for n-strings of discrete symbols, not just for the conditional independence models parametrized by tables of probabilities in S_L. Hence we integrate over S_L in the proof below merely for reasons of clarity and concreteness. We use the notation

$$P_{\phi}(V) = \int_V \phi(\underline{\theta}) \, d\underline{\theta}.$$

for any (measurable) $V \subseteq S_L$.

Proposition 3.6.2 *Under the preceding assumptions we have for any* $\epsilon > 0$

$$P_{\phi}(\underline{\theta} | P_{\underline{\theta}}(A) \leq \epsilon) \geq 1 - \epsilon, \tag{6.5}$$

where A *is given in (6.3).*

Proof: Clearly $q(\mathbf{x})$ is another probability distribution on the space of n-strings. Hence we let $q(\mathbf{x})$ take the place of $\mathbf{p}(\mathbf{x})$ in the preceding lemma and let $\mathbf{G}(X = \mathbf{x})$ be the underlying probability measure (on some probability space) replacing $\mathbf{P}$ in there. Then the preceding lemma says taking $c = -2\log\epsilon$ that

$$\epsilon^2 \geq \mathbf{G}(A). \tag{6.6}$$

But here

$$\begin{aligned} \mathbf{G}(A) &= \sum_{\mathbf{x}\in A} \int_{S_L} P(\mathbf{x} \mid \underline{\theta}) \cdot \phi(\underline{\theta})\, d\underline{\theta} \qquad (6.7) \\ &= \int_{S_L} \sum_{\mathbf{x}\in A} P(\mathbf{x} \mid \underline{\theta}) \cdot \phi(\underline{\theta})\, d\underline{\theta}, \end{aligned}$$

since the summation is finite, and by (6.4)

$$\int_{S_L} \sum_{\mathbf{x}\in A} P(\mathbf{x} \mid \underline{\theta}) \cdot \phi(\underline{\theta})\, d\underline{\theta} = \int_{S_L} P_{\underline{\theta}}(A) \cdot \phi(\underline{\theta})\, d\underline{\theta}.$$

The right hand side is greater than or equal to

$$\int_{\underline{\theta}: P_{\underline{\theta}}(A) \geq \epsilon} P_{\underline{\theta}}(A) \cdot \phi(\underline{\theta})\, d\underline{\theta} \geq \epsilon \cdot P_{\phi}(\underline{\theta} | P_{\underline{\theta}}(A) > \epsilon).$$

Thus in view of (6.6), on dividing by ϵ we obtain

$$\epsilon \geq P_{\phi}(\underline{\theta} | P_{\underline{\theta}}(A) > \epsilon).$$

Then

$$\epsilon \geq 1 - P_{\phi}(\underline{\theta} | P_{\underline{\theta}}(A) \leq \epsilon),$$

which proves the assertion in (6.5), as claimed. ∎

The interpretation of (6.5) is that for the most probable values of $\underline{\theta}$ the $P_{\underline{\theta}}$ -chance for the codelength $l(\mathbf{x})$ to be much shorter than $-\log q(\mathbf{x})$ is low. Or to make the same point in other words, the condition in the event (6.3) in the form

$$l(\mathbf{x}) \leq -\log q(\mathbf{x}) + 2 \cdot \log(\epsilon),$$

tells us that if we subtract from $-\log q(\mathbf{x})$ an arbitrarily large positive number we do not have high chance, as measured by $P_{\underline{\theta}}$, to get a number that exceeds $l(\mathbf{x})$. Some applications to microbiology and further results on stochastic complexity are found in (Gyllenberg et al. 1997, Gyllenberg and Koski 2001).

3.7 Exercises

1. **Jeffreys' prior**
   In Jeffreys (1998) the following prior density for a real parameter θ is proposed
   $$\phi(\theta) = \frac{\sqrt{J(\theta)}}{\int \sqrt{J(w)}dw},$$
   where $J(\theta)$ is the Fisher information (see Chapter 2). This presupposes that the integral exists. Determine Jeffreys' prior for $\mathrm{Be}(\theta)$.

2. If $p(\underline{\theta}|\mathbf{x})$ denotes a generic posterior density for a parameter $\underline{\theta} = (\theta_1, \ldots, \theta_J)$, then the *mean posterior estimate* for the real valued parameter θ_j is denoted by $\widehat{\theta}_{j;\mathrm{MP}}$ and given by the integral
   $$\widehat{\theta}_{j;\mathrm{MP}} = \int_{S_L} \theta_j p(\underline{\theta}|\mathbf{x})\, d\underline{\theta}.$$
   Let $p(\underline{\theta}|\mathbf{x}) = \mathrm{Dir}(n_1 + \alpha q_1, \ldots, n_L + \alpha q_L)$. Find $\widehat{\theta}_{j;\mathrm{MP}}$.

3. Consider the thumb tack experiment in the formulation by D. Heckerman and the conditional independence model for the experiment. What is the value of $P(X_{n+1} = \mathrm{head}|\mathbf{x})$?

4. Let $p(\underline{\theta}|\mathbf{x}) = \mathrm{Dir}(n_1 + \alpha q_1, \ldots, n_L + \alpha q_L)$ be the posterior density as obtained in above. What is the maximum posterior estimate $\widehat{\theta}_{\mathrm{MAP}}$ of θ in this case ? Compare $\widehat{\theta}_{\mathrm{MAP}}$ to $\widehat{\theta}_{\mathrm{MP}}$ from Exercise 2 above.

5. Let the likelihood for $\underline{\theta} = (\theta_1, \ldots, \theta_L)$ with the data $\mathbf{x}$ be given by
   $$p(\mathbf{x}|\underline{\theta}) = \prod_{j=1}^{L} \theta_j^{n_j},$$
   where n_j is the number of times the symbol x_j (in some finite alphabet with J symbols) is present in $\mathbf{x}$ and $\sum_{j=1}^{L} \theta_j = 1$. As a prior on $\underline{\theta}$ we take a finite *Dirichlet mixture* defined by
   $$p(\underline{\theta}) = \sum_{i=1}^{k} \lambda_i \mathrm{Dir}\left(\alpha^{(i)} q_1^{(i)}, \ldots, \alpha^{(i)} q_L^{(i)}\right),$$

where $\lambda_i \geq 0$, $\sum_{i=1}^{k} \lambda_i = 1$ (the mixture distribution), $\alpha^{(i)} > 0$, $q_j^{(i)} > 0$, $\sum_{i=1}^{L} q_j^{(i)} = 1$ for every i. Determine the mean posterior estimate $\widehat{\theta}_{j;\mathrm{MP}}$ for $j = 1, \ldots, L$.

6. Show that any Dirichlet mixture is closed under sampling.

7. Derive (4.2) in detail.

8. **Predictive Likelihood**
Let $X_1, X_2, \ldots, X_{m+n}$ be independent Be (θ) - distributed random variables, i.e., $\theta = P(X_i = 1 \mid \theta)$. Let

$$S = \sum_{i=1}^{n} X_i, T = \sum_{i=n+1}^{n+m} X_i.$$

(a) Compute

$$P(T = t \mid S + T = s + t).$$

(b) Take $m = 1$. Compute

$$\frac{P(X_{n+1} = 1 \mid S + X_{n+1} = s + 1)}{P(X_{n+1} = 0 \mid S + X_{n+1} = s)}$$

and compare with the result in Exercise 3 above.

9. **The Binomial Beta Distribution**
Compute for $k = 0, \ldots, n$

$$q(k) = \int_0^1 P(k \mid \theta) \cdot \phi(\theta) \, d\theta, \tag{7.1}$$

if $P(k \mid \theta)$ is Bin(n, θ) and $\phi(\theta)$ is $\mathcal{B}\mathrm{e}(\theta; \alpha_1, \alpha_2)$, see Appendix 3.8.3.

10. **Updates of Posterior Densities**
The sequential update of the posterior probability of a model M was above given as

$$P\left(M \mid \mathbf{x}^{n+1}\right) = \frac{P\left(x_{j_{n+1}} \mid \mathbf{x}^n, M\right)}{P\left(x_{j_{n+1}} \mid \mathbf{x}^n\right)} \cdot P(M \mid \mathbf{x}^n).$$

Use now similar computations to find the appropriate formula for update of the posterior density

$$\Psi\left(\theta|X_1 = x_1, \ldots, X_{n+1} = x_{n+1}\right)$$

for any multinomial process model for $X_1, \ldots, X_{n+1}$ given $\Theta = \theta$ and any prior density $\Psi(\theta)$.

11. Consider the following argument (as quoted by Neapolitan (1990)):
"It is the fatal flaw of the subjectivistic theory that it pretends that one person's probability is just as good as the other's and that no objective consideration can choose between them, when it is perfectly plain that one will lose his shirt because he is wrong about the probability. We ask more of a mathematical system than mere (coherence): we ask also that it truly describes the world."
Is this is a valid criticism of Bayesian probability?

12. Compute the evidence

$$q\left(\mathbf{x}\right) = \int_{S_L} p\left(\mathbf{x} \mid \underline{\theta}\right) \cdot \phi\left(\underline{\theta}\right) d\underline{\theta}$$

when

$$p\left(\mathbf{x}|\underline{\theta}\right) = \prod_{j=1}^{L} \theta_j^{n_j},$$

where n_j is the number of times the symbol x_j (in some finite alphabet with L symbols) is present in $\mathbf{x}$, $\sum_{j=1}^{L} \theta_j = 1$ and the prior $\phi\left(\underline{\theta}\right)$ is the symmetric Dirichlet density with $\alpha_1 = \alpha_2 = \ldots = \alpha_L = 1$. In (Boulton and Wallace 1969) $-\log q\left(\mathbf{x}\right)$ is called the information content of a population with a finite number of members. This is equal to the optimum length of a message which fully describes the members of the population.

13. **Local Compositional Complexity**

    (a) This is a continuation of Exercise 12 above. Prove that

$$-\frac{1}{n} \log q\left(\mathbf{x}\right) \geq \frac{1}{n} \log \frac{n!}{\prod_{i=1}^{L} n_i!}. \tag{7.2}$$

    Hint: Use first the lower bound in Proposition 3.5.1 and then (3.8).

(b) In (Wootton 1994) the quantity

$$\frac{1}{n} \log \frac{n!}{\prod_{i=1}^{L} n_i!}$$

in the right hand side of the inequality (7.2) is introduced as a measure of *local compositional complexity* as seen in a window covering a subsequence of length n in a longer sequence. The window is moved, e.g., in steps of one symbol along the sequence or along all in a database of sequences. This measure can be used to do *segmentation* of biological sequences, i.e., for finding segments of low complexity. The vector $(n_1, \ldots, n_L)$ ordered in decreasing order as $(n_{[1]}, \ldots, n_{[L]})$ is called the *compositional state*.

Assume that the probability distribution on the underlying alphabet is $\mathrm{U}(1, L)$. Let for $k = 1, \ldots, n$

$$r_k = \text{ the number of times the integer } k \text{ occurs among } (n_i)_{i=1}^{L}.$$

The vector $(r_1, \ldots, r_n)$ is called the *size distribution*. Show that the probability P_0 of a compositional state $(n_{[1]}, \ldots, n_{[L]})$ is

$$P_0 = \frac{1}{L^n} \frac{n!}{\prod_{i=1}^{L} n_i!} \frac{L!}{\prod_{k=1}^{n} r_k!}.$$

14. **A Weak Form of Zipf's Law**

It has been observed in many problems of biology and other fields that the *size distribution*, e.g., the frequency of biological genera (plural of genus) with a specified number of species of an empirical material follows *Zipf's law*. This is a probability density also known as the *Pareto distribution*.

A weak version of Zipf's law is that the expected proportion of genera having exactly s species is proportional to

$$\propto \frac{1}{s^{1+\alpha}}, \qquad 0 < \alpha < 1.$$

Let now Y be a random variable given by

$$Y = \theta \cdot (1 - \theta)^{j-1},$$

where θ is a random variable that has the distribution $\mathcal{B}e\,(\theta; \alpha_1, \alpha_2)$, see Appendix 3.8.3. Show that if $0 < \alpha_2 < 1$, then Y, i.e. $E(Y)$, satisfies the weak version of Zipf's law with j as variable. (Hill 1970).

15. **Consistency of the Maximum Likelihood Estimate**

    (a) Show using the law of large numbers from Chapter 1 that if $\widehat{\theta}_{\text{ML}}$ is the maximum likelihood estimate (3.3) within the appropriate model family that for every $\epsilon > 0$

$$\lim_{n\to\infty} P\left(|\widehat{\theta}_{\text{ML}} - \theta| > \epsilon | \Theta = \theta\right) = 0. \tag{7.3}$$

    (b) Show using the Chebyshev's inequality from Chapter 1 that for every $\epsilon > 0$

$$\lim_{n\to\infty} \max_{\theta} P\left(|\widehat{\theta}_{\text{ML}} - \theta| > \epsilon | \Theta = \theta\right) = 0. \tag{7.4}$$

    What is the relationship between chance and probability in this result?

16. A sequence of training strings $\mathbf{x}^1, \ldots, \mathbf{x}^t$ of equal length $= n$ is used to learn a model weight matrix $\mathcal{M}$ from section 1.2.10 in Chapter 1. Find the maximum likelihood estimates of $p_i(x_j)$ of the probabilities to be employed in $\mathcal{M}$, if

$$P\left(\mathbf{x}^1, \ldots, \mathbf{x}^t | \mathcal{M}\right) = \prod_{s=1}^{t} P\left(\mathbf{x}^s | \mathcal{M}\right). \tag{7.5}$$

    Note the constraint $\sum_{j=1}^{L} p_i(x_j) = 1$ for every i.

3.8 Appendix: Dirichlet Densities

3.8.1 Euler's Gamma

The *gamma* function $\Gamma(z)$ is defined for complex numbers z, whose real part is positive, by the definite integral

$$\Gamma(z) = \int_0^{\infty} x^{z-1} e^{-x} dx. \tag{8.1}$$

A special case, obtained by the substitution $x = u^2/2$ is

$$\Gamma\left(\frac{1}{2}\right) = \sqrt{\pi}.$$

The recursion formula is

$$\Gamma(z) = (z-1)\Gamma(z-1). \tag{8.2}$$

Hence if $z = n$, where n is a positive integer, we have the factorial

$$\Gamma(n) = (n-1)! \tag{8.3}$$

3.8.2 The Dirichlet density

Let $S_L \subset R^L$ be the *simplex*

$$S_L = \left\{ (\theta_1, \ldots, \theta_L) \,|\theta_i \geq 0, \qquad i = 1, \ldots, L, \sum_{i=1}^{L} \theta_i = 1 \right\}. \tag{8.4}$$

Let for $\alpha_i > 0$

$$\phi(\theta_1, \ldots, \theta_L) = \begin{cases} \dfrac{\prod_{i=1}^{L} \theta_i^{\alpha_i - 1}}{Z}, & \text{if } \theta_1, \ldots, \theta_L \in S_L, \\ 0 & \text{otherwise.} \end{cases} \tag{8.5}$$

Here

$$\frac{1}{Z} = \frac{\Gamma\left(\sum_{i=1}^{L} \alpha_i\right)}{\prod_{i=1}^{L} \Gamma(\alpha_i)}. \tag{8.6}$$

The density $\phi(\theta_1, \ldots, \theta_L)$ is called a *Dirichlet density*. We designate it symbolically by

$$\text{Dir}(\alpha_1, \ldots, \alpha_L). \tag{8.7}$$

If $\alpha_1 = \alpha_2 = \ldots = \alpha_L = \kappa$, then we talk about a *symmetric Dirichlet density.* For the proof of the property

$$\int_{S_L} \phi(\theta_1, \ldots, \theta_L)\, d\theta_1 \ldots d\theta_L = 1 \tag{8.8}$$

we refer to (Wilks 1962). This means also that

$$\int_{S_L} \prod_{i=1}^{L} \theta_i^{\alpha_i - 1} d\theta_1 \ldots d\theta_L = \frac{\prod_{i=1}^{L} \Gamma(\alpha_i)}{\Gamma\left(\sum_{i=1}^{L} \alpha_i\right)}. \tag{8.9}$$

(Gupta and Richards 1987) is a concise compendium of knowledge about the Dirichlet distributions. Applications of Dirichlet distributions in statistical genetics are expounded in (Lange 1997).

T.J. Santner & D.E. Duffy (1989): *The Statistical Analysis of Discrete Data.* Springer -Verlag, New York, Berlin Heidelberg et al.

S.S. Wilks (1962): *Mathematical Statistics*, John Wiley and Sons, New York and London.

- Various papers

S.F. Altschul (1991): Amino Acid Substitution Matrices from an Information Theoretic Perspective. *Journal of Molecular Biology*, 219, pp. 555–565.

S.F. Altschul, M. Boguski, W. Gish and J. Wootton (1994): Issues in searching molecular sequence databases. *Nature Genetics*, 6, pp. 119–129.

O.G. Berg and P.H. von Hippel (1987): Selection of DNA Binding Sites by Regulatory Proteins. Statistical - mechanical Theory and Application to Operators and Promoters, *Journal of Molecular Biology*, 193, pp. 723–750.

J.O. Berger (1985): *Statistical Decision Theory and Bayesian Analysis. Second Edition.* Springer–Verlag, New York, Berlin et al..

D.M. Boulton and C.S. Wallace (1969): The Information Content of a Multistate Distribution. *Journal of Theoretical Biology*, 23, pp. 269–278.

B.S. Clarke and A. R. Barron (1990): Information Theoretic Asymptotics of Bayes Methods. *IEEE Transactions on Information Theory*, 36, pp. 453–471.

A.P. Dawid (1992): Prequential Analysis, Stochastic Complexity and Bayesian Inference. J.M. Bernardo, J.O. Berger, A.P. Dawid and A.F.M. Smith (ed.s) *Bayesian Statistics*, 4, Oxford University Press, pp. 109–125.

N. Goldman (1993): Statistical Tests of DNA Substitutions. *Journal of Molecular Evolution*, 36, pp. 182–198.

I.J. Good (1967): A Bayesian Significance Test for Multinomial Distributions. *Journal of the Royal Statistical Society, Ser. B*, 29, pp. 399–431.

R.D. Gupta and D.St.P. Richards (1987): Multivariate Liouville Distributions. *Journal of Multivariate Analysis*, 23, pp. 232–256.

M. Gyllenberg and T. Koski (2001): Bayesian Predictiveness, Exchangeability and Sufficientness in Bacterial Taxonomy. Submitted to *Mathematical Biosciences.*

M. Gyllenberg, T. Koski and M. Verlaan (1997): Classification of Binary Vectors by Stochastic Complexity. *Journal of Multivariate Analysis*, 62, pp. 47–72.

B.M. Hill (1970): Zipf's Law and Prior Distributions for the Composition of a Population. *Journal of the American Statistical Association*, 65, pp. 1220–1232.

S. Karlin and S.F. Altschul (1990): Methods for assessing the statistical significance of molecular sequence features by using general scoring schemes. *Proceedings of the National Academy of Sciences USA*, 87, pp. 2264–2268

R.E. Kass and A.E. Raftery (1994): Bayes Factors. *Journal of the American Statistical Association*, 90, pp. 733–795.

J. Rissanen (1997): Stochastic complexity and learning. *Journal of Computer and System Sciences* , 55, pp. 89–95.

H.V. Roberts (1965): Probabilistic Prediction. *Journal of the American Statistical Association*, 60, pp. 50–62.

- Other textbooks and monographs used here are

P. Baldi and S. Brunak (1998): *Bioinformatics. The Machine Learning Approach.* A Bradford Book. The MIT Press. Cambridge, Massachusetts; London.

E.A. Bender (1996): *Mathematical Methods in Artificial Intelligence.* IEEE Computer Society Press, Los Alamitos, California.

J.M. Bernardo and A.F.M. Smith (1994): *Bayesian Theory.* John Wiley and Sons, Chichester, New York, Brisbane, Toronto and Singapore.

C.M. Bishop (1995): *Neural Networks for Pattern Recognition*, Clarendon Press, Oxford.

R.T. Cox (1961): *The Algebra of Probable Inference.* The Johns Hopkins Press, Baltimore, Maryland.

R.F. Doolittle (ed.) (1996): *Computer Methods for Macromolecular Sequence Analysis, Methods in Enzymology Volume 266*, Academic Press, San Diego, New York 1996.

S. Halldén (1973): *Sannolikhetens logik.* CWK Glerups Bokförlag, Lund.

R.W. Hamming (1991): *The Art of Probability for Scientists and Engineers*, Addison - Wesley Publishing Company, Reading Massachusetts etc.

J.A. Hartigan (1983): *Bayes Theory.* Springer Verlag, New York.

E.T. Jaynes: *Probability Theory: The Logic of Science.* `http://omega.albany.edu:8008/JaynesBook.html`

H. Jeffreys (1998): *Theory of Probability, Third Edition.* Oxford University Press.

A.I. Khuri (1993): *Advanced Calculus with Applications in Statistics.* John Wiley and Sons, New York.

K. Lange (1997): *Mathematical and Statistical Methods for Genetic Analysis.* Springer Verlag, New York, Berlin, Heidelberg.

R. von Mises with H. Geiringer (1964): *Mathematical Theory of Probability and Statistics.* Academic Press, New York and London.

R.E. Neapolitan (1990): *Probabilistic Reasoning in Expert Systems.* John Wiley and Sons, New York.

N.J. Nilsson (1990): *The Mathematical Foundations of Learning Machines.* Morgan Kaufmann Publishers, San Mateo, California.

F. Priest and B. Austin (1993): *Modern Bacterial Taxonomy. Second edition.* Chapman and Hall, London.

3.8.3 The Beta Density

As a special case for $L = 2$ we obtain in (8.9) the *Beta integral*

$$\int_0^1 \theta^{\alpha_1 - 1}(1-\theta)^{\alpha_2 - 1} d\theta = \frac{\Gamma(\alpha_1) \cdot \Gamma(\alpha_2)}{\Gamma(\alpha_1 + \alpha_2)}. \tag{8.10}$$

Thus

$$f(\theta) = \begin{cases} \frac{\Gamma(\alpha_1 + \alpha_2)}{\Gamma(\alpha_1) \cdot \Gamma(\alpha_2)} \theta^{\alpha_1 - 1}(1-\theta)^{\alpha_2 - 1} & 0 \le \theta \le 1 \\ 0 & \text{elsewhere.} \end{cases} \tag{8.11}$$

is a probability density called the *Beta density* and denoted by

$$\mathcal{B}\mathrm{e}(\theta; \alpha_1, \alpha_2).$$

Note the difference in the heuristic notation between Beta and Bernoulli Be(p). If $\theta = (\theta_1, \ldots, \theta_L)$ is a random variable that assumes values in S_L in (8.4) and has the symmetric Dir $(\alpha, \ldots, \alpha)$ distribution, then the marginal density of any θ_i is given by

$$\theta_i \in \mathcal{B}\mathrm{e}(\theta; \alpha, (L-1)\alpha). \tag{8.12}$$

3.9 References and Further Reading:

- The modelling discussion in the introductory sections borrows heavily from

P. Cheeseman (1988): An Inquiry into Computer Understanding. *Computational Intelligence*, **4**, 58–66.

M.A. Hearst (ed.) (1997): Banter on Bayes: debating the usefulness of Bayesian approaches to solving practical problems. *IEEE Expert,* Nov./Dec., pp. 18–25.

D. Heckerman (1997): Bayesian Networks for Data Mining. *Data Mining,* pp. 81–119.

D.V. Lindley (1987): The Probability Approach to the Treatment of Uncertainty in Artificial Intelligence and Expert Systems. *Statistical Science,* 2, pp. 17–24.

J.S. Liu and C.E. Lawrence (1999): Bayesian inference on biopolymer models. *Bioinformatics,* 15, pp. 38–52.

D.V. Lindley (1970): A non-frequentist view of probability and statistics. *The Teaching of Probability & Statistics*, L. Råde editor, Almqvist & Wicksell, Uppsala, pp. 209–222.

D.V. Lindley (1978): The Bayesian Approach. *Scandinavian Journal of Statistics*, **9**, 45–58.

J.C. Wootton (1994): Non-globular Domains in Protein Sequences: Automated Segmentation Using Complexity Measures. *Computers and Chemistry*, 18, pp. 269–285.

Chapter 4

EM Algorithm

4.1 Finite Mixtures of Distributions

4.1.1 Introduction

In this chapter we shall consider a technically more complicated instance of maximum likelihood estimation, where the models constitute a family of *finite mixture of probability distributions.* A finite mixture of distributions is a degenerate special case of a *hidden Markov model.* One well known algorithm for finding the maximum likelihood estimate of the parameters of the finite mixture is known as the EM algorithm. Therefore the principles in the EM algorithm will emerge later with hidden Markov models but then under the name 'Baum–Welch learning'. It should be emphasized that Baum–Welch learning for hidden Markov models was invented and developed independently of the discoveries of EM for other families of models.

The EM algorithm has a long history in statistical genetics, some of the relevant work and references are surveyed in (Weeks and Lange 1989, Lange 1997).

Applications of the EM algorithm in bioinformatics are many. One topic is that of finding motifs (Lawrence and Reilly 1990, Cardon and Stormo 1992 and Bailey and Elkan 1995) which will be discussed in this chapter.

4.1.2 Mixtures

Let us suppose that $\{(X_1, Y_1), \ldots, (X_n, Y_n)\} = \{(X_l, Y_l)\}_{l=1}^n$ is a sequence of independent (pairs of) random variables with the same distribution and assuming values in $\mathcal{X} \times \mathcal{Y}$. For every l and $j = 1, \ldots, L$

$$\alpha_j = P(X_l = x_j),$$

and for any $y \in \mathcal{Y}$

$$p(y|\phi_j) = P(Y_l = y|X_l = x_j),$$

where ϕ_j are some (parameter) vectors in a Euclidean vector space. The functions $p(\cdot|\phi_j)$ have for all j the same known analytic form, except that the parameters may be unknown. We take θ as the *parameter* with

$$\theta = (\alpha_1, \ldots, \alpha_L; \phi_1, \ldots, \phi_L).$$

We think initially of two data sequences $\mathbf{y} = (y_1 \ldots y_n) \in \mathcal{Y}^n$ and $\mathbf{x} = (x_{j_1} \ldots x_{j_n}) \in \mathcal{X}^n$. The assumption of pairwise independence means that

$$p(\mathbf{x}, \mathbf{y}|\theta) = \prod_{l=1}^{n} P(Y_l = y_l, X_l = x_{j_l}|\theta) = \prod_{l=1}^{n} p(y_l|\phi_{j_l}) \cdot \alpha_{j_l}. \tag{1.1}$$

The learning task is the estimation of θ in a situation where the sequence $\mathbf{x}$ is *hidden* (or *missing*) and we have only access to $\mathbf{y}$. Any practically implementable estimate must therefore rely on processing of only $\mathbf{y}$.

Using the rules for computing marginal distributions we get for any (X_l, Y_l) that

$$f(y|\theta) = \sum_{j=1}^{L} p(X_l = x_j, Y_l = y|\theta) \tag{1.2}$$

so that

$$f(y|\theta) = \sum_{j=1}^{L} \alpha_j \cdot p(y|\phi_j). \tag{1.3}$$

The probability distribution $f(y|\theta)$ is a *finite mixture*. In this context $\{\alpha_j\}_{j=1}^L$ is called a *mixing distribution*.

4.1.3 Likelihood and Missing Information

Since the sequence $\mathbf{x}$ is hidden, we phrase our model in terms of the

THE MODEL FAMILY:

CONDITIONED ON $\underline{\theta}$, THE SYMBOLS IN $\mathbf{y}$ ARE INDEPENDENT WITH THE DISTRIBUTION $f(y|\theta)$.

Hence the likelihood function for $\mathbf{y}$ with relation to θ is

$$p(\mathbf{y} \mid \theta) := f(y_1|\theta) \cdot f(y_2|\theta) \cdot \cdots \cdot f(y_n|\theta).$$

The maximum likelihood estimate is

$$\widehat{\theta}_{\mathrm{ML}} = \arg\max_{\theta} p(\mathbf{y} \mid \theta).$$

The computation of this estimate can in general only be done by a numerical method. Reviews of the various techniques for this are found in (Grim 1982, Qian and Titterington 1991, Redner and Walker 1984). The EM algorithm is a useful way of finding $\widehat{\theta}_{\mathrm{ML}}$. Some of the relevant original developments of EM algorithm are found in (Dempster et al. 1977) and (Sundberg 1974) and a recent extensive survey is (McLachlan and Krishnan 1997).

By the missing information principle stated in Chapter 1 we consider the posterior probability

$$p(\mathbf{x}|\mathbf{y},\theta) = \frac{p(\mathbf{x},\mathbf{y} \mid \theta)}{p(\mathbf{y} \mid \theta)}.$$

Owing to our assumptions we get from (1.1)

$$p(\mathbf{x}|\mathbf{y},\theta) = \prod_{l=1}^{n} \frac{p(y_l|\phi_{j_l}) \cdot \alpha_{j_l}}{f(y_l|\theta)}. \tag{1.4}$$

Hence we can write

$$\begin{aligned} \log p(\mathbf{y} \mid \theta) &= \log p(\mathbf{x},\mathbf{y} \mid \theta) - \log p(\mathbf{x}|\mathbf{y},\theta) \qquad (1.5) \\ &= \sum_{l=1}^{n} \log p(y_l|\phi_{j_l}) \cdot \alpha_{j_l} - \log p(\mathbf{x}|\mathbf{y},\theta). \end{aligned}$$

We shall continue by giving a lower bound for $\log p(\mathbf{y} \mid \theta)$ using (1.5).

4.2 An Expectation Maximization Algorithm

4.2.1 Quasi-log Likelihood

Let us suppose that we have by some means obtained an approximation $\theta^{(t)}$ to the estimate $\widehat{\theta}_{\mathrm{ML}}$ with

$$\theta^{(t)} = \left(\alpha_1^{(t)}, \ldots, \alpha_L^{(t)}; \phi_1^{(t)}, \ldots, \phi_L^{(t)}\right).$$

The general idea is to improve $\theta^{(t)}$ so as to get closer to $\widehat{\theta}_{\mathrm{ML}}$. It turns out to be useful to multiply both sides of (1.5) with $p\left(\mathbf{x}|\mathbf{y}, \theta^{(t)}\right)$ and sum this over all possible sequences $\mathbf{x}$. Since

$$\sum_{\mathbf{x}} p\left(\mathbf{x}|\mathbf{y}, \theta^{(t)}\right) = 1,$$

where $\sum_{\mathbf{x}}$ designates $\sum_{\mathbf{x}\in\mathcal{X}^n}$, it holds from (1.5) that

$$\log p\left(\mathbf{y} \mid \theta\right) = \sum_{\mathbf{x}} p\left(\mathbf{x}|\mathbf{y}, \theta^{(t)}\right) \cdot \log p\left(\mathbf{x}, \mathbf{y} \mid \theta\right) - \sum_{\mathbf{x}} p\left(\mathbf{x}|\mathbf{y}, \theta^{(t)}\right) \log p\left(\mathbf{x}|\mathbf{y}, \theta\right).$$

We introduce the auxiliary function

$$Q\left(\theta|\theta^{(t)}\right) := \sum_{\mathbf{x}} p\left(\mathbf{x}|\mathbf{y}, \theta^{(t)}\right) \cdot \log p\left(\mathbf{x}, \mathbf{y} \mid \theta\right).$$

This is nothing else but the conditional expectation

$$Q\left(\theta|\theta^{(t)}\right) = E\left[\log p\left(\left(X_1, Y_1\right), \ldots, \left(X_n, Y_n\right) |\theta\right) \mid \left(Y_1, \ldots, Y_n\right) = \mathbf{y}, \theta^{(t)}\right]$$

and in particular an estimate of the logarithm of the complete probability distribution. Using this notation we have for any θ the identity

$$\begin{aligned} \log p\left(\mathbf{y} \mid \theta\right) - \log p\left(\mathbf{y} \mid \theta^{(t)}\right) &= Q\left(\theta|\theta^{(t)}\right) - Q\left(\theta^{(t)}|\theta^{(t)}\right) \\ &\quad + \sum_{\mathbf{x}} p\left(\mathbf{x}|\mathbf{y}, \theta^{(t)}\right) \log \frac{p\left(\mathbf{x}|\mathbf{y}, \theta^{(t)}\right)}{p\left(\mathbf{x}|\mathbf{y}, \theta\right)}. \end{aligned} \tag{2.1}$$

We see here that

$$\sum_{\mathbf{x}} p\left(\mathbf{x}|\mathbf{y}, \theta^{(t)}\right) \log \frac{p\left(\mathbf{x}|\mathbf{y}, \theta^{(t)}\right)}{p\left(\mathbf{x}|\mathbf{y}, \theta\right)} \geq 0,$$

since this is the Kullback distance between the probability distributions $p\left(\cdot|\mathbf{y},\theta^{(t)}\right)$ and $p\left(\cdot|\mathbf{y},\theta\right)$. We have thus obtained

$$\log p\left(\mathbf{y}\mid\theta\right)-\log p\left(\mathbf{y}\mid\theta^{(t)}\right)\geq Q\left(\theta|\theta^{(t)}\right)-Q\left(\theta^{(t)}|\theta^{(t)}\right).$$

By this account, if we determine

$$\theta^{(t+1)}=\arg\max_{\theta}Q\left(\theta|\theta^{(t)}\right)$$

we have found an estimate $\theta^{(t+1)}$ such that

$$\log p\left(\mathbf{y}\mid\theta^{(t+1)}\right)\geq\log p\left(\mathbf{y}\mid\theta^{(t)}\right)$$

and we have improved on $\theta^{(t)}$ in the sense of increased likelihood. Therefore we call $Q\left(\theta|\theta^{(t)}\right)$ a *quasi-log likelihood.*

4.2.2 Step E and Step M

At this stage we may propose an algorithmic idea:

Expectation Maximum Algorithm (EM)

Start: An estimate $\theta^{(t)}$ is given by

$$\theta^{(t)}=\left(\alpha_1^{(t)},\ldots,\alpha_l^{(t)};\phi_1^{(t)},\ldots,\phi_L^{(t)}\right).$$

Step E: Calculate the conditional expectation

$$Q\left(\theta|\theta^{(t)}\right)=\sum_{\mathbf{x}}p\left(\mathbf{x}|\mathbf{y},\theta^{(t)}\right)\cdot\log p\left(\mathbf{x},\mathbf{y}\mid\theta\right). \tag{2.2}$$

Step M: Determine $\theta^{(t+1)}$ by

$$\theta^{(t+1)}=\arg\max_{\theta}Q\left(\theta|\theta^{(t)}\right).$$

Let $\theta^{(t+1)}\mapsto\theta^{(t)}$ and repeat from step **E**. ∎

There are some immediate questions:

- Is this really an algorithm?
- Does this converge?
- Does this converge to a global/local maximum of the likelihood function $p\left(\mathbf{y}\mid\theta\right)$?

4.2.3 An Explicit Form of Step M

Let us first find a more detailed expression for the quasi-log likelihood function. Using (1.4) and (2.2) we get

$$Q\left(\theta|\theta^{(t)}\right)=\sum_{j_1=1}^{L}\cdots\sum_{j_n=1}^{L}\prod_{l=1}^{n}\frac{p\left(y_l|\phi_{j_l}^{(t)}\right)\cdot\alpha_{j_l}^{(t)}}{f\left(y_l|\theta^{(t)}\right)}\cdot\log\prod_{l=1}^{n}p\left(y_l|\phi_{j_l}\right)\cdot\alpha_{j_l}. \tag{2.3}$$

A simple calculation (exercise) shows that

$$Q\left(\theta|\theta^{(t)}\right)=\sum_{j=1}^{L}\sum_{l=1}^{n}\log\left[p\left(y_l|\phi_j\right)\cdot\alpha_j\right]\cdot\frac{p\left(y_l|\phi_j^{(t)}\right)\cdot\alpha_j^{(t)}}{f\left(y_l|\theta^{(t)}\right)}. \tag{2.4}$$

Then it follows that

$$\begin{aligned}Q\left(\theta|\theta^{(t)}\right) &= \sum_{j=1}^{L}\left[\sum_{l=1}^{n}\frac{p\left(y_l|\phi_j^{(t)}\right)\cdot\alpha_j^{(t)}}{f\left(y_l|\theta^{(t)}\right)}\right]\log\alpha_j \\ &\quad+\sum_{j=1}^{L}\sum_{l=1}^{n}\log\left[p\left(y_l|\phi_j\right)\right]\cdot\frac{p\left(y_l|\phi_j^{(t)}\right)\cdot\alpha_j^{(t)}}{f\left(y_l|\theta^{(t)}\right)}.\end{aligned} \tag{2.5}$$

This expression yields by a simple formal differentiation using Lagrange multipliers that the **M**-step for the mixing distribution can be written as

$$\alpha_j^{(t+1)}=\frac{1}{n}\sum_{l=1}^{n}\frac{p\left(y_l|\phi_j^{(t)}\right)\cdot\alpha_j^{(t)}}{f\left(y_l|\theta^{(t)}\right)} \tag{2.6}$$

and that

$$\phi_j^{(t+1)}=\arg\max_{\phi_j}\sum_{l=1}^{n}\log\left[p\left(y_l|\phi_j\right)\right]\cdot\frac{p\left(y_l|\phi_j^{(t)}\right)\cdot\alpha_j^{(t)}}{f\left(y_l|\theta^{(t)}\right)} \tag{2.7}$$

for $j=1,\ldots,L$. Evidently the operation (2.6) is implementable enough to merit the name algorithm. In addition, it turns out that (2.7) can often be solved *uniquely* and *explicitly*. The most well known example of this involves the exponential type of family of distributions.

4.3 Exponential Type of Family of Distributions

4.3.1 Definitions and Notations:

A discrete probability distribution q on the alphabet $\mathcal{Y}$ belongs to an *exponential type of family* of distributions if

$$q\left(y|\psi\right) = \frac{e^{\psi^T \cdot h(y)} \cdot b(y)}{C\left(\psi\right)}, \tag{3.1}$$

where $b(y)$ is a real non-negative function (that depends only on y), ψ is a real $m \times 1$ vector and $h(y)$ is a $1 \times m$ -valued mapping from $\mathcal{Y}$ to a subset of R^m. Here, for any y, $\psi^T \cdot h(y)$ is a scalar product and

$$C\left(\psi\right) = \sum_{y \in \mathcal{Y}} e^{\psi^T \cdot h(y)} \cdot b(y)$$

is the necessary standardization.

We are also supposing that m *is the least integer for which* (3.1) *is true* (Andersen 1970). This means that

$$h(y) = \left(h_1(y), \ldots, h_m(y)\right)$$

is such that $h_1(y)$, ..., $h_m(y)$ are linearly independent functions on $\mathcal{Y}$. Then $h(y)$ is known as a *sufficient statistic* for the family.

The parameter ψ is known as the *canonical parameter* of the family. The *expectation parameter* ϕ corresponding to ψ is defined as the $1 \times m$ vector

$$\phi = \sum_{y \in \mathcal{Y}} h(y) \cdot q\left(y|\psi\right). \tag{3.2}$$

The correspondence $\phi \leftrightarrow \psi$ can be shown to be one-to-one and continuously differentiable. This property is tacitly assumed, if need be, in any relevant computations.

Example 4.3.1 A simple instance of these formulas is given by the Poisson distribution $\mathrm{Po}\left(\lambda\right)$, where $\mathcal{Y} = \{0, 1, 2, \ldots, \}$ and

$$b(y) = \frac{1}{y!}, \qquad h(y) = y, \qquad \psi = \ln \lambda, C\left(\psi\right) = e^{\lambda}.$$

(You may check that $\lambda = \phi$ in (3.2) holds true). ∎

Example 4.3.2 An important property of the exponential type family is that if $\mathbf{y}$ is a sequence of symbols treated as conditionally independent outcomes, then

$$q(\mathbf{y}|\psi) = \prod_{l=1}^{n} q(y_l|\psi) = \frac{e^{\psi^T \cdot \sum_{l=1}^{n} h(y_l)} \cdot \prod_{l=1}^{n} b(y_l)}{C(\psi)^n} \tag{3.3}$$

which is again of an exponential type. ■

Example 4.3.3 The *multinomial process* in Chapter 3 above is a model for n random variables Y_l, conditionally independent given $\underline{\theta} = (\theta_1, \ldots, \theta_L)$, $\sum_{j=1}^{L} \theta_j = 1$, and assuming values in the same finite alphabet with J symbols,

$$\prod_{l=1}^{n} P(Y_l = y_{j_l}|\underline{\theta}) = \prod_{j=1}^{L} \theta_j^{n_j}, \tag{3.4}$$

where n_j is the number of times the symbol y_j is present. This belongs to the exponential type, since we have

$$\prod_{l=1}^{n} P(Y_l = y_{j_l}|\underline{\theta}) = \left(1 - \sum_{j=1}^{L-1} \theta_j\right)^n \cdot e^{\sum_{j=1}^{L-1} \psi_j \cdot n_j}, \tag{3.5}$$

where the canonical parameters are

$$\psi_j = \ln\left(\frac{\theta_j}{\theta_L}\right), \qquad j = 1, \ldots, L-1.$$

Note that $m = L - 1$ in this example. ■

4.3.2 EM for Exponential Type of Family of Distributions

If we assume enough differentiability (that can actually be rigorously justified) we can calculate that for $q(\mathbf{y}|\psi)$ in (3.3)

$$\nabla_\psi \log q(\mathbf{y}|\psi) = \sum_{l=1}^{n} h(y_l) - n \cdot \phi, \tag{3.6}$$

where log now designates the natural logarithm and $\nabla_{\psi} \log q(y|\psi)$ is the gradient vector of derivatives w.r.t ψ (Khuri 1993, p. 274). Then the maximum likelihood estimate of ϕ is seen to be obtained as

$$\widehat{\phi}_{\mathrm{ML}} = \frac{1}{n} \sum_{l=1}^{n} h(y_l). \tag{3.7}$$

We plan to keep this in mind in the case where the conditional distributions $p(y|\phi_j)$ in (1.3) belong to an exponential type with ϕ_j as the expectation parameter and ψ_j as the canonical parameter

$$p(y|\phi_j) = \frac{e^{\psi_j^T(\phi_j) \cdot h_j(y)} \cdot b_j(y)}{C_j(\psi_j(\phi_j))}, \tag{3.8}$$

and

$$\phi_j = \sum_{y \in \mathcal{Y}} h_j(y) \cdot \frac{e^{\psi_j^T(\phi_j) \cdot h_j(y)} \cdot b_j(y)}{C_j(\psi_j(\phi_j))}. \tag{3.9}$$

Then we can show, using (3.6), by an appropriate optimization calculus that the M-step of the EM algorithm (2.7) becomes

$$\phi_j^{(t+1)} = \frac{\sum_{l=1}^{n} h_j(y_l) \frac{\alpha_j^{(t)} p(y_l|\phi_j^{(t)})}{f\left(y_l|\theta^{(t)}\right)}}{\sum_{l=1}^{n} \frac{\alpha_j^{(t)} p(y_l|\phi_j^{(t)})}{f\left(y_l|\theta^{(t)}\right)}}. \tag{3.10}$$

This is clearly an algorithmically feasible expression. If (3.10) is written as

$$\phi_j^{(t+1)} = \sum_{l=1}^{n} h_j(y_l)\, w\left(y_l|\alpha_j^{(t)}, \phi_j^{(t)}\right)$$

we see more clearly the structural similarity to (3.7). Here the weight is

$$w\left(y_l|\alpha_j^{(t)}, \phi_j^{(t)}\right) = \frac{1}{\sum_{l=1}^{n} \frac{\alpha_j^{(t)} p(y_l|\phi_j^{(t)})}{f\left(y_l|\theta^{(t)}\right)}} \cdot \frac{\alpha_j^{(t)} p(y_l|\phi_j^{(t)})}{f(y_l|\theta^{(t)})}.$$

Thus (3.7) is obtained for $\alpha_j^{(t)} p(y_l|\phi_j^{(t)}) = f\left(y_l|\theta^{(t)}\right)$ for all j, as is natural.

4.3.3 Summary of EM for Exponential Type of Families

The EM algorithm for mixtures of exponential type family of distributions amounts to the following explicit iterations for each $j = 1, \ldots, L$:

$$\alpha_j^{(t+1)} = \frac{1}{n} \sum_{l=1}^{n} \frac{p\left(y_l|\phi_j^{(t)}\right) \cdot \alpha_j^{(t)}}{f\left(y_l|\theta^{(t)}\right)} \tag{3.11}$$

and

$$\phi_j^{(t+1)} = \frac{\sum_{l=1}^{n} h_j\left(y_l\right) \frac{\alpha_j^{(t)} p\left(y_l|\phi_j^{(t)}\right)}{f\left(y_l|\theta^{(t)}\right)}}{\sum_{l=1}^{n} \frac{\alpha_j^{(t)} p\left(y_l|\phi_j^{(t)}\right)}{f\left(y_l|\theta^{(t)}\right)}}. \tag{3.12}$$

By Bayes' formula

$$P\left(x_l = x_j|y_l, \phi_j^{(t)}, \alpha_j^{(t)}\right) = \frac{p\left(y_l|\phi_j^{(t)}\right) \cdot \alpha_j^{(t)}}{f\left(y_l|\theta^{(t)}\right)},$$

which makes (3.11) intuitively natural.

4.4 Iterations and Convergence

Boyles (1983) and Jeff Wu (1989) established the general results on convergence of the EM algorithm. Let

$$L(\theta) = \log p(\mathbf{y}|\theta) = \sum_{l=1}^{n} \log p\left(y_l|\theta\right) \tag{4.1}$$

and $\theta^{(t)}$ be an instance of the EM algorithm. If we know that $L\left(\theta^{(t)}\right)$ is a bounded sequence, then we know that

$$L\left(\theta^{(t)}\right) \rightarrow L^*$$

as $t \rightarrow \infty$, since, as shown above, the EM algorithm forces $L\left(\theta^{(t)}\right)$ to be a monotonous sequence. This does not, however, imply in general that L^* is a local (global) maximum of $\log p(\mathbf{y}|\theta)$.

Moreover, if we know that $L\left(\theta^{(t)}\right) \rightarrow L^*$, it does not follow in general that $\theta^{(t)} \rightarrow \theta^*$ for some θ^*. However, let us assume that

(a) θ lies in a subset C of the r -dimensional Euclidean space R^r.

(b) $L(\theta)$ is continuous and differentiable in the interior of C.

(c) $\theta^{(t)}$ lies in the interior of C.

(d) the quasi-log likelihood $Q(a|b)$ is continuous in $a, b \in C$.

(e) $\{L(\theta^{(t)})\}_{t=0}^{\infty}$ is a bounded sequence for any $\theta^{(0)}$.

Then following (Boyles 1983) and (Wu 1983) we can show that convergence of the EM algorithm means convergence to a local maximum of $L(\theta)$. In particular the assumption (d) holds true for the exponential type of families in a mixture of distributions. A detailed convergence analysis of this kind is carried through in Chapter 16 below.

4.5 A Model for Fragments with Motifs

4.5.1 Pattern in a Fragment

We think here of n fragments (sequences) $\mathbf{o}^{(l)}$, all of the same length $N+1$. We assume that each of them contains a special pattern, also called a *motif* of known length W. The position of the pattern in the fragment is not known and the pattern is not always present in a sequence in an exactly unaltered form, i.e., there can be *mutations*. There are, however, no *insertions* of new symbols or *deletions* of present ones. Relaxation of this assumption is a part of modelling by hidden Markov chains in Chapter 12. Any fragment $\mathbf{o}^{(l)}$ is presumed to contain precisely one occurrence of the motif hidden in the midst of two segments of random character. The idea is simple, but the notations required in the precise formulation of the model tend to become complicated. The biological interpretation will be discussed in greater detail in later chapters.

4.5.2 Modelling

The position of occurrence of the pattern is modelled by means of the hidden random variable X, which is a integer valued with

$$\alpha_t = P(X = t), \qquad t \in (0, \ldots, N - W + 1).$$

The notation $\mathbf{o} = o_0 \dots o_N$ has the effect that $X = N - W + 1$ indicates that the pattern is inserted as a block of W last symbols.

Conditioned on $X = t$ we consider W independent random variables $\{Y_{t+i}\}_{i=0}^{W-1}$ each assuming values in a finite discrete alphabet $\mathcal{O} = \{o_1, \dots, o_K\}$. We designate a set of symbol probabilities by

$$b_i(o_l) = P\left(Y_{t+i} = o_l \mid X = t\right), \qquad l = 1, \dots, K; \qquad i = 0, \dots, W-1.$$

In words, each $b_i(o_l)$ gives the probability of the symbol o_l at the i:th the position counting from left with the position of occurrence of the motif as $i = 0$. By conditional independence we have

$$P\left(Y_t = o_{l_t}, \dots, Y_{t+W-1} = o_{l_{t+W-1}} \mid X = t\right) = \prod_{i=0}^{W-1} b_i\left(o_{l_{t+i}}\right).$$

The random sequence $= (Y_t \dots Y_{t+W-1})$ with the stated joint distribution is called a **motif** of length W. In other words the pattern or motif is defined in a probabilistic sense by the distributions $(\underline{b}_i)_{i=0}^{W-1}$, where

$$\underline{b}_i := (b_i(o_1), \dots, b_i(o_K)).$$

Every $\underline{b}_i$ is assumed to have a high *information content* in the sense introduced in Chapter 2.

Next, consider the independent, identically distributed random variables $(Z_l)_{l \geq 1}$, which assume values in $\mathcal{O}$ and with the distribution $\underline{p} = (p_1, \dots, p_K)$ on $\mathcal{O}$ for any l. The variables (Z_l) are also independent of T and $\{Y_{X+i}\}_{i=0}^{W-1}$.

The $\mathbf{p}$ is a distribution with a *smaller information content* (more random) than any of $\underline{b}_i$. The random variables Z_i are called a background sequence.

We are, in other words, modelling fragments of length $N+1$ $(W < N+1)$

$$\mathbf{o} = o_{j_0} \dots o_{j_N}$$

so that we can compute the probabilities

$$P\left(Y_0 = o_{j_0}, \dots, Y_N = o_{j_N}\right)$$

by a decomposition into three factors: two calculated by the statistics of the background sequence preceding and succeeding, respectively, the occurrence of the motif, and one factor computed using $\prod_{i=0}^{W-1} b_i\left(o_{l_{t+i}}\right)$, corresponding to

$$\mathbf{o} = \underbrace{o_{j_0}\ldots o_{j_{t-1}}}_{\text{background}}\ \underbrace{o_{j_t}\ldots o_{j_{t+W-1}}}_{\text{motif}},\ \underbrace{o_{j_{t+W}}\ldots o_{j_N}}_{\text{background}}.$$

A MODEL FAMILY FOR FRAGMENTS WITH MOTIFS:

CONDITIONED ON $X = t$ AND THE PARAMETERS $\underline{p}$ and $\{\underline{b}_i\}_{i=0}^{W-1}$,

THE SYMBOLS IN $\mathbf{o} = o_{j_0}\ldots o_{j_N}$ HAVE THE PROBABILITY

$$P\left(Y_0^{(l)},\ldots,Y_N^{(l)}|X^l = t,\underline{p},\{\underline{b}_i\}_{i=0}^{W-1}\right) = \prod_{j=0}^{t-1} p_{l_j} \prod_{i=0}^{W-1} b_i\left(o_{l_{t+i}}\right) \prod_{j=t+W}^{N} p_{l_j}. \quad (5.1)$$

Note that if each $\underline{b}_i$ is taken as column, then $\{\underline{b}_i\}_{i=0}^{W-1}$ is a *weight matrix model* in the sense of Chapter 1. If $\{\underline{b}_i\}_{i=0}^{W-1}$ is estimated from a multiple alignment of fragments, it may be called a *profile* (Gribskov and Veretnik 1996).

4.5.3 Generating Sequences with a Motif

The model family above is a generator of sequences:

1) Choose t at random from $0,\ldots,N-W+1$.

2) Choose t independent symbols $o_{j_0}\ldots o_{j_{t-1}}$ from $\underline{p}$.

3) Choose, in order, $o_{j_t}\ldots o_{j_{t+W-1}}$ from each of $\underline{b}_0,\ldots,\underline{b}_{W-1}$, respectively. This realizes the hidden motif.

4) Choose $N-t-W$ symbols $o_{j_{t+W}}\ldots o_{j_N}$ from $\underline{p}$.

5) The complete sequence is obtained by concatenation

$$\mathbf{o} = o_{j_0}\ldots o_{j_{t-1}}o_{j_t}\ldots o_{j_{t+W-1}}o_{j_{t+W}}\ldots o_{j_N}.$$

4.5.4 Learning the Model for One Motif

It is required to estimate the probabilistic model within the preceding family models. We have access to a training set of n sequences $\mathbf{o}^{(l)}$, with the length $N+1$. In addition the length W of motif is known in advance.

The starting positions X of the motif occurrences constitute the *missing information.* Thus with

$$\left(X^l, \left(Y_0^{(l)} \dots Y_N^{(l)}\right)\right)_{l=1}^{n}$$

we have the mixtures of probabilities

$$\begin{aligned} p\left(\mathbf{o}^l\right) &= P\left(Y_0^{(l)} = o_{j_0}^{(l)}, \dots, Y_N^{(l)} = o_{j_N}^{(l)} | \underline{p}, \{\underline{b}_i\}_{i=0}^{W-1}\right) \\ &= \sum_{t=0}^{N-W+1} P\left(Y_0^{(l)} = o_{j_0}^{(l)}, \dots, Y_N^{(l)} = o_{j_N}^{(l)} | X^l = t, \underline{p}, \{\underline{b}_i\}_{i=0}^{W-1}\right) \cdot \alpha_t, \end{aligned}$$

since the position of occurrence of the motif was given by $\alpha_t = P(X = t)$. This is a mixture model in the sense of section 4.1.2 above with $\mathcal{Y} = \underbrace{\mathcal{O} \times \mathcal{O} \times \ldots \times \mathcal{O}}_{N+1 \text{ times}}$ and $\mathcal{X} = \{0, 1, \ldots, N - W + 1\}$. However, the parameters, $\underline{p}$ and $\{\underline{b}_i\}_{i=0}^{W-1}$ are the same for each component distribution of the mixture.

The exponential type family for sequence models in exempel 4.3.3 and the corresponding EM algorithm above is not straightforwardly applicable, as the distribution in (5.1) obviously cannot be written in the form (3.4): in the motif model the probability of any o_j changes in accordance with the position $X^l = t$. The motif model is heterogeneous.

Various versions of EM algorithms for estimating $\underline{p}, \underline{b}_1, \ldots, \underline{b}_W$ and ultimately the values of X^l are derived in (Bailey and Elkan 1995, Cardon and Stormo 1992, Lawrence and Reilly 1990). A more explicitly Bayesian approach is found in (Liu et al. 1995). Some steps in this derivation are touched upon in an exercise of this chapter.

4.6 Exercises

1. (a) Show that

$$p(\mathbf{y}|\theta) = \sum_{\mathbf{x}} p(\mathbf{y}, \mathbf{x}|\theta) = \prod_{l=1}^{n} p(y_l|\theta), \tag{6.1}$$

where $\sum_{\mathbf{x}}$ designates the summation over all values of $\mathbf{x} \in \mathcal{X}^n$ and can thus be explicitly given as

$$\sum_{\mathbf{x}} p(\mathbf{y}, \mathbf{x}|\theta) = \sum_{j_1=1}^{L} \sum_{j_2=1}^{L} \cdots \sum_{j_n=1}^{L} p(\mathbf{y}, x_{j_1}, \ldots, x_{j_n}|\theta).$$

(b) Show that under the assumptions introduced above one detailed expression of the auxiliary function $Q\left(\theta|\theta^{(t)}\right)$ introduced in the EM algorithm for finite mixtures is

$$Q\left(\theta|\theta^{(t)}\right) = \sum_{j_1=1}^{L} \cdots \sum_{j_n=1}^{L} \prod_{l=1}^{n} \frac{p\left(y_l|\phi_{j_l}^{(t)}\right) \cdot \alpha_{j_l}^{(t)}}{f\left(y_l|\theta^{(t)}\right)} \cdot \log \prod_{l=1}^{n} p\left(y_l|\phi_{j_l}\right) \cdot \alpha_{j_l}. \tag{6.2}$$

(c) Show finally from (b) that

$$Q\left(\theta|\theta^{(t)}\right) = \sum_{j=1}^{L} \sum_{l=1}^{n} \log\left[p\left(y_l|\phi_j\right) \cdot \alpha_j\right] \cdot \frac{p\left(y_l|\phi_j^{(t)}\right) \cdot \alpha_j^{(t)}}{f\left(y_l|\theta^{(t)}\right)}. \tag{6.3}$$

2. Show using the expression (6.3) in Exercise 1 above that if

$$\theta^{(t+1)} = \left(\alpha_1^{(t+1)}, \ldots, \alpha_K^{(t+1)}; \phi_1^{(t+1)}, \ldots, \phi_K^{(t+1)}\right)$$

is an estimate obtained by the E-step and by the M-step, respectively, then

$$Q\left(\theta^{(t+1)}|\theta^{(t)}\right) - Q\left(\theta^{(t)}|\theta^{(t)}\right) \geq n \cdot D\left(\underline{\alpha}^{(t+1)}|\underline{\alpha}^{(t)}\right), \tag{6.4}$$

where $D\left(\underline{\alpha}^{(t+1)}|\underline{\alpha}^{(t)}\right)$ is the relative entropy or the Kullback distance between the probability distribution

$$\underline{\alpha}^{(t+1)} = \left(\alpha_1^{(t+1)}, \ldots, \alpha_K^{(t+1)}\right)$$

and the probability distribution $\underline{\alpha}^{(t)}$ (defined analogously).

Show using this that if

$$L\left(\theta^{(t)}\right) \to L^*$$

as $t \to \infty$, then $\underline{\alpha}^{(t)}$ converges to a value $\underline{\alpha}^*$, as $t \to \infty$.

8. **Hidden Binomial Experiments**
$X|Y \in \text{Bin}(n,p)$, with $\text{Bin}(n,p)$ defined in Example 1.2.7 in Chapter 1. X is hidden and Y is some function of X that is actually observed. Show that the EM algorithm for p is

$$p^{(t+1)} = \frac{E[X|Y,p^{(t)}]}{n}. \tag{6.8}$$

(Weeks and Lange 1989).

9. **The exact estimate**
In (Höglund 1974) the following method of estimation (a competitor to maximum likelihood) is proposed and analyzed. The distribution of a string $\mathbf{x}$ belongs to an exponential type of family. The string $\mathbf{x}$ itself is hidden, but we observe the value of a sufficient statistic $t = T(\mathbf{x})$. Then $t = T(\mathbf{x})$ has the probability distribution

$$q(t|\psi) = \frac{e^{\psi^T \cdot t} \cdot f(t)}{C(\psi)}. \tag{6.9}$$

Let us consider the set of values of the parameter given by

$$\check{\Psi}(t) = \left\{ \psi | q(t|\psi) = \max_s q(s|\psi) \right\}.$$

In other words, for each ψ in $\check{\Psi}(t)$ the value t is the most probable value of T. Hereby $\check{\Psi}(t)$ is called *the set of exact estimates.*

(a) Let $\mathbf{x} = x_1 \ldots x_n$ be an n-string of outcomes of independent $\text{Be}(p)$ - distributed random variables. Show the set of exact estimates based on

$$t = x_1 + \ldots + x_n$$

is

$$\check{\Psi}(t) = \left[\frac{t}{n+1}, \frac{t+1}{n+1} \right].$$

(b) Let $\mathbf{x} = x_1 \ldots x_n$ be an n-string of outcomes of independent $\text{Ge}(p)$ - distributed random variables. Show that

$$t = x_1 + \ldots + x_n$$

where $\theta = P(X = 1|\Theta = \theta)$. Set

$$T = X_1 + \ldots + X_n.$$

The distribution of Θ is given by a prior density $\Psi(\theta)$.

(a) Show that the posterior density for Θ given $X_1, \ldots, X_n$ is the same as the posterior density for Θ given T or

$$\Psi(\theta|X_1 = x_1, \ldots, X_n = x_n) = \Phi(\theta|T = x_1 + \ldots + x_n)$$

for all θ. This means that the posterior distribution based on the full data is the same as the posterior distribution based on the summary T. We call T a *sufficient statistic.*

(b) Show that

$$P(X_1 = x_1, \ldots, X_n = x_n|\Theta = \theta)$$

belongs to an exponential type of family.

6. Consider n random variables X_l, conditionally independent given that $\underline{\Theta} = \underline{\theta} = (\theta_1, \ldots, \theta_J)$, $\sum_{j=1}^{J} \theta_j = 1$, and assuming values in the same finite alphabet with J symbols,

$$\prod_{l=1}^{n} P(X_l = x_{j_l}|\underline{\theta}) = \prod_{j=1}^{J} \theta_j^{n_j}, \tag{6.7}$$

where n_j is the number of times the symbol x_j is present. The distribution of $\underline{\Theta}$ is given by a prior density $\Psi(\underline{\theta})$. Show that the posterior density for $\underline{\Theta}$ given $X_1, \ldots, X_n$ is the same as the posterior density for Θ given $T = (n_1, \ldots, n_J)$ or

$$\Psi(\underline{\theta}|X_1 = x_1, \ldots, X_n = x_n) = \Phi(\underline{\theta}|T = (n_1, \ldots, n_J))$$

for all $\underline{\theta}$. *Hint:* Find the distribution of T.

7. The **factorization criterion** for a sufficient statistic says that T is a sufficient statistic for θ if and only if we have

$$P(X_1 = x_1, \ldots, X_n = x_n|\Theta = \theta) = g(T(\underline{x})|\Theta = \theta) \cdot h(\underline{x}),$$

where $\underline{x} = x_1, \ldots, x_n$, and the function $h(\underline{x})$ is the same for all θ. Show that the sufficient statistics in Exercises 5 and 6 above are properly called sufficient statistics also in the sense of this criterion.

3. Let $y = (y_1, \ldots, y_d)$, where $y_i \in \{0,1\}$, $i = 1, \ldots, d$, let

$$\phi_j = \left(\phi_{j|1}, \ldots, \phi_{j|d}\right),$$

where $0 \leq \phi_{j|i} \leq 1$ for every $i = 1, \ldots, d$, $j = 1, \ldots, L$ and

$$p\left(y|\phi_j\right) = \prod_{i=1}^{d} \phi_{j|i}^{y_i} \cdot \left(1 - \phi_{j|i}\right)^{1-y_i}, \qquad j = 1, \ldots, L. \tag{6.5}$$

The model for n binary vectors $y_1, \ldots, y_n$ is thus that they are independent samples of

$$f\left(y|\theta\right) = \sum_{j=1}^{L} \alpha_j \cdot p\left(y|\phi_j\right), \tag{6.6}$$

where estimation of θ involves estimation of $\phi_{j|1}, \ldots, \phi_{j|d}$ for all j.
Show using $Q\left(\theta|\theta^{(t)}\right)$ for the mixture (6.6) in the form given in (2.5) that an explicit expression for the M-step of the EM algorithm with respect to the parameters $\phi_{j|i}$ can be obtained without differentiation by invoking the *Gibbs' inequality* (5.8) from Chapter 2.
Hint: Set

$$w\left(j|y^{(l)}\right) = \frac{p\left(y_l|\phi_j^{(t)}\right) \cdot \alpha_j^{(t)}}{f\left(y_l|\theta^{(t)}\right)}.$$

Equivalently we can thus maximize the standardized sum

$$\frac{1}{\sum_{l=1}^{n} w\left(j|y^{(l)}\right)} Q_j = \sum_{l=1}^{n} \ln\left[p\left(y_l|\phi_j\right)\right] \cdot \frac{\frac{p\left(y_l|\phi_j^{(t)}\right) \cdot \alpha_j^{(t)}}{f\left(y_l|\theta^{(t)}\right)}}{\sum_{l=1}^{n} w\left(j|y^{(l)}\right)}.$$

Here consider first each term in the sum separately and then separate with regard to each bit in $y = (y_1, \ldots, y_d)$.

4. Prove (3.10).

5. **Sufficiency**
Assume that $X_n \in \{0,1\}$ for $n = 1, 2, \ldots$, are conditionally independent given Θ so that

$$P\left(X_1 = x_1, \ldots, X_n = x_n|\Theta = \theta\right) = \prod_{i=1}^{n} P\left(X_i = x_i|\Theta = \theta\right),$$

is an outcome of the sufficient statistic. Show that the set of exact estimates based on t is

$$\check{\Psi}(t) = \left[\frac{t}{n-1+t}, \frac{t+1}{n+t}\right].$$

(c) Let $\mathbf{x} = x_1 \ldots x_n$ be an n-string of outcomes of independent Po(λ) - distributed random variables. Show that

$$t = x_1 + \ldots + x_n$$

is sufficient for $\psi = \log \lambda$ and determine the set of exact estimates based on t, which is

$$\check{\Psi}(t) = \left[\frac{t}{n}, \frac{t+1}{n}\right].$$

10. **An EM Algorithm for Motifs**
Write the hidden variables X^l of occurrence positions of motifs as the equivalent indicator variables

$$U_{l,t} = \begin{cases} 1, & \text{if } X^l = t \text{ in sequence } \mathbf{o}^{(l)}, \\ 0 & \text{otherwise.} \end{cases} \tag{6.10}$$

By this we can take the simultaneous probability of the fragments and of the variables X^l as

$$\prod_{l=1}^{n} p\left(\mathbf{o}^l, X^l\right) = \prod_{l=0}^{n} \prod_{t=0}^{N-W+1} \left[\prod_{j=0}^{t-1} p_{l_j} \cdot \prod_{i=0}^{W-1} b_i\left(o_{l_{t+i}}^{(l)}\right) \cdot \prod_{j=t+W}^{N} p_{l_j} \cdot \alpha_t\right]^{U_{l,t}}.$$

Introduce in addition

$$I_{o_j}\left(\mathbf{o}_i^l\right) = \begin{cases} 1, & \text{if } \mathbf{o}_i^l = o_j \text{ in sequence } \mathbf{o}^{(l)}, \\ 0 & \text{otherwise.} \end{cases} \tag{6.11}$$

Then one can write

$$\prod_{l=1}^{n} p\left(\mathbf{o}^l, X^l\right) = \prod_{l=0}^{n} \prod_{t=0}^{N-W+1} \left[\alpha_t \prod_{j=1}^{K} C_j \prod_{i=0}^{W-1} b_i\left(o_j\right)^{I_{o_j}\left(\mathbf{o}_{t+i}^l\right)}\right]^{U_{l,t}},$$

where

$$C_j = \prod_{r \in \Delta(l,t)} p_j^{I_{o_j}\left(o_r^l\right)}$$

and $\Delta(l,t)$ are the non-motif positions in $\mathbf{o}^l$.

Find the estimate of $U_{l,t}$ in the E-step by computing

$$E\left[\ln \prod_{l=1}^{n} p\left(\mathbf{o}^l, X^l\right) \mid \left(\mathbf{o}^l,\right)_{l=1}^{n}, \underline{p}^{(s)} \left(b_i^{(s)}\right)_{i=0}^{W-1}\right],$$

where $\underline{p}^{(s)}$ and $\left(b_i^{(s)}\right)_{i=0}^{W-1}$ denote the current estimates of the symbol probabilities. (Bailey and Elkan 1995).

4.7 References and Further Reading:

- EM algorithm and mixtures

R. Boyles (1983): On the Convergence of the EM algorithm. *Journal of the Royal Statistical Society Ser. B*, 45, pp. 47–50.

A.P. Dempster, N.M. Laird and D.B. Rubin (1977): Maximum-likelihood from incomplete data via the EM algorithm. *Journal of the Royal Statistical Society Ser. B*, pp. 1–38.

J. Grim (1982): On Numerical Evaluation of Maximum Likelihood Estimates for Finite Mixtures of Distributions. *Kybernetika* (Prague), 18, pp. 173–191.

A.I. Khuri (1993): *Advanced Calculus with Applications in Statistics.* John Wiley and Sons, New York.

G.J. McLachlan and T. Krishnan (1997): *The EM Algorithm and Extensions.* John Wiley & Sons, New York.

W. Qian and D.M. Titterington (1991): Estimation of Parameters in hidden Markov models. *Philosophical Transactions of the Royal Society of London, A*, 337, pp. 407 – 428.

R. Sundberg (1974): Maximum Likelihood Theory for Incomplete Data from an Exponential Family. *Scandinavian Journal of Statistics*, pp. 49–58.

C.F. Jeff Wu (1983): On the Convergence Properties of the EM Algorithm. *The Annals of Statistics*, pp. 95–103.

- Statistical genetics

  K. Lange (1997): *Mathematical and Statistical Methods for Genetic Analysis.* Springer Verlag, New York, Berlin, Heidelberg.

  D. Weeks and K. Lange (1989): Trials, Tribulations and Triumphs of the EM Algorithm in Pedigree Analysis. *IMA Journal of Mathematics Applied in Medicine and Biology*, 6, pp. 209–232.

- Profiles and Motifs and Binding Sites

  T. L. Bailey (1995): *Discovering motifs in DNA and protein sequences: The approximate common substring problem.* Ph.D. Thesis, Department of Computer Science and Engineering, University of California, San Diego.

  T. L. Bailey and C. Elkan (1995): Unsupervised Learning of Multiple Motifs in Biopolymers using EM, *Machine Learning*, 21(1-2): pp. 51–80.

  L.R. Cardon and G.D. Stormo (1992): Expectation Maximization Algorithm for the Identifying Protein Binding Sites with Variable Lengths from Unaligned DNA Fragments. *Journal of Molecular Biology*, 223, pp. 159–170.

  M. Gribskov and S. Veretnik (1996): Identification of Sequence Patterns with Profile Analysis. *Methods in Enzymology*, 183, pp. 146–159.

  C.E. Lawrence and A.A. Reilly (1990): An Expectation Maximization (EM) Algorithm for the Identification and Characterization of Common Sites in Unaligned Biopolymer Sequences. *Proteins: Structure, Function, and Genetics*, 7, pp. 41–51.

  J.S. Liu, A.F. Neuwald and C.E. Lawrence (1995): Bayesian Models for Multiple Local Sequence Alignment and Gibbs Sampling Strategies. *Journal of the American Statistical Association*, 90, pp. 1156–1170.

- The exponential type of family and the exact estimate

  E.B. Andersen (1970): Sufficiency and Exponential Families for Discrete Sample Spaces. *Journal of the American Statistical Association*, 65, pp. 1248–1255.

  Th. Höglund (1974): The Exact Estimate - A Method of Statistical Estimation. *Zeitschrift für Wahrscheinlichkeitstheorie und verwandte Gebiete.* 29, pp. 257–271.

Chapter 5

Alignment and Scoring

5.1 Summary and Overview

5.1.1 Background

One important problem in bioinformatics is the comparison of sequences. By this one hopes to find segments of similarity, e.g., motifs in two or more sequences. The formats of sequences making the desired comparison possible are called *alignments*. In an alignment sequences are arranged as rows in a scheme, where each column should contain as many matches of symbols as possible. Clearly this is in general not possible without a judicious insertion of spaces into the sequences. The *scoring function* quantifies the degree of agreement or disagreement between the aligned sequences and counts the penalties for the inserted spaces. Alignments are either *global* or *local* depending on whether every symbol in the sequences has to participate in the comparison and scoring or not, respectively.

Some rudimentary knowledge about the machinery of aligning two or more sequences is essential in understanding, e.g., the probabilistic modelling of sequence families and some basic notions are for this reason included, although the material in this chapter involves no probabilistic concepts. There are a number of specialist treatments of alignments like (Gusfield 1997, Waterman 1995).

It is not difficult to guess that there are many possible scoring functions, but three considerations limit the number of reasonable choices. The first is biological: does the score function correspond to ideas of biological relatedness and molecular evolution (Buneman 1974, Waterman et al. 1976). The second is algorithmic: The number of alternative alignments is growing exponentially in sequence length, can an *optimal* alignment be found in reasonable time? Thirdly one may desire the criterion to have a probabilistic rationale. An interesting probabilistic interpretation is found in (Bucher and Hoffmann 1996). This and other probabilistically motivated alignment systems arising from hidden Markov models are discussed in the later chapters. There are two different types of *scoring method* in an alignment system:

- Minimum distance methods of pairwise alignments
- Maximum similarity methods of pairwise alignments

In this chapter we shall recapitulate the results of (Seller 1974) about the minimum distance methods for pairs of sequences.

5.2 Pairwise Alignments

We restrict ourselves to the purely formal and mathematically technical aspects of defining and computing the best alignment between two sequences using algorithmic methods based on *dynamic programming.* The dynamic programming method was introduced to biological sequence comparison by (Needleman and Wunsch 1970).

We shall treat mainly *global alignments.* Local alignment (Smith – Waterman) is presented as an exercise. Heuristic methods of alignment are not a natural part of this presentation.

5.2.1 Definitions

It is important to set clearly the definitions for describing sequence alignments. Some of these may seem pedantic, but will hopefully add to clarity of presentation.

For any sequence $\mathbf{x}$ $\mathbf{x}[i;j]$ is the contiguous subsequence $\mathbf{x}$ that starts at position i and ends at position j of $\mathbf{x}$. In particular, $\mathbf{x}[1;i]$ is the *prefix* of sequence $\mathbf{x}$ which ends at position i, and if $|\mathbf{x}|$ is the length of $\mathbf{x}$ then $\mathbf{x}[j;|\mathbf{x}|]$ is the *suffix* of $\mathbf{x}$ that begins at position j.

Definition 5.2.1 A **(global) alignment** of two sequences $\mathbf{x}$ and $\mathbf{y}$ *from the same alphabet* is obtained by first inserting chosen spaces ('-'), either into or at ends of $\mathbf{x}$ and $\mathbf{y}$, and then placing the two resulting sequences one above the other so that every symbol or space in either sequence is opposite to a unique symbol or a unique space in the other sequence. We denote the aligned sequences by $\mathbf{x}^*$ and $\mathbf{y}^*$. ∎

By the qualifier *global* we mean that all of the symbols in $\mathbf{x}$ and $\mathbf{y}$ are participating in the alignment. In a *local* alignment we search for substrings that are well matched to each other.

Example 5.2.1 Let $\mathbf{x} = ATAAGC$, $\mathbf{y} = GAAAACC$. An alignment in the sense of this definition is

$$\begin{matrix} \mathbf{x}^* \\ \mathbf{y}^* \end{matrix} = \begin{matrix} -ATAAGC- \\ GAAAA-CC \end{matrix} \tag{2.1}$$

∎

Often one refers to, e.g., $\frac{A}{A}$ as a *column*. The following terminology is of importance

> Two symbols arranged over one another are called **matched**. This differs from the use of the concept of match in an exercise in Chapter 1. There is no match between two ('-'). We shall in the sequence talk about *pseudo-alignments*, where $\frac{-}{-}$ are allowed as alignment columns.
>
> If two matched symbols are are equal, the match is called an **identity**, otherwise the match is called a **substitution**.
>
> An **insertion** or **deletion**, together referred to as *indels*, is one or more symbols aligned against $-$.

The difference between an indel and a substitution is in the resulting length of the string. In example 5.2.1 there are four identities, the column $\frac{A}{A}$ occurs three times and the column $\frac{C}{C}$ occurs once, there is one substitution $\frac{T}{A}$ and there are three indels. $y_1^* = G$ is said to be inserted into the first sequence or deleted from the second, depending on the point of view.

5.2.2 More Terminology: Gaps

We introduce a formal definition (Gusfield 1997).

Definition 5.2.2 A *gap* is any consecutive run of $'-'$ in a single row of a given alignment. ∎

An illustration is

$$\begin{array}{l} C\,T\,T\,T\,A\,C - - A - C \\ C - - - C\,C\,C\,C\,A\,T\,C \end{array}.$$

This is an alignment with six spaces distributed into three gaps.

5.3 An Evolutionary Distance

Our idea is now to define an *evolutionary distance* between **x** and **y** using the standard notions of a *metric* or distance as known in mathematics. The evolutionary distance is at the same time one criterion for the *best (global) pairwise alignment.* In this section we take the basic alphabet as

$$\mathcal{X}^* = \{x_1, \ldots, x_J, '-'\},$$

i.e., the alphabet $\mathcal{X} = \{x_1, \ldots, x_J\}$ extended by the symbol for insertion and deletion (indel).

5.3.1 Metric Spaces

Next we recall the definition of a metric space (cf. Exercise 10 in Chapter 1 and Exercise 9 in Chapter 2).

Definition 5.3.1 Suppose that there is a real valued function $\rho(x, y)$ defined on $\mathcal{X}^* \times \mathcal{X}^*$ with the following properties.

(a) $\rho(x, y) \geq 0$;

(b) $\rho(x, y) = 0$ is and only if $x = y$;

(c) $\rho(x, y) = \rho(y, x)$;

(d) $\rho(x, z) \leq \rho(x, y) + \rho(y, z)$ for all $x, y, z \in \mathcal{X}^*$.

Then we call ρ a *metric* (or *distance function*) for $\mathcal{X}^*$. The pair $(\mathcal{X}^*, \rho)$ is called a metric space. ∎

The inequality (d) is known as the *triangle inequality* for ρ.

Example 5.3.1 By

$$\rho(x,y) = \begin{cases} 0 & \text{if } x = y, \\ \\ 1 & \text{if } x \neq y \end{cases} \tag{3.2}$$

we define a metric on (a finite set) $\mathcal{X}^* \times \mathcal{X}^*$. This is known as the *discrete metric.* ■

Example 5.3.2 Let $\mathbf{x}$ and $\mathbf{y}$ be binary vectors with d bits, i.e., x_i and y_i = 0 or 1, $i = 1, \ldots, d$. Then we define

$$\rho_{\mathrm{H}}(\mathbf{x}, \mathbf{y}) := \sum_{i=1}^{d} |x_i - y_i|. \tag{3.3}$$

Hence $\rho_{\mathrm{H}}(\mathbf{x}, \mathbf{y})$ is the number of positions where $\mathbf{x}$ and $\mathbf{y}$ differ. This is seen to be a metric on $\{0,1\}^d$ and is known as the *Hamming metric.* ■

5.3.2 Auxiliary Terminology

The following auxiliary terminology seems helpful in some of the proofs in the sequel, but will be invoked only in this chapter.

An **element** is a symbol in the alphabet $\mathcal{X}^*$.

A **letter** is an element other than $'-'$ i.e. a symbol from $\mathcal{X}$.

A **sequence** is a finite string of letters, i.e., a sequence in the sense of a previous definition.

A **padded sequence** is a string of elements. $\mathcal{X}^{*\infty}$ designates the set of all padded sequences with a finite number of letters.

A **pseudo-alignment** of the sequences $\mathbf{x}$ and $\mathbf{y}$ is alignment of the padded sequences $\mathbf{x}^*$ and $\mathbf{y}^*$ of equal length such that removing all $'-'$ from $\mathbf{x}^*$ or $\mathbf{y}^*$ restores the sequence $\mathbf{x}$ or $\mathbf{y}$, respectively.

An **alignment** is a pseudo alignment when each column contains a letter.

Note that the earlier definitions of sequence and alignment are still valid but have been imbedded in a new terminology.

5.3.3 The Sequence Metric and the Evolutionary Distance

Let us now suppose that we have a metric ρ defined on $\mathcal{X}^* \times \mathcal{X}^*$. The discrete metric (5.3.1) is often used. This can be interpreted as defining a *cost* $\rho(x,y)$ for a substitution between x and y and for an indel

$$\rho(x,-) = \rho(-,x) = g(x) \geq 0, \qquad x \in \mathcal{X}^*. \tag{3.4}$$

Then we can extend this metric for padded sequences $\mathbf{x}^*$ and $\mathbf{y}^* \in \mathcal{X}^{*\infty}$ by

$$\overline{\rho}\left(\mathbf{x}^*, \mathbf{y}^*\right) = \sum_{i=1}^{\infty} \rho\left(x_i^*, y_i^*\right), \tag{3.5}$$

where $\mathcal{X}^{*\infty}$ designates the set of all padded sequences with a finite number of letters.

Hence the sum $\sum_{i=1}^{\infty} \rho\left(x_i^*, y_i^*\right)$ is well defined, since there are only a finite number of terms $\rho\left(x_i^*, y_i^*\right)$ differing from zero, if we always extend the padded sequences to be equally long by padding with more of '$-$'.

Note that $\overline{\rho}$ is readily verified to be a metric on $\mathcal{X}^{*\infty} \times \mathcal{X}^{*\infty}$ in the sense of definition 5.3.1 and hence $(\mathcal{X}^{*\infty}, \overline{\rho})$ is a metric space.

Definition 5.3.2 Let $\mathbf{x}$ and $\mathbf{y}$ be two sequences of symbols from $\mathcal{X}$ and let $\mathbf{x}^*$ and $\mathbf{y}^*$ be two corresponding padded sequences. Then the *evolutionary distance* $D(\mathbf{x}, \mathbf{y})$ between $\mathbf{x}$ and $\mathbf{y}$ is defined as

$$D(\mathbf{x}, \mathbf{y}) = \min_{\mathbf{x}^*, \mathbf{y}^*} \overline{\rho}\left(\mathbf{x}^*, \mathbf{y}^*\right), \tag{3.6}$$

where the minimum is computed over the set of all pseudoalignments of $\mathbf{x}$ with $\mathbf{y}$. ■

The evolutionary distance is a metric for strings, the proof from (Sellers 1974) is given in an appendix 5.7 below. Since the number of different alignments and pseudo alignments is finite[1], the minimum in (3.6) exists always. Note that alignments of $\mathbf{x}$ with $\mathbf{y}$ are special cases of pseudoalignments. In fact, since $\rho(-,-) = 0$, the minimum in (3.6) is always obtained for an alignment.

[1]But extremely large, if $|\mathbf{x}| = |\mathbf{y}| = n$, then the number of alignments grows asymptotically as $\left(1+\sqrt{2}\right)^{2n+1} n^{-1/2}$, see (Waterman 1995, p. 187)

5.4 Dynamic Programming

Definition 5.4.1 For any two sequences $\mathbf{x}, \mathbf{y}$ with symbols from a common alphabet $D_{i,j}$ is the evolutionary distance between the prefixes $\mathbf{x}[1;i]$ and $\mathbf{y}[1;j]$. ■

With this notation, if $|\mathbf{x}| = n, |\mathbf{y}| = m$, then $D(\mathbf{x}, \mathbf{y}) = D_{n,m}$. We will compute $D_{n,m}$ by solving the more general problem of computing $D_{i,j}$ for all combinations of i and j, where i ranges from zero to n and j ranges from zero to m. The technique for doing this is another instance of *dynamic programming* or of the *Bellman optimality principle* and has here three explicit components:

- *recurrence relation*
- *tabular computation*
- *traceback*

We shall explain each of these in turn. Dynamic programming, in the sequel reappearing as the *Viterbi algorithm*, is capable of computing solutions to a number of optimization problems that involve exponential growth in the number of alternatives. The word 'programming' does not refer to creating a piece of software code for the optimization but to the tabular mode of computation.

We shall establish an algorithm for pairwise alignment by showing how to compute the evolutionary distance.

5.4.1 The Recurrence Relation

We denote by $\mathbf{1}$ the sequence containing only the element $'-'$ repeated any suitable number of times. The start conditions for the evolutionary distance are

$$D_{0,0} = 0, \qquad D_{i,\mathbf{1}} = \sum_{h=1}^{i} \rho(x_h, -), \qquad D_{\mathbf{1},j} = \sum_{h=1}^{j} \rho(-, y_h).$$

The formulas for $D_{i,\mathbf{1}}$ and $D_{\mathbf{1},j}$ given here agree with (3.6) in definition 5.3.2, except that there should be a minimization. Here minimization is not necessary, since there is but one number which to find a minimum of in each set. We claim that there is *recurrence relation for the evolutionary distance*, given by

$$D_{i,j} = \min\left[D_{i-1,j} + \rho(x_i, -), D_{i-1,j-1} + \rho(x_i, y_j), D_{i,j-1} + \rho(-, y_j)\right],$$

We shall now prove this assertion (stated formally in Proposition 5.4.3 below) a by imitating the proof of the analogous recursion relation for the *edit distance* given in (Gusfield 1997). The first lemma uses an idea known as the Bellman optimality principle in a very transparent manner.

Lemma 5.4.1 *The value of $D_{i,j}$ must be $D_{i-1,j}+\rho(x_i,-)$, $D_{i-1,j-1}+\rho(x_i,y_j)$ or $D_{i,j-1} + \rho(-, y_j)$. There are no other possibilities.*

Proof: Let us suppose that the sequences $\mathbf{x}[1;i]$ and $\mathbf{y}[1;j]$ have with respect to $D_{i,j}$ the minimizing padded sequences $\mathbf{x}^*[1;k]$ and $\mathbf{y}^*[1;k]$, respectively. Then

$$\begin{aligned} D_{i,j} &= \overline{\rho}\left(\mathbf{x}^*[1;k], \mathbf{y}^*[1;k]\right) = \sum_{i=1}^{k} \rho\left(x_i^*, y_i^*\right) \\ &= \sum_{i=1}^{k-1} \rho\left(x_i^*, y_i^*\right) + \rho\left(x_k^*, y_k^*\right). \end{aligned} \qquad (4.1)$$

But

$$\sum_{i=1}^{k-1} \rho\left(x_i^*, y_i^*\right) = \overline{\rho}\left(\mathbf{x}^*[1;k-1], \mathbf{y}^*[1;k-1]\right).$$

Here we must have that $\overline{\rho}\left(\mathbf{x}^*[1;k-1], \mathbf{y}^*[1;k-1]\right)$ is minimal in the sense of providing the least value for $\overline{\rho}$ over all pseudo alignments of sequences with the same letters as $\mathbf{x}^*[1;k-1]$ and $\mathbf{y}^*[1;k-1]$. The reason is that if this was not true, then $\overline{\rho}\left(\mathbf{x}^*[1;k], \mathbf{y}^*[1;k]\right)$ would not be minimal either.

We can always arrange the optimal pseudo alignment so that both x_k^* and y_k^* are not both equal to '$-$' at the same time. Hence there are three alternatives for the expression in the right hand side of (4.1):

(i) $D_{i-1,j} + \rho\left(x_i^*, -\right)$, if $x_k^* \neq -$ and $y_k^* = -$:

(ii) $D_{i-1,j-1} + \rho\left(x_i^*, y_j^*\right)$, if $x_k^* \neq -$ and $y_k^* \neq -$;

(iii) $D_{i,j-1} + \rho\left(-, y_j^*\right)$, if $x_k^* = -$ and $y_k^* \neq -$.

Hence one of these is always equal to $D(i, j)$. ∎

Lemma 5.4.2

$$D_{i,j} \leq \min\left\{D_{i-1,j} + \rho(x_i, -), D_{i-1,j-1} + \rho(x_i, y_j), D_{i,j-1} + \rho(-, y_j)\right\}. \quad (4.2)$$

Proof: The objective is now to show it is possible to align $\mathbf{x}[1; i]$ and $\mathbf{y}[1; j]$ so that each and every one of the numbers on the right hand side of (4.2) are assumed. First of all, it *is* possible to align $\mathbf{x}[1; i]$ and $\mathbf{y}[1, j]$ so that

$$\overline{\rho}\left(\mathbf{x}^*[1; i], \mathbf{y}^*[1; j]\right) = D_{i-1,j} + \rho(\mathbf{x}_i, -).$$

To do this we align $\mathbf{x}[1; i-1]$ and $\mathbf{y}[1; j]$ so that their distance is $D_{i-1,j}$ and then use the column $\frac{x_i}{-}$ as alignment. But then the sum of these terms is larger than or equal to $D_{i,j}$. The same argument carries through in the other cases. ∎

Proposition 5.4.3 *When both* $\mathbf{x}^*$ *and* $\mathbf{y}^*$ *are different from the sequence* **1**, *then*

$$D_{i,j} = \min\left[D_{i-1,j} + \rho(x_i, -), D_{i-1,j-1} + \rho(x_i, y_j), D_{i,j-1} + \rho(-, y_j)\right]. \quad (4.3)$$

Proof: Lemma 5.4.1 says that $D_{i,j}$ must be one of the three values in the right hand side of (4.3). Lemma 5.4.2 says $D_{i,j}$ is smaller than or equal to the smallest of these values. It follows that $D_{i,j}$ is equal to the smallest of these and we have obtained the recursion for $D_{i,j}$ as claimed. ∎

5.4.2 Tabular Computation

It is perhaps tempting to use the recurrence relation (4.3) to compute the value of the evolutionary distance. The problem is that the number of recursive calls grows exponentially with n and m, as can be shown. But there are only $(n+1) \times (m+1)$ combinations of i and j. Efficient computation of the evolutionary distance is therefore done by a tabular bottom−up method. There are more efficient versions (Ukkonen 1985).

In the tabular method we work with a *dynamic programming table.* The table recapitulates all the values of $D_{i,j}$. The sequence $\mathbf{x}$ corresponds to the vertical axis of the table and the sequence $\mathbf{y}$ to the horizontal axis. Because of the start conditions the table has a column and row for $\mathbf{1}$ at the beginning of the table. The values of this row and column are directly filled in from the initial conditions.

After the first table has been set up the table is suitably filled one row at a time in the order of increasing i. In each row the cells are filled in order of increasing j.

To see how this is done let us note that by the general recurrence relation for $D_{i,j}$ all the values needed for the computation of $D_{1,1}$ are contained once $D_{\mathbf{1},\mathbf{1}}$, $D_{\mathbf{1},1}$, and $D_{1,\mathbf{1}}$ have been computed. Hence $D_{1,1}$ can be computed as soon as the first row and the first column have been filled in. The computation up to this stage is shown in Figure 5.1, in which the metric in the initial conditions is the discrete metric in (3.2). Then again, by the recur-

	1	y_1	y_2	y_3	y_4
1	0	1	2	3	4
x_1	1	D_{11}			
x_2	2				
x_3	3				
x_4	4				
x_5	5				$D(x,y)$

Figure 5.1: Table for pairwise alignment

rence relation, after $D_{1,1}$ has been computed all the values needed for the computation of $D_{1,2}$ are at hand. Following this description we see that the values for row one can be computed in order of increasing j. After that all the values needed needed to compute the values in row two are known, and that row can be filled in in the order of increasing j. By extension all of the table can be filled one row at a time, in order of increasing j. The value of the evolutionary distance is found in the final cell as $D_{n,m} = D(\mathbf{x}, \mathbf{y})$.

5.4.3 The Traceback: Optimal Alignment

Once the value of evolutionary distance has been computed by tabular bottom-up computation, we would be interested in knowing how the best alignment itself is to be found. One computational technique for doing this is the trace-back carried out by *pointers.*

At each cell pointers are saved and show which of $(i-1,j),(i-1,j-1)$, or $(i,j-1)$ was responsible for giving the optimal $D_{i,j}$. Then when $D_{n,m}$ has been found the pointers are followed to produce alignments. The traceback works by building the alignment in reverse starting from the final cell. At each step of the traceback we move back from the current cell (i,j) to one of the cells singled out by the pointer. At the same time we add a column of elements onto the front of the current alignment.

The column added is $\begin{smallmatrix}x_i\\y_j\end{smallmatrix}$ if the pointer gives us the step $(i-1,j-1)$. The column added it is $\begin{smallmatrix}x_i\\-\end{smallmatrix}$ if the step was $(i-1,j)$. If the step was $(i,j-1)$ we take $\begin{smallmatrix}-\\y_j\end{smallmatrix}$. In Figure 5.2 the optimal alignment has $\begin{smallmatrix}x_5\\y_4\end{smallmatrix}$ as its final column.

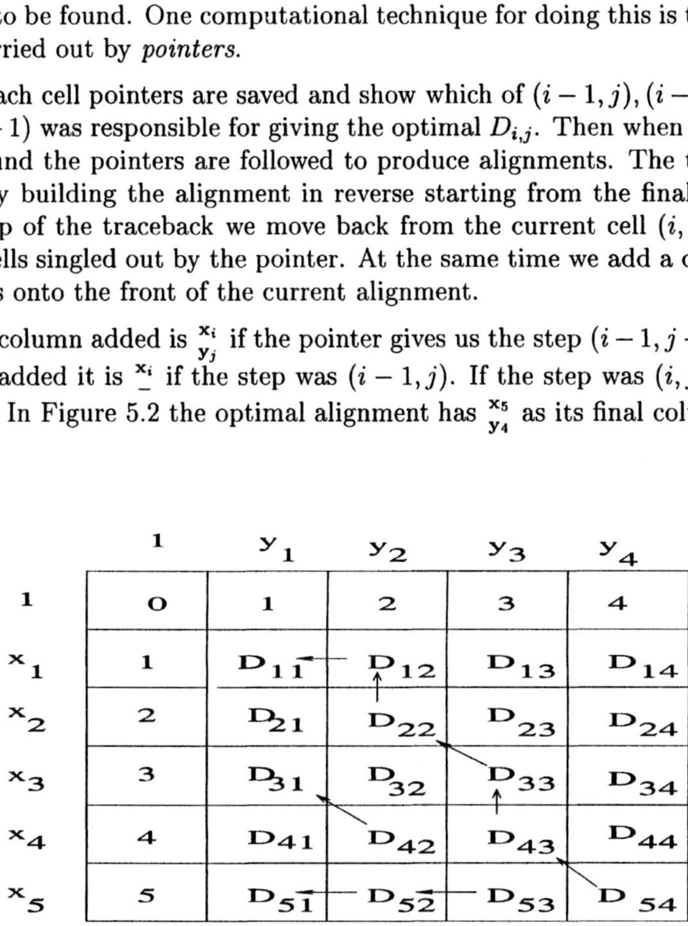

Figure 5.2: Traceback for pairwise alignment

5.5 Global Similarity Alignment, Gap Penalties

A *similarity measure* between $\mathbf{x}$ and $\mathbf{y}$ is

$$S(\mathbf{x}, \mathbf{y}) = \max_{\mathbf{x}^*, \mathbf{y}^*} \bar{s}(\mathbf{x}^*, \mathbf{y}^*), \tag{5.1}$$

where the maximum is computed over the set of all (pseudo) alignments of $\mathbf{x}$ with $\mathbf{y}$ and where

$$\bar{s}(\mathbf{x}^*, \mathbf{y}^*) = \sum_{i=1}^{L} s(x_i^*, y_i^*).$$

In this we have a similarity function $s(x^*, y^*)$ on $\mathcal{X}^* \times \mathcal{X}^*$. The similarity function $s(x, y)$ is defined on pairs of letters. An example is

$$s(x, y) = \begin{cases} +1 & \text{if } x = y \\ \\ -\mu & x \neq y. \end{cases} \tag{5.2}$$

A biologically important source of similarity functions is the *substitution matrices* for exchange of residues in amino acid sequences. A survey of substitution matrices is found in (Johnson and Overington 1993).

Frequently the insertions and deletions (indels) of adjacent symbols are not thought to be the sum of single deletions or insertions but the result of one event in evolution of biological sequences. Therefore it is sometimes required to weight these multiple indels in minimum distance or maximum similarity alignments differently from merely summing single indel weights.

A given set of biological sequences is likely to possess both fairly conserved regions and highly variable regions. Hence it is not reasonable to assign identical penalties for all regions of the sequences. Substitutions and indels in a region of high conservation should be penalized differently in one position compared to another. Also, the gaps are more or less likely in different regions of the sequence. This information is captured by giving position sensitive gap penalties. In Chapter 12 it will turn out that there are hidden Markov models that automatically provide the user with a coherent set of position sensitive gap penalties.

Next we make a formalization of the weighting of the gaps. If $\mathbf{1}$ is the sequence containing only repetitions of the element $'-'$, then the initial conditions for the similarity measure are

$$E_{0,0} = S_{0,0} = 0, \qquad E_{i,1} = S_{i,1} = -\widehat{g}(i),$$

$$F_{j,1} = S_{1,j} = -\widehat{g}(j).$$

Here we take the **affine gap penalty**

$$\widehat{g}(n) = \alpha + \beta \cdot (n-1). \tag{5.3}$$

The coefficient α corresponds to the **gap open** penalty and the coefficient β to the **gap extension** penalty. The affine function is chosen to make the running time of the algorithm reasonable.

Let $S_{i,j}$ denote the similarity measure as defined in (5.1) between the prefixes $\mathbf{x}[1;i]$ and $\mathbf{y}[1;j]$.

Proposition 5.5.1 *If*

$$E_{i,j} = \max\left\{S_{i,j-1} - \alpha, E_{i,j-1} - \beta\right\} \tag{5.4}$$

and

$$F_{i,j} = \max\left\{S_{i-1,j} - \alpha, F_{i-1,j} - \beta\right\}, \tag{5.5}$$

then

$$S_{i,j} = \max\left[S_{i-1,j-1} + s(x_i, y_j), E_{i,j}, F_{i,j}\right]. \tag{5.6}$$

Proof: As in all proofs concerned with the properties of pairwise alignments, we consider the situation in the final column of the pairwise alignment assuming that the preceding columns of the alignment have been established. For alignment of $\mathbf{x}[1;i]$ and $\mathbf{y}[1;j]$ there are the three alternatives for the final column in the pairwise alignment:

$$\begin{matrix} x_i & x_i & - \\ - & y_j & y_j \end{matrix}.$$

There are, in fact, only three possible values of $S_{i,j}$ corresponding to these three alternatives, respectively, in a certain way.

If $\begin{matrix} x_i \\ y_j \end{matrix}$ is the rightmost column then we obtain the value

$$S_{i-1,j-1} + s(x_i, y_j)$$

as one of the three possible values of $S_{i,j}$. This is owed to the observation that $S_{i,j}$ cannot be maximal unless the alignment of the prefixes $\mathbf{x}[1;i-1]$ and $\mathbf{y}[1;j-1]$ has the similarity $S_{i-1,j-1}$.

For $\genfrac{}{}{0pt}{}{x_i}{-}$

$$\max_{\mathbf{x}^*,\mathbf{y}^*} \overline{s}\left(\mathbf{x}^*,\mathbf{y}^*\right) = \sum_{l=1}^{L-1} s\left(x_l^*, y_l^*\right) + s\left(x_i, -\right) \tag{5.7}$$

and for $\genfrac{}{}{0pt}{}{-}{y_j}$

$$\max_{\mathbf{x}^*,\mathbf{y}^*} \overline{s}\left(\mathbf{x}^*,\mathbf{y}^*\right) = \sum_{l=1}^{L-1} s\left(x_l^*, y_l^*\right) + s\left(-, y_j\right). \tag{5.8}$$

In both cases we must decide whether we are optimally opening a gap or extending an existing gap. Let us consider (5.7) first. Here the optimality principle implies that

$$\sum_{l=1}^{L-1} s\left(x_l^*, y_l^*\right) + s\left(x_i, -\right) = \max_{1\leq k\leq j} \left\{S_{i,j-k} - \widehat{g}(k)\right\},$$

since we are choosing a gap in $\mathbf{y}^*$ with k elements optimally and align $\mathbf{x}[1;i]$ and $\mathbf{y}[1;j-k]$ optimally, too. For the same reason

$$\sum_{l=1}^{L-1} s\left(x_l^*, y_l^*\right) + s\left(-, y_j\right) = \max_{1\leq k\leq i} \left\{S_{i-k,j} - \widehat{g}(k)\right\}.$$

We prove next that for $E_{i,j}$ in (5.4)

$$E_{i,j} = \max_{1\leq k\leq j} \left\{S_{i,j-k} - \widehat{g}(k)\right\} \tag{5.9}$$

and for $F_{i,j}$ in (5.5)

$$F_{i,j} = \max_{1\leq k\leq i} \left\{S_{i-k,j} - \widehat{g}(k)\right\}. \tag{5.10}$$

The idea is to prove that the right hand side of (5.9) satisfies the recursion (5.4). For this note that it holds by construction of maximum that

$$E_{i,j}^* = \max_{1\leq k\leq j} \left\{S_{i,j-k} - \widehat{g}(k)\right\} = \max\left\{\max_{2\leq k\leq j} \left[S_{i,j-k} - \widehat{g}(k)\right],\ S_{i,j-1} - \widehat{g}(1)\right\}$$

and since $\widehat{g}(n) = \widehat{g}(n-1) + \beta$, we obtain in the right hand side, by a change of variable,

$$E^*_{i,j} = \max\left\{ \max_{1 \le k-1 \le j-1} \left\{ S_{i,(j-1)-(k-1)} - \widehat{g}(k-1) \right\} - \beta, S_{i,j-1} - \alpha \right\} =$$

$$= \max\left\{ E^*_{i,j} - \beta,\ S_{i,j-1} - \alpha \right\}.$$

But this is nothing else but the recursion for $E_{i,j}$, as was to be proved. We handle $F_{i,j}$ in a similar manner.

Hence we have proved that $S_{i,j}$ can have only three possible values $S_{i-1,j-1} + s(x_i, y_j)$, $E_{i,j}$ or $F_{i,j}$. It is clear that these values are possible for all $\mathbf{x}[1;i]$ and $\mathbf{y}[1;j-k]$. Hence $S_{i,j}$ is equal to the maximal of these three values. ∎

5.6 Exercises

1. The alphabet is $\mathcal{X}^* = \{a, b, c, d, '-'\}$ and the metric on $\mathcal{X}^*$ is the discrete metric in (5.3.1). Find the optimal pairwise alignment $\mathbf{x} = acbba$ and $\mathbf{y} = abca$ with regard to the evolutionary distance (Seller 1974).

2. **Local Alignment by Smith–Waterman algorithm**
By pairwise local alignment we mean finding optimally aligned substrings of two sequences. Let $\mathbf{x}$ and $\mathbf{y}$ be two strings with n and m letters, respectively. For any string $\mathbf{x}$, $\mathbf{x}[i;j]$ is the the contiguous subsequence $\mathbf{x}$ that starts at position i and ends at position j of $\mathbf{x}$. The objective function of local alignment is

$$H(\mathbf{x}, \mathbf{y}) = \max_{1 \le i \le j \le n, 1 \le k \le l \le m} [0, S(\mathbf{x}[i;j], \mathbf{y}[k;l])],$$

where $S(\mathbf{x}[i;j], \mathbf{y}[k;l])$ is the similarity for the best alignment of $\mathbf{x}[i;j]$ with $\mathbf{y}[k;l]$ corresponding to the global formulation

$$S(\mathbf{x}, \mathbf{y}) = \max_{\mathbf{x}^*, \mathbf{y}^*} \overline{s}(\mathbf{x}^*, \mathbf{y}^*), \tag{6.1}$$

where the maximum is computed over the set of all (pseudo) alignments of $\mathbf{x}$ with $\mathbf{y}$ and where

$$\overline{s}(\mathbf{x}^*, \mathbf{y}^*) = \sum_{i=1}^{L} s(x^*_i, y^*_i).$$

The Smith–Waterman algorithm (Smith and Waterman 1981) comes in terms of the quantity $H_{i,j}$ defined as

$$H_{i,j} = \max_{1 \le r \le i \le n, 1 \le s \le j \le m} \{0, S(\mathbf{x}[r;j], \mathbf{y}[s;j])\}.$$

This is the best score of any substrings of contiguous letters ending at x_i and y_j, or it is 0 if no such alignments have a positive score. Then the exercise is to show that the following proposition is true.

Proposition 5.6.1 *If*

$$\delta = s(x, -) = s(-, x) > 0,$$

then

$$H_{i,j} = \max\{H_{i-1,j} - \delta, H_{i,j-1} - \delta, H_{i-1,j-1} + s(x_i, y_j), 0\}. \qquad (6.2)$$

■

Hint: Just as in the proof in the preceding, we think of possible alignments ending at the column $\begin{smallmatrix} x_i^* \\ y_j^* \end{smallmatrix}$. Then these must yield the recursion with three terms. If this recursion ends with negative values, then there are no positively scoring substrings ending at (i, j) and the value of $H_{i,j}$ is equal to zero.

5.7 Appendix: On the Evolutionary Distance

5.7.1 The Metric Property

We prove that the evolutionary distance is a metric for strings, which is a theorem due to (Sellers 1974, Waterman 1995 pp. 192 - 193), a generalization is found in (Waterman et al. 1976). The advantages of having a metric as the evolutionary distance are many. For example this property may aid in construction of phylogenetic trees, which are the biologically most natural way of describing evolution.

Proposition 5.7.1 *If*

$$\overline{\rho}(\mathbf{x}^*, \mathbf{y}^*) = \sum_{i=1}^{\infty} \rho(x_i^*, y_i^*), \tag{7.1}$$

where ρ is a metric on $\mathcal{X}^*$, then

$$D(\mathbf{x}, \mathbf{y}) = \min_{\mathbf{x}^*, \mathbf{y}^*} \overline{\rho}(\mathbf{x}^*, \mathbf{y}^*) \tag{7.2}$$

is a metric on the set of sequences (finite strings of symbols of $\mathcal{X}$).

Proof: We start by changing the domain of $D(\mathbf{x}, \mathbf{y})$. For every sequence $\mathbf{x}$ there is a class of padded sequences $\mathbf{x}^*$ obtained by inserting '$-$' in $\mathbf{x}$. We denote this class by $X(\mathbf{x})$ ($\subseteq \mathcal{X}^{*\infty}$). The sequence $\mathbf{x}$ belongs itself to $X(\mathbf{x})$. We can interpret (7.2) as a function of pairs of such classes $X(\mathbf{x})$,$X(\mathbf{y})$. For since every padded sequence in $X(\mathbf{x})$ contains the same finite number of letters, we have that

$$D(\mathbf{x}, \mathbf{y}) = D(X(\mathbf{x}), X(\mathbf{y})),$$

as we can define $D(X(\mathbf{x}), X(\mathbf{y}))$ as the distance between the the closest padded sequences in the respective classes which is

$$D(X(\mathbf{x}), X(\mathbf{y})) = \min_{\mathbf{x}^*, \mathbf{y}^*} \overline{\rho}(\mathbf{x}^*, \mathbf{y}^*),$$

The minimum in (3.6) is by definition computed over all the pseudoalignments of $\mathbf{x}$ with $\mathbf{y}$ and this means just searching in $X(\mathbf{x})$ and in $X(\mathbf{y})$. Also,

$$D(X(\mathbf{x}), X(\mathbf{x})) = 0$$

since we can here align any padded equivalent of $\mathbf{x}$ with itself using identities. Thus we can just as well regard D as a function of equivalence classes of padded sequences in $\mathcal{X}^*$ with finitely many letters.

Since $\overline{\rho}$ inherits the property of being a metric from ρ, it is obvious that the defining properties (a)-(c) of a metric are transferred to $D(X(\mathbf{x}), X(\mathbf{y}))$. We need only to prove the triangle inequality (d). We consider

$$D(X(\mathbf{x}), X(\mathbf{y})) + D(X(\mathbf{y}), X(\mathbf{z})) = \overline{\rho}(\mathbf{x}^*, \mathbf{y}_1^*) + \overline{\rho}(\mathbf{y}_2^*, \mathbf{z}^*),$$

where we have used the definition (7.2) for some representatives $\mathbf{x}^*$, $\mathbf{y}_1^*$, $\mathbf{y}_2^*$ and $\mathbf{z}^*$ of the pertinent equivalence classes. Next we prove that we can modify

$\mathbf{y}_1^*$ and $\mathbf{y}_2^*$ so that they become identical padded sequences without changing the values of $\overline{\rho}(\mathbf{x}^*, \mathbf{y}_1^*)$ and $\overline{\rho}(\mathbf{y}_2^*, \mathbf{z}^*)$.

For this we proceed as follows. The two padded sequences $\mathbf{y}_1^*$ and $\mathbf{y}_2^*$ are different only with respect to the appearance of the elements $'-'$. The letters in $\mathbf{y}_1^*$ and $\mathbf{y}_2^*$ are the same. Hence we add $'-'$ to both of them to augment them so that they sit in the same padded sequence, say $\mathbf{y}^*$.

Every time we add $'-'$ to $\mathbf{y}_1^*$ we add the $'-'$ to the same position in $\mathbf{x}^*$ and wherever $'-'$ is put to $\mathbf{y}_2^*$ we put $'-'$ in the same position in $\mathbf{z}^*$, too. This does not change the values of the metric $\overline{\rho}$. Here we recall that columns $\begin{smallmatrix}-\\-\end{smallmatrix}$ are permitted in pseudo alignments. Assuming that this procedure has been carried out and denoting the modified padded sequences again with $\mathbf{x}^*$ and $\mathbf{z}^*$ we have

$$\begin{aligned}\overline{\rho}(\mathbf{x}^*, \mathbf{y}_1^*) + \overline{\rho}(\mathbf{y}_2^*, \mathbf{z}^*) &= = \overline{\rho}(\mathbf{x}^*, \mathbf{y}^*) + \overline{\rho}(\mathbf{y}^*, \mathbf{z}^*) \\ &\geq \overline{\rho}(\mathbf{x}^*, \mathbf{z}^*)\end{aligned}$$

since $\overline{\rho}$ is a metric and thus obeys the triangle inequality. But then by the definition of D we have

$$\overline{\rho}(\mathbf{x}^*, \mathbf{z}^*) \geq D(X(\mathbf{x}), X(\mathbf{z})),$$

which proves the assertion, that is,

$$D(X(\mathbf{x}), X(\mathbf{y})) + D(X(\mathbf{y}), X(\mathbf{y})) \geq D(X(\mathbf{x}), X(\mathbf{z})). \qquad \blacksquare$$

5.7.2 Edge Lengths of a Tree and the Related Metrics

The notion of a metric is of importance in constructing (taxonomic) trees from similarity data or in phylogenetic alignment (Gusfield 1997). A phylogenetic alignment involves the evolutionary relationships (or models of them) between the sequences to be aligned. In these applications there are additional properties required form a metric.

An undirected graph is called a *tree* if it is connected and contains no cycles. The edge lengths are a set of positive weights u_e associated with each edge of a tree. In biological terms they often indicate the distance in time of evolutionary events between the nodes.

If $E(x, y)$ is a path between the nodes x and y in a tree, then

$$\delta(x, y) = \sum_{e \in E(x,y)} u_e \tag{7.3}$$

is *metric* between nodes in a tree. If x, y, z, t are nodes in a tree this metric satisfies the condition

$$\delta(x,y) + \delta(z,t) \leq \max\left(\delta(x,z) + \delta(y,t), \delta(x,t) + \delta(y,z)\right). \qquad (7.4)$$

This is a stronger requirement than the triangle inequality.

Conversely, if we have a metric on a set of points V that satisfies (7.4), then there is a tree with V amongst its nodes and with edge lengths λ_e that give $\delta(x,y)$ as $\delta(x,y) = \sum_{e \in E(x,y)} u_e$. These statements are owed to (Buneman 1974). The property

$$\delta(x,y) \leq \max\left(\delta(x,z), \delta(z,y)\right). \qquad (7.5)$$

implies (7.4) and is known as the *ultra metric property*.

5.8 References and Further Reading:

P. Bucher and K. Hoffmann (1996): A Sequence Similarity Search Algorithm Based on a Probabilistic Interpretation of an Alignment Scoring System. *International Conference on Intelligent Systems for Molecular Biology*, pp. 44–51.

P. Buneman (1974): A Note on the Metric Properties of Trees. *Journal of Combinatorial Theory* (**B**), pp. 48–50.

D. Gusfield (1997): *Algorithms on Strings, Trees and Sequences. Computer Science and Computational Biology.* Cambridge University Press. Cambridge.

M.S. Johnson and J.P. Overington (1993): A Structural Basis for Sequence Comparisons. An Evaluation of Scoring Methodologies. *Journal of Molecular Biology*, 233, pp. 716–738.

S.B. Needleman and C.D. Wunsch (1970): A General Method Applicable to the Search for Similarities in the Amino Acid Sequence of Two Proteins. *Journal of Molecular Biology*, 48, pp. 443–453.

P.H. Sellers (1974): On the Theory and Computation of Evolutionary Distances. *SIAM Journal Applied Mathematics*, 26, pp. 787–793.

T.F. Smith and M.S. Waterman (1981): Identification of Common Molecular Subsequences. *Journal of Molecular Biology*, 147, pp. 645–656.

E. Ukkonen (1985): Algorithms for Approximate String Matching. *Information and Control*, 64, p. 100–118.

M.S. Waterman, T.F. Smith and W.A. Beyer (1976): Some Biological Sequence Metrics. *Advances in Mathematics*, 20, pp. 367–387.

M.S. Waterman (1995): *Introduction to Computational Biology. Maps, sequences and genomes.* Chapman and Hall, London, Glasgow, Weinheim, New York, Melbourne, Madras.

Chapter 6

Mixture Models and Profiles

6.1 Summary and Overview

6.1.1 Prior Information for a Weight Matrix

A key idea in any probabilistic analysis is the idea of modelling. It is models (theories) of the world that determine what information to condition on (and what possible outcomes are) in making probabilistic estimates. The rules of probability show how different statements of uncertainty have to fit together (coherence).

Chapter 1 introduced simple probability models for families of sequences like the weight matrices

$$\mathcal{M}: \quad \begin{array}{ccc} p_1(o_1) & \ldots & p_N(o_1) \\ p_1(o_2) & \ldots & p_N(o_2) \\ \vdots & \vdots & \vdots \\ p_1(o_K) & \ldots & p_N(o_K) \end{array} . \tag{1.1}$$

Each column is a probability distribution $\underline{p}_j = (p_j(o_l))_{l=1}^{K}, \quad j = 1, \ldots, N$, on the alphabet $\mathcal{O} = \{o_1, \ldots, o_K\}$. The purpose here is to regard $\underline{p}_j$ as random variables with a common prior density $\phi(\underline{p})$. The prior $\phi(\underline{p})$ will be modeled as a mixture of Dirichlet densities. The parameters of the mixture are estimated using a multiple alignment of sequences. In these terms $\mathcal{M}$ in (1.1) corresponds to the statistics of the symbols in a match segment of the alignment.

The introduction of this prior density based on the background knowledge in the form of multiple alignments is, by the well known formulas of mean posterior probabilities with Dirichlet priors, also related to the capacity for generalization, when using an estimated weight matrix. A non-probabilistic account of generalization and background knowledge in molecular biology is (Hirsh and Noordewier 1994).

6.1.2 Family and Superfamily Representations

By a *family* we mean a set of strings defined by *biological similarity*. Often one wants to find common properties (like substrings) that characterize or represent the family.

Next we quote the statement about protein classification by the group (Murzin et al. 1995)[1] responsible for the upkeep and development of **scop** (Structural Classification of Proteins) database. In **scop** proteins are classified to reflect both structural and evolutionary relatedness. The major levels in the hierarchy are family, superfamily, and fold, described below.

Quote 6.1.1 We recapitulate a slight modification of (Murzin et al. 1995).

- Family: Proteins clustered together into families are clearly evolutionarily related or **homologous**. Generally this means that pairwise symbol identities between the protein sequences are 30% and greater.

- Superfamily: Proteins that have low sequence identities, but whose structural and functional features suggest that a common evolutionary origin is probable are placed together in superfamilies.

- Fold: Proteins are defined as having a common fold if they have same major secondary structures in same arrangement and with the same topological connections. Proteins placed together in the same fold category may not have a common evolutionary origin: the structural similarities could arise just from the physics and chemistry of proteins favoring certain packing arrangements and chain topologies.

- Class: Convenient clustering of folds.

[1]See also `http://scop.mrc-lmb.cam.ac.uk/scop/`

In the course of molecular evolution protein sequences can divert beyond simple sequence comparison and therefore methods detecting distant homologies are needed. Here statistical methods like profiles and Hidden Markov Modelling (HMM) have turned out to work well.

6.1.3 Multiple Alignment and Priors of Mixtures of Dirichlet Densities

One source of prior knowledge useful for modelling in bioinformatics is found in trusted multiple alignments of sequences belonging to a family. If the family is thought to include subfamilies it may feel too restrictive to assume only *one* prior distribution for all columnwise probability distributions.

In (Sjölander et al. 1996) alignment information is used to estimate a prior mixture distribution for profile probability distributions. The mixture distribution used in parameter estimation is, in fact, a mixture of model-averaged or predictive multinomial distributions for the frequency counts in some chosen columns of a multiple alignment. The model averaging is done by integrating a multinomial distribution for the frequency counts with respect to a mixture of Dirichlet densities. Tatusov et al. (1994) have reported good empirical results in detection of conserved segments using mixtures of Dirichlet priors. Another application of mixtures of Dirichlet densities is protein classification in (Wang et al. 1996).

6.1.4 On a General Theory of Mixtures of Dirichlet Densities

There is a mathematical theory of continuous mixtures of Dirichlet densities developed initially in (Antoniak 1974, Ferguson 1973) for extension of Bayesian methods to non-parametric problems in statistics.

Ferguson considered a random probability measure μ, to be called a Dirichlet process, on a separable metric space S such that for any measurable partition $s_1, \ldots, s_K$ of S the vector

$$\mu(s_1), \ldots, \mu(s_K)$$

has a Dirichlet distribution $\mathrm{Dir}(\alpha(s_1), \ldots, \alpha(s_K))$, where α is a non-negative, finite measure on S.

Next a random probability measure μ on S is called a *mixture of Dirichlet processes* if

$$\mathbf{P}\left(\mu\left(s_1\right) \leq y_1, \mu\left(s_2\right) \leq y_2, \ldots, \mu\left(s_K\right) \leq y_K\right)$$
$$= \int_U F_{\mathrm{Dir}}\left(y_1, y_2, \ldots, y_K | \alpha\left(s_1, u\right), \ldots, \alpha\left(s_K, u\right)\right) dH(u),$$

where $F_{\mathrm{Dir}}\left(y_1, y_2, \ldots, y_K | \alpha\left(s_1, u\right), .., \alpha\left(s_K, u\right)\right)$ is the distribution function of a density $\mathrm{Dir}\left(\alpha_1, \ldots, \alpha_K\right)$ and U is an index space and H is a probability distribution on U.

One most interesting fact is that μ is almost surely a discrete probability on S. Then Antoniak shows that the size distribution of frequencies of distinct values in a sample (of values in S) of size n from μ is the *Ewens distribution* (Tavaré and Ewens 1998) known in statistical genetics. These facts are not needed in the sequel.

6.2 A Finite Mixture of Prior Dirichlets

6.2.1 Multiple Alignment and Prior Information

The prior knowledge to be used with the mixture modelling is a set of training sequences

$$\left\{\mathbf{o}^{(1)}, \ldots, \mathbf{o}^{(T)}\right\} \tag{2.1}$$

with symbols in

$$\mathcal{O} = \left\{o_1, \ldots, o_K\right\},$$

and *multiple alignments* for them. A multiple alignment is constructed in principle by insertion of '$-$' into (or at either end of) each of the strings so that the resulting sequences have the same length. Then the strings are arrayed in T rows of L columns each, so that each element of each extended string is in a unique column. Some scoring criterion is required for an algorithmic solution. The algorithms for multiple alignment are treated in (Gusfield 1997). In theory the dynamic programming algorithms in Chapter 5 can be extended to find optimal multiple alignments, but in practice this is not computationally feasible. Let us suppose that an algorithm has produced a multiple alignment with at least one *match segment*, a bit vaguely defined as a segment of N adjacent columns such that there are many symbol identities over each of the columns (recall the definition of a sequence family above).

For presentation we extend, as shown earlier, the alphabet $\mathcal{O}$ by adding the symbol $'-'$ for insertion and deletion

$$\mathcal{O}^{*} = \{o_1, \ldots, o_K, '-'\},$$

and write a match segment $\mathcal{A}$ in a multiple alignment as strings of symbols for this alphabet or

$$\mathcal{A}: \quad \begin{array}{ccc} o_1^{(1)\,*} & \ldots & o_N^{(1)\,*} \\ o_1^{(2)\,*} & \ldots & o_N^{(2)\,*} \\ \vdots & \vdots & \vdots \\ o_1^{(T)\,*} & \ldots & o_N^{(T)\,*}. \end{array} \tag{2.2}$$

The purpose of multiple alignment can perhaps be said to be to find an array $\mathcal{A}$ such that regions of common motifs in $\mathbf{o}^{(l)}, l = 1, \ldots, T$ are made visible.

Formally, for the alignment in $\mathcal{A}$ in (2.2) we set for $l = 1, \ldots, K$ and for $t = 1, \ldots, N$

$$n_{tl} = \sum_{i=1}^{T} I_{\{o_t^{(i)\,*} = o_l\}}, \tag{2.3}$$

where

$$I_{\{o_t^{(i)\,*} = o_l\}} = \begin{cases} 1 & \text{if } o_t^{(i)\,*} = o_l, \\ 0 & \text{otherwise.} \end{cases}$$

Note that the spaces $'-'$ are not counted here. Each n_{tl} is simply the number of times (i.e., frequency) the symbol o_l occurs in column t in $\mathcal{A}$ in (2.2). We obtain the counts of symbol frequencies corresponding to column t as

$$\underline{n}_t = (n_{t1}, \ldots, n_{tK}). \tag{2.4}$$

The effective data set is thus N count vectors $\underline{n}_1, \underline{n}_2, \ldots, \underline{n}_N$ of the symbol frequencies from N match columns in some multiple alignments. Many multiple alignments of different (protein) families are included so that N may become large. We set

$$n(t) := n_{t,1} + n_{t,2} + \ldots + n_{t,K}.$$

A piece of expert knowledge about proteins quoted in (Sjölander et al. 1996) is that amino acids can be grouped according to similarity along several properties and hence be naturally representable by a mixture. As the family of models for distributions (Sjölander et al. 1996) suggest mixtures $\phi(\underline{p})$ of Dirichlet distributions ϕ_j on the simplex S_K defined in (7.1),

$$\phi(\underline{p}) = q_1\phi_1(\underline{p}) + \ldots + q_j\phi_j(\underline{p}) + \ldots + q_M\phi_M(\underline{p}),$$

where

$$\phi_j(\underline{p}) = \frac{\Gamma\left(\sum_{i=1}^{K}\alpha_{j,i}\right)}{\prod_{i=1}^{K}\Gamma(\alpha_{j,i})}\prod_{i=1}^{K}p_i^{\alpha_{j,i}-1}. \tag{2.5}$$

By this we look for several underlying *prototypical distributions* $\phi_j(\underline{p})$ that can be seen as generating the observed symbol frequency counts $\underline{n}$:

1) First a component ϕ_j is selected at random according to the mixing distribution $(q_j)_{j=1}^{M}$.

2) Then a probability distribution $\underline{p} = (p_1, \ldots, p_K)$ is chosen at random according to ϕ_j.

3) Finally, $\mathbf{o}^{(l)} = o_{j_1}^{(l)} \ldots o_{j_l}^{(l)}$ for $l = 1, \ldots, T$ are generated independently according to the distribution $\underline{p}$. Or, alternatively, $\underline{n}_t$ are generated from a multinomial distribution with the parameters $\underline{p}$.

The parameters are the mixing coefficients q_j and the hyperparameters of the components ϕ_j and are to be estimated using the frequency count data (2.4). The estimation is done by a maximum likelihood method. First we derive the likelihood function for the frequency counts.

6.2.2 A Likelihood Function for the Frequency Counts

It turns out to be useful to expand the notation. We designate the hyperparameters by

$$\underline{\alpha}_j = (\alpha_{j,1}, \ldots, \alpha_{j,K})$$

and set

$$\Theta = \{\underline{\alpha}_j, q_j; j = 1, \ldots, M\},$$

and from (2.5)

$$\phi(\underline{p}; \underline{\alpha}_j) = \phi_j(\underline{p}).$$

Let also a multinomial distribution for the frequency counts (see Section 1.2.9 in Chapter 1) with the parameter $\underline{p}$ be denoted by $P(\underline{\mathbf{n}} = \underline{n} | \underline{\mathbf{p}} = \underline{p})$.

As we want to estimate the hyperparameters for the whole model family, model averaging is done as in Chapter 3. Using the mixture of Dirichlet densities model averaging gives

$$\begin{aligned} P\left(\underline{\mathbf{n}}=\underline{n};n,\Theta\right) &= \int_{S_K} P\left(\underline{\mathbf{n}}=\underline{n}|\underline{\mathbf{p}}=\underline{p}\right)\phi\left(\underline{p}\right)d\underline{p} \\ &= \sum_{j=1}^{M} q_j \int_{S_K} P\left(\underline{\mathbf{n}}=\underline{n}|\underline{\mathbf{p}}=\underline{p}\right)\phi\left(\underline{p};\underline{\alpha}_j\right)d\underline{p}, \end{aligned} \tag{2.6}$$

where S_K is given in (7.1) below. We write (2.6) as

$$P\left(\underline{\mathbf{n}}=\underline{n};n,\Theta\right) = \sum_{j=1}^{M} q_j P\left(\underline{\mathbf{n}}=\underline{n};n,\underline{\alpha}_j\right), \tag{2.7}$$

where

$$P\left(\underline{\mathbf{n}}=\underline{n};n,\underline{\alpha}_j\right) := \int_{S_K} P\left(\underline{\mathbf{n}}=\underline{n}|\underline{\mathbf{p}}=\underline{p}\right)\phi\left(\underline{p};\underline{\alpha}_j\right)d\underline{p}$$

is a *predictive multinomial distribution* or *compound multinomial distribution* (Mosimann 1962). Here we have

$$P\left(\underline{\mathbf{n}}=\underline{n};n,\underline{\alpha}_j\right) = \frac{\Gamma(n+1)\Gamma(\sum_{i=1}^{K}\alpha_{j,i})}{\Gamma(n+\sum_{i=1}^{K}\alpha_{j,i})} \prod_{i=1}^{K} \frac{\Gamma(n_i+\alpha_{j,i})\Gamma(\alpha_{j,i})}{\Gamma(n_i+1)}. \tag{2.8}$$

The details behind this formula are given in Section 6.8.2 below. We take $P\left(\underline{\mathbf{n}}=\underline{n};n,\Theta\right)$ as the likelihood function for Θ on basis of the data $\underline{n}$. The log likelihood function $\mathcal{L}\left(\Theta\right)$ for Θ with the N count sequences

$$\underline{n}_1, \underline{n}_2, \ldots, \underline{n}_N$$

is equal to the sum of the log likelihoods (multiplied by minus one) of the columnwise frequency counts

$$\mathcal{L}\left(\Theta\right) = -\sum_{t=1}^{N} \log P\left(\underline{\mathbf{n}}=\underline{n}_t;n(t),\Theta\right). \tag{2.9}$$

Since $P\left(\underline{\mathbf{n}}=\underline{n}_t;n(t),\Theta\right)$ contains an averaging over the parameters in a family of models, the minimization of $\mathcal{L}\left(\Theta\right)$ corresponds to the minimization of the sum of Bayesian codelengths of $\underline{n}_t, t=1,\ldots,N$, as presented in Chapter

3. In the terminology of (Berger 1985, Section 3.5) this would amount to the 'ML–II approach to prior selection'.

Empirically and theoretically the estimation of a mixture involves the problem of *estimation of M, the number of distributions in the mixture.* A survey of this topic is (Fraley and Raftery 1998). In the applications to protein sequences different practical judgements and heuristics (Sjölander et al. 1996, R. Hughey and A. Krogh 1995) are used.

The asymptotics of the maximum likelihood estimate for mixtures are discussed in (Redner 1981) under the assumption that the data are actually generated by one of the models in the family of mixtures. (Redner 1981) gives also a method for estimating M.

6.2.3 The Mixture Estimation Algorithm

There would be a straightforward procedure for the iterative estimates, if the distributions $P\left(\underline{\mathbf{n}} = \underline{n}; n, \underline{\alpha}_j\right)$ in (8.7) were of the exponential type of family treated in 4. We give another related algorithm, the idea of which is present in (Duda and Hart 1973), as applied in (Sjölander et al. 1996).

Estimation of the Hyper Parameters $\underline{\alpha}_j$

First we reparametrize using

$$\alpha_{j,i} = e^{w_{j,i}},$$

which has the advantage that $w_{j,i}$ is an unconstrained real number. Then from (2.9)

$$\frac{\partial}{\partial w_{j,i}} \mathcal{L}(\Theta) = -\sum_{t=1}^{N} \frac{\partial}{\partial w_{j,i}} \log P\left(\underline{\mathbf{n}} = \underline{n}_t; n(t), \Theta\right) \cdot \frac{\partial}{\partial w_{j,i}} \alpha_{j,i}.$$

Let now $P\left(\xi = j | \underline{n}_t, n(t), \Theta\right)$ be a posterior probability which is an update of q_j by Bayes' rule as

$$P\left(\xi = j | \underline{n}_t, n(t), \Theta\right) = \frac{q_j P\left(\underline{\mathbf{n}}_t = \underline{n}_t; n(t), \underline{\alpha}_j\right)}{\sum_{r=1}^{M} q_r P\left(\underline{\mathbf{n}}_t = \underline{n}_t; n(t), \underline{\alpha}_r\right)}, \tag{2.10}$$

By Lemma 6.9.1 in the Appendix we obtain

$$\frac{\partial}{\partial \alpha_{j,i}} \log P\left(\underline{\mathbf{n}} = \underline{n}_t; n, \Theta\right) = P\left(\xi = j | \underline{n}_t, n(t), \Theta\right) \cdot \frac{\partial}{\partial \alpha_{j,i}} \log P\left(\underline{\mathbf{n}} = \underline{n}_t; n(t), \underline{\alpha}_j\right).$$

Thus

$$\frac{\partial}{\partial w_{j,i}}\mathcal{L}(\Theta) = -\sum_{t=1}^{N} \frac{\partial}{\partial \alpha_{j,i}} \log P\left(\mathbf{\underline{n}} = \underline{n}_t; n(t), \underline{\alpha}_j\right) \cdot P\left(\xi = j | \underline{n}_t, n(t), \Theta\right) \cdot \frac{\partial}{\partial w_{j,i}} \alpha_{j,i}.$$

Furthermore, from Lemma 6.9.2 in the Appendix

$$\begin{aligned}\frac{\partial}{\partial \alpha_{j,i}} \log P\left(\mathbf{\underline{n}} = \underline{n}_t; n(t), \underline{\alpha}_j\right) &= \Psi\left(\sum_{i=1}^{K} \alpha_{j,i}\right) - \Psi\left(n(t) + \sum_{i=1}^{K} \alpha_{j,i}\right) \\ &\quad + \Psi\left(n_{t,i} + \alpha_{j,i}\right) - \Psi\left(\alpha_{j,i}\right).\end{aligned}$$

Here Ψ is the *digamma* function, the logarithmic derivative of $\Gamma(z)$,

$$\Psi(z) = \frac{d}{dz} \log \Gamma(z) = \frac{\Gamma'(z)}{\Gamma(z)}. \tag{2.11}$$

Ψ is easily available for computation, for example, as a standard expression in Mathematica^R (Ruskeepää 1999). Thus

$$\frac{\partial}{\partial w_{j,i}}\mathcal{L}(\Theta) = 0$$

yields

$$\sum_{t=1}^{N} \alpha_{j,i} P\left(\xi = j | \underline{n}_t, n(t), \Theta\right) \cdot \Phi(t) = 0, \tag{2.12}$$

where

$$\Phi(t) = \Psi\left(\sum_{i=1}^{K} \alpha_{j,i}\right) - \Psi\left(n(t) + \sum_{i=1}^{K} \alpha_{j,i}\right) + \Psi\left(n_{t,i} + \alpha_{j,i}\right) - \Psi\left(\alpha_{j,i}\right). \tag{2.13}$$

A Newton–Raphson algorithm for estimating the parameters of a single Dirichlet density is given in (Narayanan 1991), who thus needs a special case of (2.12).

Estimation of the Mixing Coefficients

First we reparametrize by

$$q_j = \frac{Q_j}{\sum_{j=1}^{K} Q_j}, \qquad j = 1, \ldots, K.$$

Then from (2.9)

$$\frac{\partial}{\partial Q_j}\mathcal{L}(\Theta) = -\sum_{t=1}^{N}\frac{\partial}{\partial Q_j}\log P\left(\underline{\mathbf{n}} = \underline{n}_t; n(t), \Theta\right).$$

Taking

$$P\left(\xi = j|\underline{n}_t, n(t), \Theta\right) = \frac{q_j P\left(\underline{\mathbf{n}}_t = \underline{n}_t; n(t), \underline{\alpha}_j\right)}{\sum_{r=1}^{M} q_r P\left(\underline{\mathbf{n}}_t = \underline{n}_t; n(t), \underline{\alpha}_r\right)}$$

and invoking Lemma 6.9.4 we obtain (exercise) from $\frac{\partial}{\partial Q_j}\mathcal{L}(\Theta) = 0$

$$\widehat{q}_j = \frac{Q_j}{\sum_{j=1}^{K} Q_j} = \frac{1}{N}\sum_{t=1}^{N} P\left(\xi = j|\underline{n}_t, n(t), \Theta\right), \tag{2.14}$$

which is an intuitively natural result.

Summary and Iteration

The equations (2.12) and (2.14) constitute a system of coupled non-linear equations that are solved by an iteration of EM - type or by a gradient descent algorithm. We denote the resulting estimates by

$$q_j^{\mathrm{ML}}, \qquad \underline{\alpha}_j^{\mathrm{ML}}, \qquad j = 1, \ldots, M.$$

6.3 Estimates of the Symbol Probability Distributions

Given an estimate of Θ the Bayesian estimate of $\underline{p}_j$ in the weight matrix (1.1) (derived from the multiple alignment $\mathcal{A}$ in (2.2)) can be evaluated. We use (8.12) in section 6.8.4.

This means that we estimate the probability of symbol o_i at a specific position in the alignment as

$$\widehat{p}_i = \sum_{j=1}^{M}\widehat{q}_j \cdot \frac{\sum_{t=1}^{N} n_{t,i} + \widehat{\alpha}_{j,i}^{\mathrm{ML}}}{\sum_{t=1}^{N} n(t) + \sum_{i=1}^{K}\widehat{\alpha}_{j,i}^{\mathrm{ML}}}. \tag{3.1}$$

Here $\sum_{t=1}^{N} n_{t,i}$ is equal to the number of times the letter o_i appears in an arbitrarily chosen match column in the pertinent multiple alignment.

In (3.1) the mixing probability $\widehat{q}_j$ can be either the estimate of $\widehat{q}_j = q_j^{\mathrm{ML}}$ given by the algorithm or as the estimate obtained as

$$\frac{q_j^{\mathrm{ML}} \prod_{t=1}^{N} P\left(\underline{\mathbf{n}}_t = \underline{n}_t; n(t), \underline{\alpha}_j^{\mathrm{ML}}\right)}{\sum_{r=1}^{M} q_r^{\mathrm{ML}} \prod_{t=1}^{N} P\left(\underline{\mathbf{n}}_t = \underline{n}_t; n(t), \underline{\alpha}_r^{\mathrm{ML}}\right)}.$$

6.4 Aligning a Sequence to a Profile

One classification or discrimination problem in bioinformatics is to consider a protein for membership in an established cluster at some level of a taxonomic hierarchy, or searching a protein database for a new candidate member of a protein family (Wang et al. 1996). The profile analysis (Gribskov et al. 1987, 1990, 1996) for this problem is based on the idea that for discrimination it is usually much more effective to align the candidate string to representation of the known family members rather than to align to single members of the family.

6.5 Profile

We consider again the match segment (2.2) in a multiple alignment.

Definition 6.5.1 A *profile* for a given multiple alignment of strings $\mathcal{A}$ specifies for each column the *relative frequency* that each symbol appears in that column. ∎

We set for $l = 1, \ldots, K$

$$p_{tl} = \frac{n_{tl}}{T} = \frac{1}{T} \sum_{i=1}^{T} I_{\{o_t^{(i)*} = o_l\}}, \qquad \text{for } o_l \in \mathcal{O}, \qquad 1 \le t \le N, \tag{5.1}$$

Thus p_{tl} is the relative frequency of the symbol o_l in column t in $\mathcal{A}$.

We complete this by augmenting with the relative frequency for the space $'-'$ and making an abuse of the notation we set

$$n_{t0} = \sum_{i=1}^{T} I_{\{o_t^{(i)*} = '-'\}}, \qquad p_{t0} = \frac{n_{t0}}{T}.$$

We then set

$$\mathbf{p}_t = (p_{t0}, p_{t1} \ldots, p_{tK}). \tag{5.2}$$

The *profile* **Pr** in the sense of definition 6.5.1 is then

$$\mathbf{Pr} = (\mathbf{p}_1, \mathbf{p}_2, \ldots, \mathbf{p}_N). \tag{5.3}$$

A profile seen in this way is an $(K+1) \times N$ matrix. Each column in the matrix corresponds to the relative frequencies of the elements in the alphabet in that position (column) in $\mathcal{A}$.

6.5.1 Optimal Fit of a String to a Profile

Given a profile **Pr** one frequently asks how to find the *best similarity fit* of a new string $\mathbf{o}^*$ with the profile **Pr**. A similarity function $\widehat{s}$ or *score* between a sequence and a profile is to be maximized. First we have the log likelihood weighting

$$\widehat{s}(\mathbf{p}_i, o) = -\sum_{k=1}^{K} s(o_k, o) \log \widehat{p}_{ik}, \tag{5.4}$$

where $s(o_k, o)$ is a similarity measure on $\mathcal{O}$ (like in (5.2) in Chapter 6). Here we can think of averaging with respect to Dirichlet mixtures or regularization for obtaining $\widehat{p}_{ik}$ from the column $\mathbf{p}_j$ in profile and hence log is well defined. Also, the variant

$$\widehat{s}(\mathbf{p}_i, o) = -\sum_{k=1}^{K} s(o_k, o) \log \frac{\widehat{p}_{ik}}{\rho_k}$$

is useful, where ρ_k is a probability (background probability) for o_k established from the entire database including the family or other cluster we are interested in.

Let $\mathbf{o}^* = o_1^* \ldots o_j^* \ldots o_{n+L-1}^*$ be a sequence the membership of which in a family we are interested in. What is sought is the segment in this sequence where the profile **Pr** in (5.3) would fit best. Formally this means that we want to find

$$\max_{1 \le j \le n} T_{L,j},$$

where

$$T_{L,j} = \sum_{i=1}^{L} \widehat{s}\left(\mathbf{p}_i, o_{i+j-1}^*\right), \qquad j = 1, \ldots, n.$$

Two different gap penalty functions will be be required. For the gaps in a sequence $\mathbf{o}$ we take the affine function

$$\widehat{g}_{seq}(n) = \alpha + \beta \cdot (n-1). \tag{5.5}$$

For the gaps in the profile $\mathbf{Pr}$ we introduce the penalty $-\gamma_i$ for opening a gap at position $\mathbf{p}_i$ and the penalty $-\delta_i$ for extending a gap through $\mathbf{p}_i$.

Then we can establish the following algorithm just as the alignment algorithms in Chapter 5 were established.

Proposition 6.5.1 *We initialize*

$$T_{0,j} = 0, \qquad 0 \leq j \leq n + L - 1, \qquad T_{i,0} = \kappa_1 + \sum_{k=2}^{i} \delta_k$$

$$E_{i,j} = \max \left\{ T_{i,j-1} - \alpha, E_{i,j-1} - \beta \right\}, \tag{5.6}$$

and

$$F_{i,j} = \max \left\{ T_{i-1,j} - \gamma_i, F_{i-1,j} - \delta_i \right\}, \tag{5.7}$$

then

$$T_{i,j} = \max \left[T_{i-1,j-1} + \widehat{s}\left(\mathbf{p}_i, o^*_{i+j-1}\right), E_{i,j}, F_{i,j} \right]. \tag{5.8}$$

Thereby we have obtained the best alignment with the score

$$T\left(\mathbf{Pr}, \mathbf{o}^{(r+1)}\right) = \max\{T_{L,j} | 1 \leq j \leq n\}.$$

■

This can be easily modified to give a local alignment of the sequence with the profile, i.e., to determine where in the sequence the entire profile fits best. The indication of similarity or good fit of $\mathbf{o}^*$ with the profile is that the score $T(\mathbf{Pr}, \mathbf{o}^*)$ will be high. An extreme value theory for the significance of the maximal score is derived in (Goldstein and Waterman 1994).

For discrimination between alternative families of sequences for $\mathbf{o}^*$ we may repeat the prodecure for match segments of each family, respectively, and $\mathbf{o}^*$ and then choose the family that gives the best score $T(\mathbf{Pr}, \mathbf{o}^*)$.

6.6 Exercises

1. **Beta Mixtures**
   Let X be a Bernoulli random variable $Be(\theta)$, that is

$$P(X = 1 \mid \theta) = \theta.$$

   Let the prior density ϕ for θ be a mixture of two *Beta* densities or

$$\phi(\theta) = \frac{1}{2}\mathcal{B}\mathrm{e}(\theta; 1, 2) + \frac{1}{2}\mathcal{B}\mathrm{e}(\theta; 2, 2).$$

   Show by a straightforward calculation that the posterior density for θ given $X = 1$ is

$$\phi(\theta \mid X = 1) = \frac{2}{5}\mathcal{B}\mathrm{e}(\theta; 2, 2) + \frac{3}{5}\mathcal{B}\mathrm{e}(\theta; 3, 2).$$

   What is the interpretation of this result?

2. Show (2.14), i.e.,

$$\widehat{q}_j = \frac{1}{N}\sum_{t=1}^{N} P\left(\xi = j | \underline{n}_t, n(t), \Theta\right). \tag{6.1}$$

   Hint: Use the reparametrization and Lemma 6.9.4.

3. **Product Moments of Dirichlet Distributions**
   Let

$$\underline{\mathbf{p}} = (\mathbf{p}_1, \ldots, \mathbf{p}_K)$$

   be a random variable that assumes values in the simplex S_K defined in (7.1). If E denotes expectation with respect to the Dirichlet density $\mathrm{Dir}(\alpha_1, \ldots, \alpha_K)$, then the expectation of the ith component in $\underline{\mathbf{p}}$ is

$$E[\mathbf{p}_i] = \frac{\alpha_i}{\sum_{i=1}^{K} \alpha_i}. \tag{6.2}$$

   This gives an interpretation of the hyperparameters α_i.

(a) Show that for $i \neq j$ the product moment is

$$E[\mathbf{p}_i \cdot \mathbf{p}_j] = \frac{\alpha_i \alpha_j}{\left(\sum_{i=1}^K \alpha_i\right) \cdot \left(\sum_{i=1}^K \alpha_i + 1\right)} \tag{6.3}$$

and

$$E\left[\mathbf{p}_i^2\right] = \frac{\alpha_i (\alpha_i + 1)}{\left(\sum_{i=1}^K \alpha_i\right) \cdot \left(\sum_{i=1}^K \alpha_i + 1\right)}. \tag{6.4}$$

(b) Show that the variance of $\mathbf{p}_i$ is

$$\mathrm{Var}[\mathbf{p}_i] = \frac{\alpha_i \left(\sum_{i=1}^K \alpha_i - \alpha_i\right)}{\left(\sum_{i=1}^K \alpha_i\right)^2 \cdot \left(\sum_{i=1}^K \alpha_i + 1\right)}. \tag{6.5}$$

(c) Show that if the *covariance* between the random variables X and Y is defined as

$$\mathrm{cov}(X, Y) := E[(X - E(X)) \cdot (Y - E(Y))],$$

then

$$\mathrm{cov}(\mathbf{p}_i, \mathbf{p}_j) = \frac{-\alpha_i \alpha_j}{\left(\sum_{i=1}^K \alpha_i\right)^2 \cdot \left(\sum_{i=1}^K \alpha_i + 1\right)}. \tag{6.6}$$

4. Let

$$\underline{\mathbf{p}} = (\mathbf{p}_1, \ldots, \mathbf{p}_K)$$

be a random variable that assumes values in the simplex S_K defined in (7.1). The mixture

$$\phi = q_1\phi_1 + \ldots + q_j\phi_j + \ldots + q_M\phi_M \tag{6.7}$$

is the distribution of $\underline{\mathbf{p}}$ and ϕ_j is the Dirichlet density $\mathrm{Dir}(\alpha_{j,1}, \ldots, \alpha_{j,K})$.

(a) Compute

$$E[\mathbf{p}_i].$$

(*Hint*: Use a suitable iterated expectation

$$E[X] = E[E[X|Y]]$$

and Exercise 3 above.)

(b) Compute

$$\mathrm{Var}\left[\mathbf{p}_i\right].$$

(Hint: Here the formula

$$\mathrm{Var}(X) = E\left[\mathrm{Var}(X|Y)\right] + \mathrm{Var}[E\left[X|Y\right]]$$

may be found useful.)

5. **The Megaprior Heuristic** is owed to (Bailey 1996, Bailey and Gribskov 1996) and imposes in a more strict way a biological background knowledge about what constitutes a reasonable probability distribution for the symbols in a match state. The **megaprior** is the Dirichlet density obtained by scaling the hyperparameters as

$$\mathrm{Dir}\left(\tau \cdot \alpha_{j,1}, \ldots, \tau \cdot \alpha_{j,K}\right), \tag{6.8}$$

where

$$\tau = \frac{\kappa \cdot N}{\alpha_o},$$

where κ and N are parameters to be chosen for some purposes and

$$\alpha_o = \sum_{j=1}^{M} \sum_{i=1}^{K} \alpha_{j,i}.$$

The megaprior is no longer a prior density in a strictly Bayesian sense, if N denotes the size of the training set.

(a) Let

$$\underline{\mathbf{p}} = (\mathbf{p}_1, \ldots, \mathbf{p}_K)$$

be a random variable that assumes values in the simplex S_K defined in (7.1) and has the megaprior density in (6.8). Explain the effect, as a function of κ and N, of the scaling $\tau = \frac{\kappa N}{\alpha_o}$ on the expectation and variance of $\mathbf{p}_i$ by comparison to the results in Exercise 3 in the preceding.

(b) Scale each component ϕ_j of the Dirichlet mixture prior

$$\phi = q_1\phi_1 + \ldots + q_j\phi_j + \ldots + q_M\phi_M.$$

as in (6.8). Here ϕ_j is the Dirichlet density $\mathrm{Dir}\left(\alpha_{j,1}, \ldots, \alpha_{j,K}\right)$. What is the effect of scaling on the expectation and variance of $\mathbf{p}_i$?

6.7 Appendix A: Dirichlet Mixtures

6.7.1 Definition of Dirichlet Mixtures

A mixture of Dirichlet densities is based on a collection of individual Dirichlet densities that operate jointly to assign probabilities to (probability) distributions. The last mentioned distributions are thought to be sampled to generate the strings in the background knowledge represented in the multiple alignment.

We let q_j, $j = 1, \ldots, M$, be a probability distribution, which will play the role of a mixing distribution. A Dirichlet density ϕ_j is a probability density for a random variable $\underline{\mathbf{p}}$ that assumes values in a simplex of probability distributions

$$S_K = \left\{ \underline{p} = (p_1, \ldots, p_K) \,|p_i \geq 0, i = 1, \ldots, K, \sum_{i=1}^{K} p_i = 1 \right\} \tag{7.1}$$

and has the expression

$$\phi_j\left(\underline{p}\right) = \phi\left(\underline{p}; \underline{\alpha}_j\right) = \frac{\Gamma\left(\sum_{i=1}^{K} \alpha_{j,i}\right)}{\prod_{i=1}^{K} \Gamma\left(\alpha_{j,i}\right)} \prod_{i=1}^{K} p_i^{\alpha_{j.i}-1}, \tag{7.2}$$

where the hyperparameter is

$$\underline{\alpha}_j = \left(\alpha_{j,1}, \ldots, \alpha_{j,K}\right). \tag{7.3}$$

Definition 6.7.1 A *Dirichlet mixture* is a probability density

$$\phi = q_1\phi_1 + \ldots + q_j\phi_j + \ldots + q_M\phi_M \tag{7.4}$$

on S_K (and zero outside). ■

We introduce the symbol Θ to designate the parameters of the Dirichlet mixture

$$\Theta = \{\underline{\alpha}_j, q_j; j = 1, \ldots, M\}. \tag{7.5}$$

Then we can write ϕ in (7.4) as

$$\phi\left(\underline{p}; \Theta\right) = q_1\phi\left(\underline{p}; \underline{\alpha}_1\right) + \ldots + q_j\phi\left(\underline{p}; \underline{\alpha}_j\right) + \ldots + q_M\phi\left(\underline{p}; \underline{\alpha}_M\right). \tag{7.6}$$

6.8 Appendix B: The Predictive Multinomial Mixture

Here we present a number of formulas derived by averaging a multinomial distribution with regard to a mixture of Dirichlet densities. The frequency counts are designated for ease of writing by $\underline{n}$ and n, whereas the formulas are applied in the text to the columnwise frequency counts $\underline{n}_t$ and $n(t)$.

6.8.1 The Predictive Multinomial Process

Let $\mathbf{x}$ be a string with symbols that are modeled independent conditioned on the random variable $\underline{\mathbf{p}}$ assuming the value $\underline{p} = (p_1, \ldots, p_K)$. With n_i as the frequency counts

$$\underline{n} = (n_1, \ldots, n_K)$$

the probability of obtaining $n_1, \ldots, n_K$ is equal to

$$P\left(\underline{\mathbf{n}} = \underline{n} | \underline{\mathbf{p}} = \underline{p}\right) = \Gamma(n+1) \prod_{i=1}^{K} \frac{p_i^{n_i}}{\Gamma(n_i + 1)}, \tag{8.1}$$

by the multinomial theorem. If we average (8.1) over all models $\underline{p}$ in the family we get the *(prior) predictive multinomial distribution*

$$\int_{S_K} P\left(\underline{\mathbf{n}} = \underline{n} | \underline{\mathbf{p}} = \underline{p}\right) \phi\left(\underline{p}; \underline{\alpha}\right) d\underline{p}$$

$$= \Gamma(n+1) \int_{S_K} \prod_{i=1}^{K} \frac{p_i^{n_i}}{\Gamma(n_i + 1)} \frac{\Gamma\left(\sum_{i=1}^{K} \alpha_{j,i}\right)}{\prod_{i=1}^{K} \Gamma\left(\alpha_i\right)} \prod_{i=1}^{K} p_i^{\alpha_i - 1} dp_1 \ldots dp_K . \tag{8.2}$$

We denote this integral by $P\left(\underline{\mathbf{n}} = \underline{n}; n, \underline{\alpha}\right)$ and then we have by known formulas for evaluating integrals of Dirchlet densities

$$P\left(\underline{\mathbf{n}} = \underline{n}; n, \underline{\alpha}\right) := \frac{\Gamma(n+1)\Gamma(\sum_{i=1}^{K} \alpha_i)}{\Gamma(n + \sum_{i=1}^{K} \alpha_i)} \prod_{i=1}^{K} \frac{\Gamma(n_i + \alpha_i)}{\Gamma(n_i + 1) \cdot \Gamma(\alpha_i)}. \tag{8.3}$$

In (Mosimann 1962), where the expression perhaps appears for the first time, this is called is the *compound multinomial distribution.* Here we include $n = n_1 + \ldots + n_K$ in the notation to emphasize that the length of the sequence is fixed in advance.

6.8.2 The Predictive Mixture

It will also be convenient to introduce a random variable ξ assuming values in $\{1, 2, \ldots, J\}$ such that

$$P(\xi = j) = q_j, \qquad j = 1, 2, \ldots, M. \tag{8.4}$$

This is a random variable whose outcomes are used in step 1) of the generative scheme in 6.7.1 above to pick one ϕ_j. Then let us average the multinomial distribution with the Dirichlet mixture ϕ in (7.4). Thus we compute the **predictive multinomial mixture distribution** $P(\underline{\mathbf{n}} = \underline{n}; n, \Theta)$, which is given by

$$\begin{aligned} P(\underline{\mathbf{n}} = \underline{n}; n, \Theta) &= \int_{S_K} P(\underline{\mathbf{n}} = \underline{n} | \underline{\mathbf{p}} = \underline{p})\, \phi(\underline{p}; \Theta)\, d\underline{p} \\ &= \sum_{j=1}^{M} q_j \int_{S_K} P(\underline{\mathbf{n}} = \underline{n} | \underline{\mathbf{p}} = \underline{p})\, \phi(\underline{p}; \underline{\alpha}_j)\, d\underline{p} \\ &= \sum_{j=1}^{M} q_j \cdot \Gamma(n+1) \int_{S_K} \prod_{i=1}^{K} \frac{p_i^{n_i}}{\Gamma(n_i+1)} \phi(\underline{p}; \underline{\alpha}_j)\, d\underline{p} \end{aligned}$$

and from (8.2) and (8.3) this becomes

$$= \sum_{j=1}^{M} q_j \cdot \frac{\Gamma(n+1)\Gamma(\sum_{i=1}^{K} \alpha_{j,i})}{\Gamma(n + \sum_{i=1}^{K} \alpha_{j,i})} \prod_{i=1}^{K} \frac{\Gamma(n_i + \alpha_{j,i})}{\Gamma(n_i+1)\Gamma(\alpha_{j,i})}. \tag{8.5}$$

It is, however, more natural to express this using the form in (8.3) so that (8.5) becomes

$$P(\underline{\mathbf{n}} = \underline{n}; n, \Theta) = \sum_{j=1}^{M} q_j P(\underline{\mathbf{n}} = \underline{n}; n, \underline{\alpha}_j), \tag{8.6}$$

where

$$P(\underline{\mathbf{n}} = \underline{n}; n, \underline{\alpha}_j) := \frac{\Gamma(n+1)\Gamma(\sum_{i=1}^{K} \alpha_{j,i})}{\Gamma(n + \sum_{i=1}^{K} \alpha_{j,i})} \prod_{i=1}^{K} \frac{\Gamma(n_i + \alpha_{j,i})}{\Gamma(n_i+1)\Gamma(\alpha_{j,i})}. \tag{8.7}$$

6.8.3 The Posterior Mixture Density

We wish to find the posterior density of $\underline{p} = p_1, \ldots, p_K$, given $\underline{n}$ and n expressed by means of the parameters of the Dirichlet mixture. By marginalization

$$\begin{aligned}\psi\left(\underline{p}|\underline{n}, n, \Theta\right) &= \sum_{j=1}^{M} g\left(\underline{p}, \xi = j|\underline{n}, n, \Theta\right) \\ &= \sum_{j=1}^{M} g\left(\underline{p}|\xi = j, \underline{n}, n, \Theta\right) P\left(\xi = j|\underline{n}, n, \Theta\right). \end{aligned} \tag{8.8}$$

Hence we have by the analysis above $g\left(\underline{p}|\xi = j, \underline{n}, n, \Theta\right)$ as another notation for the posterior density

$$g\left(\underline{p}|\xi = j, \underline{n}, n, \Theta\right) = \phi\left(\underline{p}|\underline{n}, n; \underline{\alpha}_j\right) = \frac{\Gamma\left(n + \sum_{i=1}^{K} \alpha_{j,i}\right)}{\prod_{i=1}^{K} \Gamma\left(\alpha_{j,i} + n_i\right)} \prod_{i=1}^{K} p_i^{n_i + \alpha_{j,i} - 1}. \tag{8.9}$$

Furthermore $P\left(\xi = j|\underline{n}, n, \Theta\right)$ is a posterior probability which is an update of (8.4) by Bayes' rule,

$$P\left(\xi = j|\underline{n}, n, \Theta\right) = \frac{q_j P\left(\underline{\mathbf{n}} = \underline{n}; n, \underline{\alpha}_j\right)}{\sum_{r=1}^{M} q_r P\left(\underline{\mathbf{n}} = \underline{n}; n, \underline{\alpha}_r\right)}. \tag{8.10}$$

Thus we have obtained a *mixture representation of the posterior density* of $\underline{p}$ as

$$\psi\left(\underline{p}|\underline{n}, n, \Theta\right) = \sum_{j=1}^{M} \frac{q_j P\left(\underline{\mathbf{n}} = \underline{n}; n, \underline{\alpha}_j\right)}{\sum_{r=1}^{M} q_r P\left(\underline{\mathbf{n}} = \underline{n}; n, \underline{\alpha}_r\right)} \phi\left(\underline{p}|\underline{n}, n; \underline{\alpha}_j\right). \tag{8.11}$$

6.8.4 Mean Posterior Probability of Symbols

Using the model averaging and the result in (8.11) we obtain

$$\begin{aligned}\widehat{p}_i &= \int_{S_K} p_i \psi\left(\underline{p}|\underline{n}, n, \Theta\right) d\underline{p} \\ &= \sum_{j=1}^{M} \frac{q_j P\left(\underline{\mathbf{n}} = \underline{n}; n, \underline{\alpha}_j\right)}{\sum_{r=1}^{M} q_r P\left(\underline{\mathbf{n}} = \underline{n}; n, \underline{\alpha}_r\right)} \int_{S_K} p_i \cdot \phi\left(\underline{p}|\underline{n}, n; \underline{\alpha}_j\right) d\underline{p} \\ &= \sum_{j=1}^{M} \left[\frac{q_j P\left(\underline{\mathbf{n}} = \underline{n}; n, \underline{\alpha}_j\right)}{\sum_{r=1}^{M} q_r P\left(\underline{\mathbf{n}} = \underline{n}; n, \underline{\alpha}_r\right)}\right] \cdot \frac{n_i + \alpha_{j,i}}{n + \sum_{i=1}^{K} \alpha_{j,i}}, \end{aligned} \tag{8.12}$$

where we have used

$$\frac{\Gamma\left(n+\sum_{l=1}^{K}\alpha_l\right)}{\prod_{l=1}^{K}\Gamma\left(n_l+\alpha_l\right)}\int_{S_K} p_i\prod_{l=1}^{K} p_l^{n_l+\alpha_l-1}dp_1\ldots dp_K=\frac{n_i+\alpha_i}{n+\sum_{i=1}^{K}\alpha_i} \tag{8.13}$$

recalling the formulas for the mean posterior estimate in Chapter 3.

6.9 Appendix C: The Lemmas for Mixture Estimation

Using $P\left(\xi=j|\underline{n},n,\Theta\right)$ from (8.10) we have:

Lemma 6.9.1

$$\frac{\partial}{\partial\alpha_{j,i}}\log P\left(\underline{\mathbf{n}}=\underline{n};n,\Theta\right)=P\left(\xi=j|\underline{n},n,\Theta\right)\cdot\frac{\partial}{\partial\alpha_{j,i}}\log P\left(\underline{\mathbf{n}}=\underline{n};\underline{\alpha}_j,n\right). \tag{9.1}$$

Proof: From (8.6) we obtain

$$\frac{\partial}{\partial\alpha_{j,i}}\log P\left(\underline{\mathbf{n}}=\underline{n};n,\Theta\right)=\frac{\frac{\partial}{\partial\alpha_{j,i}}\sum_{l=1}^{M}q_lP\left(\underline{\mathbf{n}}=\underline{n};n,\underline{\alpha}_l\right)}{P\left(\underline{\mathbf{n}}=\underline{n};n,\Theta\right)}.$$

In the right hand side we recall that $\alpha_{j,i}$ is included in only one of the terms in the sum and that the mixture distribution q_j does not depend on $\alpha_{j,i}$. Hence we obtain in the right hand side

$$\frac{q_j}{P\left(\underline{\mathbf{n}}=\underline{n};n,\Theta\right)}\cdot\frac{\partial}{\partial\alpha_{j,i}}P\left(\underline{\mathbf{n}}=\underline{n};n,\underline{\alpha}_j\right).$$

This can be rearranged, using Bayes formula (8.10), as

$$\frac{P\left(\xi=j|\underline{n},n,\Theta\right)}{P\left(\underline{\mathbf{n}}=\underline{n};n,\underline{\alpha}_j\right)}\cdot\frac{\partial}{\partial\alpha_{j,i}}P\left(\underline{\mathbf{n}}=\underline{n};n,\underline{\alpha}_j\right).$$

The rule for differentiation of logarithms gives the lemma as claimed. ∎

Lemma 6.9.2

$$\begin{aligned}\frac{\partial}{\partial\alpha_i}\log P\left(\underline{\mathbf{n}}=\underline{n};n,\underline{\alpha}\right) &= \Psi\left(\sum_{i=1}^{K}\alpha_i\right)-\Psi\left(n+\sum_{i=1}^{K}\alpha_i\right)\\ &\quad+\Psi\left(n_i+\alpha_i\right)-\Psi\left(\alpha_i\right).\end{aligned} \tag{9.2}$$

Proof: This is follows by logarithmic differentiation of (8.3) or

$$P\left(\underline{\mathbf{n}}=\underline{n};n,\underline{\alpha}\right)=\frac{\Gamma(n+1)\Gamma(\sum_{i=1}^{K}\alpha_i)}{\Gamma(n+\sum_{i=1}^{K}\alpha_i)}\prod_{i=1}^{K}\frac{\Gamma(n_i+\alpha_i)}{\Gamma(n_i+1)\Gamma(\alpha_i)}.$$

Since the terms which do not involve α_i vanish when the logarithms are differentiated, we have

$$\begin{aligned}\frac{\partial}{\partial\alpha_i}\log P\left(\underline{\mathbf{n}}=\underline{n};n,\underline{\alpha}\right) &= \frac{\partial}{\partial\alpha_i}\log\Gamma\left(\sum_{i=1}^{K}\alpha_i\right)-\frac{\partial}{\partial\alpha_i}\log\Gamma\left(n+\sum_{i=1}^{K}\alpha_i\right)\\ &\quad+\frac{\partial}{\partial\alpha_i}\log\Gamma\left(n_i+\alpha_i\right)-\frac{\partial}{\partial\alpha_i}\log\Gamma\left(\alpha_i\right).\end{aligned}$$

Then we change to the notation in (2.11). ■

Next we use the reparametrization $q_j=\dfrac{Q_j}{\sum_{j=1}^{K}Q_j}$, where $Q_j>0$.

Lemma 6.9.3

$$\frac{\partial}{\partial Q_j}P\left(\underline{\mathbf{n}}=\underline{n};n,\Theta\right)=\frac{P\left(\underline{\mathbf{n}}=\underline{n}|\alpha_j,n\right)-P\left(\underline{\mathbf{n}}=\underline{n};n,\Theta\right)}{\sum_{j=1}^{K}Q_j}. \tag{9.3}$$

Proof: In (8.6) we obtain

$$\frac{\partial}{\partial Q_j}P\left(\underline{\mathbf{n}}=\underline{n};n,\Theta\right)=\sum_{l=1}^{M}\left[\left\{\frac{\partial}{\partial Q_j}\frac{Q_l}{\sum_{j=1}^{K}Q_j}\right\}P\left(\underline{\mathbf{n}}=\underline{n};n,\underline{\alpha}_l\right)\right],$$

since $\frac{\partial}{\partial Q_j}P\left(\underline{\mathbf{n}}=\underline{n};n,\underline{\alpha}_l\right)=0$ for all j. The partial derivative of the fraction is

$$\frac{\partial}{\partial Q_j}\frac{Q_l}{\sum_{j=1}^{K}Q_j}=-\frac{Q_j}{\left(\sum_{j=1}^{K}Q_j\right)^2}$$

if $l\neq j$ and

$$\frac{\partial}{\partial Q_j}\frac{Q_j}{\sum_{j=1}^{K}Q_j}=\frac{1}{\sum_{j=1}^{K}Q_j}-\frac{Q_l}{\left(\sum_{j=1}^{K}Q_j\right)^2}.$$

if $j=l$. Thus we have obtained

$$\frac{\partial}{\partial Q_j}P\left(\underline{\mathbf{n}}=\underline{n};n,\Theta\right)=\frac{P\left(\underline{\mathbf{n}}=\underline{n};n,\underline{\alpha}_j\right)}{\sum_{j=1}^{K}Q_j}-\sum_{l=1}^{M}P\left(\underline{\mathbf{n}}=\underline{n};n,\underline{\alpha}_l\right)\frac{Q_l}{\left(\sum_{j=1}^{K}Q_j\right)^2}.$$

Since $q_j = \frac{Q_j}{\sum_{j=1}^{K} Q_j}$ this is found to be equal to

$$= \frac{P\left(\underline{\mathbf{n}} = \underline{n}; n, \underline{\alpha}_j\right) - \sum_{l=1}^{M} q_l P\left(\underline{\mathbf{n}} = \underline{n}; n, \underline{\alpha}_l\right)}{\sum_{j=1}^{K} Q_j}.$$

This is the assertion as claimed in view of (8.6). ■

Recalling $P\left(\xi = j | \underline{n}, n, \Theta\right)$ from (8.10) we have

Lemma 6.9.4

$$\frac{\partial}{\partial Q_j} \log P\left(\underline{\mathbf{n}} = \underline{n}; n, \Theta\right) = \frac{P\left(\xi = j | \underline{n}, n, \Theta\right)}{Q_j} - \frac{1}{\sum_{j=1}^{K} Q_j}. \tag{9.4}$$

Proof: By logarithmic differentiation we obtain

$$\frac{\partial}{\partial Q_j} \log P\left(\underline{\mathbf{n}} = \underline{n}; n, \Theta\right) = \frac{1}{P\left(\underline{\mathbf{n}} = \underline{n}; n, \Theta\right)} \frac{\partial}{\partial Q_j} P\left(\underline{\mathbf{n}} = \underline{n}; n, \Theta\right).$$

Using the preceding lemma we obtain in the right hand side

$$\frac{1}{P\left(\underline{\mathbf{n}} = \underline{n}; n, \Theta\right)} \cdot \left[\frac{P\left(\underline{\mathbf{n}} = \underline{n} | \underline{\alpha}_j, n\right) - P\left(\underline{\mathbf{n}} = \underline{n}; n, \Theta\right)}{\sum_{j=1}^{K} Q_j}\right]$$

$$= \frac{\frac{P\left(\underline{\mathbf{n}} = \underline{n} | \underline{\alpha}_j, n\right)}{P\left(\underline{\mathbf{n}} = \underline{n}; n, \Theta\right)} - 1}{\sum_{j=1}^{K} Q_j}.$$

Here we recall the update in (8.10)

$$P\left(\xi = j | \underline{n}, n, \Theta\right) = \frac{q_j P\left(\underline{\mathbf{n}} = \underline{n}; n, \underline{\alpha}_j\right)}{\sum_{r=1}^{M} q_r P\left(\underline{\mathbf{n}} = \underline{n}; n, \underline{\alpha}_r\right)}$$

and obtain in the right hand side

$$\frac{\frac{P\left(\xi = j | \underline{n}, n, \Theta\right)}{q_j} - 1}{\sum_{j=1}^{K} Q_j}.$$

Then we use the identity $q_j = \frac{Q_j}{\sum_{j=1}^{K} Q_j}$, and this gives the assertion in the lemma. ■

6.10 References and Further Reading:

- Mixture priors for sequence families

R. Hughey and A. Krogh (1995): *SAM: Sequence alignment and modelling software system.* Technical Report UCSC-CRL-95-7, University of California, Santa Cruz, CA, (Regularly updated.) The SAM documentation. `http://www.cse.ucsc.edu/research/compbio/sam.html`

R. Hughey and A. Krogh (1996): Hidden Markov models for sequence analysis: extensions and analysis of the basic method. *Computer Applications in Biological Sciences* (CABIOS), 12, pp. 95–107.

A. Krogh, M. Brown, I. Saira Mian, K. Sjölander and D. Haussler (1994): Hidden Markov Models in Computational Biology: Applications to Protein Modelling. *Journal of Molecular Biology,* 235, pp. 1501–1531.

K. Sjölander, K. Karplus, M. Brown, R. Hughey, A. Krogh, I. Saira Mian and D. Haussler (1996): Dirichlet Mixtures: a method for improved detection of weak but significant sequence homology. *Computer Applications in Biological Sciences* (CABIOS), 12, pp. 327–345.

- The continuous Dirichlet Mixtures

C.E. Antoniak (1974): Mixtures of Dirichlet processes with applications to Bayesian nonparametric problems.*The Annals of Statistics,* 2, pp. 1152–1174.

T.S. Ferguson (1973): A Bayesian Analysis of Some Nonparametric Problems. *The Annals of Statistics,* 1, pp. 209–230.

S. Tavaré and W.J. Ewens (1998): The Ewens Sampling Formula. `http://www-hto.usc.edu/papers/abstracts/esfrep.html`. Also in N.L. Johnson, S. Kotz and N. Balakrishnan: *Multivariate Discrete Distributions.* J. Wiley, New York.

- Background knowledge, multiple alignment and classification

L. Goldstein and M.S Waterman (1994): Approximations to profile score distributions. *Journal of Computational Biology,* 1, pp. 93–104.

M. Gribskov, A. McLachlan and D. Eisenberg (1987): Profile Analysis: Detection of similarity of distantly related proteins. *Proceedings of the National Academy of Science of USA.* 84, February, pp. 4355–4358.

M. Gribskov, R. Lüthy and D. Eisenberg (1990): Profile Analysis. *Methods in Enzymology,* 266, pp. 198–212.

M. Gribskov, S. Veretnik (1996): Identification of Sequence Patterns with Profile Analysis. *Methods in Enzymology,* 183, pp. 146–159.

D. Gusfield (1997): *Algorithms on Strings, Trees and Sequences. Computer Science and Computational Biology.* Cambridge University Press. Cambridge.

H. Hirsh and M.Noordewier (1994): Using Background Knowledge to Improve Inductive Learning. A Case Study in Molecular Biology. *IEEE Expert*, October, pp. 3–6.

A.G. Murzin, S.E. Brenner, T. Hubbard and C. Chothia (1995): **scop**: a structural classification of proteins database for the investigation of sequences and structures. *Journal of Molecular Biology*, 247, pp. 536–540.

R.L. Tatusov, S.F. Altschul and E.V. Koonin (1994): Detection of conserved segments of proteins: Iterative scanning of sequence databases with alignment blocks. *Proceedings of the National Academy of Science, USA*, 91, pp. 12091–12095.

J.T.L. Wang, Th.G. Marr, D. Shasha, B.A. Shapiro, G-W. Chirn and T.Y. Lee (1996): Complementary Classification Approaches for Protein Sequences. *Protein Engineering*, 9, pp. 381–386.

- Megapriors

T.L. Bailey (1996): Separating Mixtures Using Megapriors. *UCSD Technical Report CS96 - 471*, January.

T.L. Bailey and M. Gribskov (1996): The megaprior heuristic for discovering protein sequence patterns. *Proceedings of the Fourth International Conference on Intelligent Systems for Molecular Biology*, AAAI Press.

- Various mathematical auxiliaries

J.O. Berger (1985): *Statistical Decision Theory and Bayesian Analysis. Second Edition.* Springer–Verlag, New York, Berlin et al..

R.O. Duda and P.E. Hart (1974): *Pattern Classification and Scene Analysis*, A Wiley-Interscience Publication, John Wiley and Sons, New York, London, Sydney, Toronto.

C. Fraley and A.E. Raftery (1998): How Many Clusters? Which Clustering Method? Answers via Model-Based Cluster Analysis. *Technical Report No. 329, Department of Statistics, University of Washington, Seattle.*

J.E. Mosimann (1962): On the compound multinomial distribution, the multivariate β-distribution, and correlation among proportions. *Biometrika*, 49, pp. 65–82.

A. Narayanan (1991): Algorithm AS 266: The maximum likelihood estimation of parameters of the Dirichlet distribution. *Applied Statistics*, 40, pp. 365–374.

R. Redner (1981): Note on the Consistency of the Maximum Likelihood Estimate for Non-identifiable Distributions. *The Annals of Mathematical Statistics*, 9, pp. 225–228.

H. Ruskeepää (1999): *Mathematica Navigator. Graphics and Methods of Applied Mathematics*, Academic Press, New York, London, etc. .

Chapter 7

Markov Chains

7.1 Introduction

7.1.1 Contents

This chapter gives a brief outline of the elementary properties of Markov chains. Main attention is given to ergodic Markov chains. Ergodic Markov chains are useful, e.g., in modelling the segmentation properties of DNA sequences, as will be shown in Chapter 11.

7.1.2 Definitions and Notations

Consider an alphabet $S = \{s_1, s_2, \ldots, s_J\}$ and a sequence of random variables $X_0, X_1, \ldots, X_n, \ldots$, assuming values in S. The symbols s_j in the alphabet are called *states* and S is also called the state space. We can encode the states by giving the state s_j the label j and take for convenience of typing $S = \{1, 2, \ldots, J\}$.

Definition 7.1.1 *A sequence of random variables* $\{X_n\}_{n=0}^{\infty}$ *is called a* **Markov chain,** **(MC)**, *if for all* $n \geq 1$ *and* $j_0, j_1, \ldots, j_n \in S$,

$$P(X_n = j_n | X_0 = j_0, \ldots, X_{n-1} = j_{n-1}) = P(X_n = j_n | X_{n-1} = j_{n-1}). \quad (1.1)$$

■

The significance of an MC lies in that if $X_n = j_n$ is a future event, then the conditional probability of this event given the past history $X_0 = j_0, X_1 =$

$j_1, \ldots, X_{n-1} = j_{n-1}$ depends only upon the immediate past $X_{n-1} = j_{n-1}$ and not upon the remote past $X_0 = j_0, X_1 = j_1, \ldots, X_{n-2} = j_{n-2}$.

The condition in (1.1) is known as the *Markov property.* The Markov property yields a simple and mathematically tractable alternative to the multinomial process from Chapter 3 as a model family for $\{X_n\}_{n\geq 0}$.

Let $\{X_n\}_{n=0}^{\infty}$ be Markov chain. If $X_n = j$, we say that the *the chain is in state j at time n* or that *the chain visits the state at time n.* The conditional probabilities

$$p_{i|j} = P\left(X_n = j | X_{n-1} = i\right), \qquad n \geq 1, \qquad i, j \in S \tag{1.2}$$

are assumed to be independent of n and are called *(stationary) one step transition probabilities.* A Markov chain with stationary transition probabilities is called *homogeneous.* If the conditional probability is not defined, we put $p_{i|j} = 0$.

The numbers $p_{i|j}$ are for many good purposes displayed in matrix form

$$P = \left(p_{i|j}\right)_{i=1,j=1}^{J,J} \tag{1.3}$$

or

$$P = \begin{pmatrix} p_{1|1} & p_{1|2} & \cdots & p_{1|J} \\ p_{2|1} & p_{2|2} & \cdots & p_{2|J} \\ \vdots & \vdots & \ddots & \vdots \\ p_{J|1} & p_{J|2} & \cdots & p_{J|J} \end{pmatrix}. \tag{1.4}$$

Thus P is an $J \times J$ matrix to be called a *transition matrix.* The ith row of P is the conditional probability distribution of X_n given that $X_{n-1} = i$. Therefore the following properties hold true:

$$p_{i|j} \geq 0, \qquad \sum_{j=1}^{J} p_{i|j} = 1. \tag{1.5}$$

A matrix with this property is called a *stochastic matrix.*

7.1.3 Examples

Example 7.1.1 [A binary Markov chain] A binary Markov chain is a sequential mechanism producing zeros (0) and ones (1). It can, e.g., be used modelling a 'regulatory region' (of DNA) or 'not a regulatory region'. A regulatory region is a part of chromosomal DNA that serve as signals for turning the gene expression off or on. If at some stage 0 is produced, then at the next stage 1 will be produced with probability p and 0 will be produced with probability $1 - p$. If a 1 is produced, then at a next stage 0 will be produced with probability q and 1 will be produced with probability $1 - q$. This corresponds to the transition matrix

$$P = \begin{pmatrix} 1-p & p \\ q & 1-q \end{pmatrix}. \tag{1.6}$$

Figure 7.1 is a graphical representation of the state space and the transition matrix of this chain. Figure 7.2 shows (a segment of) the *trellis* for the sample paths of the Markov chains in this example. Each path from left to right through the trellis following at each step one of the arrows in the diagram is one possible sample path of the chain. For $n = 0, 1, 2$ there are eight ($= 2^3$) possible sample paths.

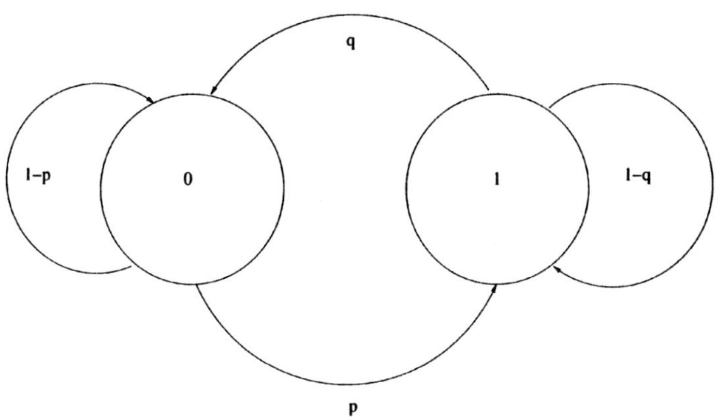

Figure 7.1: A state space graph

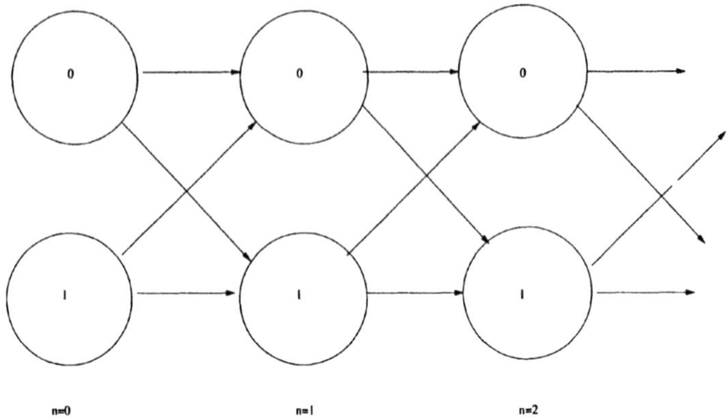

Figure 7.2: A trellis for the binary source

Example 7.1.2 (Repeaters) Binary digits 0 and 1 are transmitted over a sequence of k processors (repeaters) with noise. Let the probability that a digit which enters the lth processor is unchanged be p and let the probability that a digit that enters the lth processor is reversed be $1-p$. These probabilities are the same for all processors. In addition we assume that the processors operate independently of each other, as far as the transmission of digits, correctly or not, is concerned.

Let X_0 denote the digit that enters the first processor and X_l, $l = 1, \ldots, k$ be the digit that leaves the lth processor. This can also be graphically depicted by the *trellis* in Figure 7.2. We now show that $X_0, X_1, \ldots, X_k$ is a Markov chain by verifying the Markov property. By definition of conditional probability

$$\begin{aligned} & P\left(X_n = j_n | X_0 = j_0, \ldots, X_{n-1} = j_{n-1}\right) \\ & = \frac{P\left(X_0 = j_0, \ldots, X_{n-1} = j_{n-1}, X_n = j_n\right)}{P\left(X_0 = j_0, \ldots, X_{n-1} = j_{n-1}, X_{n-1} = j_{n-1}\right)}. \end{aligned} \tag{1.7}$$

Since we know that $X_0 = j_0$, then we can always recover the sequence of states $j_1, \ldots, j_{n-1}, j_{n-1}$ from the sequence of digits $j_1 - j_0, \ldots, j_{n-1} - j_{n-2}$ and vice versa. Hence we have that

$$P\left(X_0 = j_0, X_1 = j_1, \ldots, X_{n-1} = j_{n-1}\right)$$

$$= P(X_0 = j_0, X_1 - X_0 = j_1 - j_0, \ldots, X_{n-1} - X_{n-2} = j_{n-1} - j_{n-2}).$$

But $X_l - X_{l-1}$ is by assumption independent of $X_h - X_{h-1}$, $h \neq l$. Hence

$$P(X_0 = j_0, X_1 - X_0 = j_1 - j_0, \ldots, X_{n-1} - X_{n-2} = j_{n-1} - j_{n-2})$$
$$= P(X_0 = j_0) \cdot \prod_{l=1}^{n-1} P(X_l - X_{l-1} = j_l - j_{l-1})$$

and similarly

$$P(X_0 = j_0, X_1 = j_1, \ldots, X_{n-1} = j_{n-1}, X_n = j_n)$$
$$P(X_0 = j_0) \cdot \prod_{l=1}^{n} P(X_l - X_{l-1} = j_l - j_{l-1})$$

But by (1.7)

$$P(X_n = j_n | X_0 = j_0, X_1 = j_1, .., X_{n-1} = j_{n-1}) = P(X_n - X_{n-1} = j_n - j_{n-1}).$$

However,

$$P(X_n - X_{n-1} = j_n - j_{n-1}) = \begin{cases} p & \text{if } j_n - j_{n-1} = 0, \\ 1-p & \text{if } j_n - j_{n-1} = 1 \text{ or } j_n - j_{n-1} = -1. \end{cases}$$

Also,

$$P(X_n = j_n \mid X_{n-1} = j_{n-1}) = \begin{cases} p & \text{if } j_n - j_{n-1} = 0, \\ 1-p & \text{if } j_n - j_{n-1} = 1 \text{ or } j_n - j_{n-1} = -1. \end{cases}$$

Hence

$$P(X_n = j_n | X_0 = j_0, X_1 = j_1, \ldots, X_{n-1} = j_{n-1}) = P(X_n = j_n \mid X_{n-1} = j_{n-1})$$

as was to be proved. ■

7.1.4 Markov Chains of kth Order

Definition 7.1.2 *A sequence of random variables* $\{X_n\}_{n=0}^{\infty}$ *is called a* **kth order Markov chain,** *if for all* $n \geq 1$ *and* $j_0, j_1, \ldots, j_n \in S$,

$$P(X_n = j_n | X_0 = j_0, X_1 = j_1, \ldots, X_{n-1} = j_{n-1})$$
$$= P(X_n = j_n | X_{n-k} = j_{n-k}, \ldots, X_{n-1} = j_{n-1}), \tag{1.8}$$

for a positive integer k. ■

In this context the MC in the basic definition 7.1.1 is called a first order Markov chain. A multinomial process $\{X_n\}_{n=0}^{\infty}$ assuming values in S would then be called a Markov chain of zeroth order. MCs of orders higher than first order are often used in modelling of genomes, as will be outlined in Chapters 9 and 11. One difficulty with (1.8) is the number of parameters involved, more parsimonious high order Markov models are discussed in Exercise 13 below and in Chapter 9.

7.1.5 Joint Probability Distribution of an MC

Next we assign a probability to the sequence $j_0 \dots j_n$ using a Markov chain as model. By successive iterations of the definition of conditional probability and by successive uses of the Markov property (1.1)

$$\begin{aligned}
& P\left(X_0 = j_0, .., X_{n-1} = j_{n-1}, X_n = j_n\right) \\
& = P\left(X_n = j_n | X_0 = j_0, .., X_{n-1} = j_{n-1}\right) P\left(X_0 = j_0, .., X_{n-1} = j_{n-1}\right) \\
& = P\left(X_n = j_n | X_{n-1} = j_{n-1}\right) P\left(X_0 = j_0, .., X_{n-1} = j_{n-1}\right) \\
& = p_{j_{n-1}|j_n} P\left(X_{n-1} = j_n | X_0 = j_0, .., X_{n-2} = j_{n-2}\right) P\left(X_0 = j_0, .., X_{n-2} = j_{n-2}\right) \\
& \vdots \\
& = p_{j_{n-1}|j_n} \cdot p_{j_{n-2}|j_{n-1}} \cdots p_{j_0|j_1} \cdot p_{X_0}\left(j_0\right) \\
& = p_{X_0}\left(j_0\right) \cdot p_{j_0|j_1} \cdots p_{j_{n-2}|j_{n-1}} \cdot p_{j_{n-1}|j_n}.
\end{aligned}$$

If there is some $p_{j_{l-1}|j_l} = 0$ then the result is still true, since both sides of the equation are then zero. Hence we have proved

Proposition 7.1.3 *If $\{X_n\}_{n=0}^{\infty}$ is a Markov chain with stationary transition probabilities, then*

$$P\left(X_0 = j_0, X_1 = j_1, \dots, X_n = j_n\right) = p_{X_0}\left(j_0\right) \prod_{l=1}^{n} p_{j_{l-1}|j_l}. \tag{1.9}$$

■

The property above can be said to fully specify the family of finite-dimensional distributions of $\{X_n\}_{n=0}^{\infty}$ in a consistent manner and thus makes $\{X_n\}_{n=0}^{\infty}$ a stochastic process (Gray 1988, pp. 56–58) in the sense of the Kolmogorov–Daniell theorem. The details of this statement are beyond the scope of this presentation. Hence we can write the above as

$$\{X_n\}_{n=0}^{\infty} \in \text{Markov}\left(P, p_{X_0}\right).$$

7.2 Chapman–Kolmogorov Equations

Let $\{X_n\}_{n=0}^{\infty} \in$ Markov (P, p_{X_0}). The conditional probabilities

$$p_{i|j}(n) = P\left(X_{m+n} = j | X_m = i\right), \qquad n \geq 1, i, j \in S \tag{2.1}$$

are independent of m, as can be shown (an Exercise at the end of the chapter). The probabilities $p_{i|j}(n)$ are called the *n step transition probabilities.* Then

$$P(n) = \left(p_{i|j}(n)\right)_{i=1,j=1}^{J,J} \tag{2.2}$$

is the *n step transition matrix.* We define

$$p_{i|j}(0) = \begin{cases} 1 & \text{if } j = i, \\ 0 & \text{if } j \neq i. \end{cases}$$

Proposition 7.2.1 (Chapman–Kolmogorov equations) *For all* $m, n \geq 1$ *and* $i, j \in S$

$$p_{i|j}(m+n) = \sum_{k=1}^{J} p_{i|k}(m) \cdot p_{k|j}(n). \tag{2.3}$$

Proof: We can assume that $p_{X_0}(i) > 0$, since otherwise both sides of (2.3) will be zero. We observe that

$$\begin{aligned} p_{i|j}(m+n) &= P\left(X_{m+n} = j | X_0 = i\right) \\ &= \sum_{k=1}^{J} \frac{P\left(X_{m+n} = j, X_m = k, X_0 = i\right)}{P\left(X_0 = i\right)}. \end{aligned}$$

We try now to express $P\left(X_{m+n} = j, X_m = k, X_0 = i\right)$ in a suitable way. We consider the following identity obtained by definition of conditional probability

$$P\left(X_0, \ldots, X_m, \ldots X_n\right) = P\left(X_{m+1}, \ldots, X_n | X_0, \ldots, X_m\right) P\left(X_0, \ldots, X_m\right)$$

where we for economy of space have omitted the states visited, e.g.,

$$P\left(X_0, \ldots, X_m, \ldots, X_n\right) = P\left(X_0 = j_0, \ldots, X_m = j_m, \ldots, X_n = j_n\right).$$

Using next (1.9) we obtain (an exercise)

$$P\left(X_{m+1},\ldots,X_n|X_0,\ldots,X_m\right)=P\left(X_{m+1},\ldots,X_n|X_m\right).$$

Thus we have obtained

$$P\left(X_0,\ldots,X_m,\ldots,X_n\right)=P\left(X_{m+1},\ldots,X_n|X_m\right)P\left(X_1,\ldots,X_m|X_0\right)P\left(X_0\right).$$

If we sum in this identity over $j_1,\ldots,j_{m-1},j_{m+1}\cdots j_{n-1}$ we obtain

$$\begin{aligned}&P\left(X_0=j_0,X_m=j_m,X_n=j_n\right)\\&=P\left(X_n=j_n|X_m=j_m\right)P\left(X_m=j_m|X_0=j_0\right)P\left(X_0=j_0\right).\end{aligned} \quad (2.4)$$

Next we recall

$$\begin{aligned}p_{i|j}(m+n)&=P\left(X_{m+n}=j|X_0=i\right)\\&=\sum_{k=1}^{J}\frac{P\left(X_{m+n}=j,X_m=k,X_0=i\right)}{P\left(X_0=i\right)}.\end{aligned}$$

Replacing n by $n+m$, j_0 by i, j_m by k and j_n by j in eq. (2.4), we obtain from this

$$\begin{aligned}p_{i|j}(m+n)&=\sum_{k=1}^{J}P\left(X_{m+n}=j|X_m=k\right)P\left(X_m=k|X_0=i\right)\\&=\sum_{k=1}^{J}p_{k|j}(n)\cdot p_{i|k}(m).\end{aligned}$$

This is the assertion as claimed. ■

In other elementary texts the Chapman–Kolmogorov equation is often demonstrated by neglecting many of the details. The preceding proof is adapted from (Helms 1997). In a matrix notation the Chapman–Kolmogorov equation (2.3) can be written by aid of (2.2) as the following matrix multiplication

$$P(n+m)=P(m)\cdot P(n). \quad (2.5)$$

But we have

Proposition 7.2.2

$$P(n)=P^n. \quad (2.6)$$

Proof: This is easily proved by induction, since where $P^0 = I$ (the $J \times J$ identity matrix), $P^1 = P$, $P^2 = P \cdot P$, $P^3 = P \cdot P^2$ and so on. ∎

Corollary 7.2.3 *Chapman – Kolmogorov equation* (2.3) *can be written as*

$$P^{n+m} = P^m \cdot P^n. \tag{2.7}$$

∎

7.3 Stationarity and Invariant Distributions

Let the distribution of X_0 be denoted by $\pi(0)$. In other words,

$$\pi(0) = \left(p_{X_0}(1), \ldots, p_{X_0}(J)\right). \tag{3.1}$$

This will be called the *initial distribution.* Let us denote by

$$\pi(n) = \left(p\left(X_n = 1\right), \ldots, p\left(X_n = J\right)\right) \tag{3.2}$$

the $1 \times J$ vector of the probabilities that the chain is in state j at time n. Then by the basic properties

$$p\left(X_n = j\right) = \sum_{k=1}^{J} p_{k|j} \cdot p\left(X_{n-1} = k\right).$$

This we write using a matrix notation as

$$\pi(n) = \pi(n-1)P. \tag{3.3}$$

Definition 7.3.1 A Markov chain $\{X_n\}_{n=0}^{\infty}$ is called *stationary* if the probability $p\left(X_n = j\right)$ is independent of n for all j in the state space. Furthermore, we call a distribution π *an invariant distribution*, with

$$\pi = \left(\pi_1, \ldots, \pi_J\right),$$

if $p\left(X_0 = j\right) = \pi_j$ implies that $p\left(X_1 = j\right) = \pi_j$. ∎

Proposition 7.3.1 *Let* $\{X_n\}_{n=0}^{\infty} \in$ Markov $(P, \pi(0))$. *Every invariant distribution satisfies the equation*

$$\pi = \pi P \tag{3.4}$$

with the constraints

$$\sum_{j=1}^{J} \pi_j = 1, \qquad \pi_j \geq 0.$$

Proof: Assume first that π is an invariant distribution. Then $\sum_{j=1}^{J} \pi_j = 1$ and $\pi_j \geq 0$ are clear. Since π is invariant, by the definition 7.3.1 above we must have $\pi(0) = \pi$ and $\pi(1) = \pi$. But because $\pi(n) = \pi(n-1)P$ we obtain $\pi = \pi P$.

Assume now that π satisfies $\pi = \pi P$ and the other restrictions. Let $\pi(0) = \pi$. Then

$$\pi(1) = \pi(0)P = \pi P = \pi$$

and π is an invariant distribution. ■

Proposition 7.3.2 *Every MC with a finite state space has* at least one *invariant distribution*

Proof: We give only an outline of the proof which is based on unpublished lecture notes by Bengt Rosén. Let p be an arbitrary probability distribution on S. Set

$$p^{(n)} = \frac{1}{n}\left(p + pP + pP^2 + \ldots + pP^{n-1}\right).$$

This is a sequence of probability distributions, i.e., vectors with components with values between zero and one. Thus the well known theorem of Bolzano and Weierstrass (Khuri 1993, p. 11) shows that we can pick a convergent subsequence $p^{(n_v)}$ which converges componentwise to the vector π. We can show that π is a probability distribution.

By our construction we have the recursion relations

$$p^{(n+1)} = \frac{n}{n+1}p^{(n)} + \frac{1}{n+1}pP^n \tag{3.5}$$

and

$$p^{(n+1)} = \frac{n}{n+1}p^{(n)}P + \frac{1}{n+1}p. \tag{3.6}$$

From recursion (3.5) we obtain

$$p^{(n_v+1)} \to \pi$$

and then from (3.6) we obtain

$$\pi = \pi P,$$

which proves the claim. The omitted finer details of argument are left for the reader. ∎

Example 7.3.3 (Binary Markov chain) In example 7.1.1 above with the transition matrix (1.6) we have uniquely that

$$\pi = \left(\frac{q}{q+p}, \frac{p}{q+p} \right). \tag{3.7}$$

The invariant distribution can be interpreted as the asymptotic percentages of time the chain spends in each of the states. ∎

7.4 Ergodic Markov Chains

7.4.1 Definition and Properties

When does a Markov chain have a unique invariant distribution? Is there convergence to an invariant distribution? These are in general difficult issues, but for finite state spaces some of the answers can be stated with relative ease. The following type of Markov chains is of interest in these studies.

Definition 7.4.1 Ergodic Markov Chain The Markov chain $\{X_n\}_{n=0}^{\infty}$ is said to be *ergodic* if there is a probability distribution $a = (a_1, \ldots, a_J)$ on the state space S such that

$$\lim_{n \to \infty} \pi(n) = a \tag{4.1}$$

for *any* initial distribution $\pi(0)$. ∎

Clearly the limiting distribution a must be unique. Next result shows that for an ergodic Markov chain the limiting distribution and the invariant distribution are identical.

Proposition 7.4.1 *Let* $\{X_n\}_{n=0}^{\infty} \in \text{Markov}\,(P, p_{X_0})$. *Let us assume that*

$$\lim_{n\to\infty} \pi(n) = a, \tag{4.2}$$

where $a = (a_1, \ldots, a_J)$ *is a probability distribution. Then* a *is an invariant distribution.*

Proof: Taking limits in (3.3) yields

$$\begin{aligned} a &= \lim_{n\to\infty} \pi(n) = \lim_{n\to\infty} \pi(n+1) \\ &= \lim_{n\to\infty} (\pi(n)P) = \left(\lim_{n\to\infty} \pi(n)\right) P = aP. \end{aligned}$$

Here the interchange of order between matrix multiplication and limit is permissible, since the state space is finite. ∎

Hence we can say that an ergodic Markov chain will asymptotically become stationary.

7.4.2 A Sufficient Condition for Ergodicity

Let us assume now that $\lim_{n\to\infty} p_{i|j}(n) = a_j$ for all j and independently of i. Taking $n = 1$ and letting $m \to \infty$ in (2.3) we obtain

$$\begin{aligned} a_j &= \lim_{m\to\infty} p_{i|j}(m+1) = \sum_{k=1}^{J} \lim_{m\to\infty} p_{i|k}(m) \cdot p_{k|j}(1) \\ &= \sum_{k=1}^{J} a_k \cdot p_{k|j}(1) = \sum_{k=1}^{J} a_k \cdot p_{k|j}, \end{aligned}$$

since $p_{k|j}(1) = p_{k|j}$ for one step transitions. In other words

$$a_j = \sum_{k=1}^{J} a_k \cdot p_{k|j}, j = 1, \ldots, J, \tag{4.3}$$

which can be written as $a = aP$. Hence we see that the limiting properties of the n step transition matrices are also related to the question of limiting distributions.

There are many different proofs of the result to be presented next. A powerful proof using the coupling argument (Lindvall 1992) and a simulation method is found in (Häggström 2001). The version presented here is owed to Kullback (1966) and Csiszár (1963) and requires (4.4) below. This restrictive condition is, however, also needed for the asymptotics of maximum likelihood estimation of hidden Markov models in Chapter 17. We choose to present this particular proof also for the reason that it is based on the properties of the Kullback distance.

Theorem 7.4.2 *If a Markov chain with stationary transition probabilities and a finite state space is such that*

$$p_{i|j} > 0, \tag{4.4}$$

for all i and j, then the limit

$$\lim_{n\to\infty} p_{i|j}(n) = \pi_j \tag{4.5}$$

exists for all j and is independent of i.

Proof: The computation above shows that if the limit in (4.5) exists, the limiting vector satisfies (4.3). Hence we take $\pi = (\pi_1, \ldots, \pi_J)$, which is a solution of

$$\pi_j = \sum_{k=1}^{J} \pi_k \cdot p_{k|j}, \qquad j = 1, \ldots, J. \tag{4.6}$$

We know by Proposition 7.3.2 above that at least one such solution exists.

The idea is to investigate the behaviour, as a function of n, of the Kullback distance $D\left(P_i^{(n)}|\pi\right)$ between a row in the n step transition matrix

$$P_i^{(n)} = \left(p_{i|1}(n), p_{i|2}(n), \ldots, p_{i|J}(n)\right)$$

and the chosen invariant distribution π. In fact we show first that

$$D\left(P_i^{(n)}|\pi\right) \geq D\left(P_i^{(n+1)}|\pi\right). \tag{4.7}$$

Let us write using (4.6) and $p_{i|j}(n+1) = \sum_{k=1}^{J} p_{i|k} \cdot p_{k|j}(n) = \sum_{k=1}^{J} p_{i|k}(n) p_{k|j}$ and (3.4)

$$\begin{aligned} D\left(P_i^{(n+1)}|\pi\right) &= \sum_{j=1}^{J} p_{i|j}(n+1) \log \frac{p_{i|j}(n+1)}{\pi_j} \\ &= \sum_{j=1}^{J} \sum_{k=1}^{J} p_{i|k}(n) \cdot p_{k|j} \log \frac{\sum_{k=1}^{J} p_{i|k}(n) \cdot p_{k|j}}{\sum_{k=1}^{J} \pi_k \cdot p_{k|j}}. \end{aligned}$$

Then by the log sum inequality 2.5.4 from Chapter 2 with $a_k = p_{i|k}(n) \cdot p_{k|j}$ and $b_k = \pi_k \cdot p_{k|j}$ we obtain

$$\begin{aligned} D\left(P_i^{(n+1)}|\pi\right) &\leq \sum_{j=1}^{J}\sum_{k=1}^{J} p_{i|k}(n) \cdot p_{k|j} \log \frac{p_{i|k}(n) \cdot p_{k|j}}{\pi_k \cdot p_{k|j}} \\ &= \sum_{k=1}^{J} p_{i|k}(n) \log \frac{p_{i|k}(n)}{\pi_k} = D\left(P_i^{(n)}|\pi\right), \end{aligned}$$

as was desired. There is equality if and only if $p_{i|k}(n) = \pi_k$, as is readily seen. We know that $D\left(P_i^{(n+1)}|\pi\right) \geq 0$ and thus

$$D\left(P_i^{(1)}|\pi\right) \geq D\left(P_i^{(2)}|\pi\right) \geq \ldots \geq D\left(P_i^{(n+1)}|\pi\right) \geq 0. \tag{4.8}$$

We now assume that this sequence is strictly decreasing. The opposite case gives the desired result

$$\lim_{n\to\infty} p_{i|j}(n) = \pi_j \tag{4.9}$$

immediately. Now we prove (4.9) in the case when the sequence in (4.8) is strictly monotonous. We note first that

$$\begin{aligned} D\left(P_i^{(n)}|\pi\right) &= \sum_{h=1}^{J} p_{i|h}(n) \log \frac{p_{i|h}(n)}{\pi_h} \\ &= \sum_{j=1}^{J}\sum_{h=1}^{J} p_{i|h}(n) p_{i|j} \log \frac{p_{i|h}(n)}{\pi_h}, \end{aligned}$$

since $\sum_{j=1}^{J} p_{i|j} = 1$. Thus we obtain

$$\begin{aligned} &D\left(P_i^{(n)}|\pi\right) - D\left(P_i^{(n+1)}|\pi\right) \\ &= \sum_{j=1}^{J}\sum_{h=1}^{J} p_{i|h}(n) \cdot p_{h|j} \log \frac{p_{h|j} p_{i|h}(n)}{p_{h|j} \frac{\pi_h}{\pi_j} p_{i|j}(n+1)}. \end{aligned} \tag{4.10}$$

Here we observe that

$$\sum_{j=1}^{J}\sum_{h=1}^{J} p_{i|h}(n) \cdot p_{h|j} = \sum_{h=1}^{J} p_{i|h}(n) \sum_{j=1}^{J} p_{h|j} = \sum_{h=1}^{J} p_{i|h}(n) = 1$$

and that

$$\begin{aligned}\sum_{j=1}^{J}\sum_{h=1}^{J} p_{h|j}\frac{\pi_h}{\pi_j}p_{i|j}(n+1) &= \sum_{j=1}^{J}\frac{p_{i|j}(n+1)}{\pi_j}\sum_{h=1}^{J} p_{h|j}\pi_h \\ &= \sum_{j=1}^{J}\frac{p_{i|j}(n+1)}{\pi_j}\pi_j = \sum_{j=1}^{J} p_{i|j}(n+1) = 1.\end{aligned}$$

Therefore the difference $D\left(P_i^{(n)}|\pi\right) - D\left(P_i^{(n+1)}|\pi\right)$ as found in (4.10) is another Kullback distance between two probability distributions. On the other hand we know by (4.8) that the sequence $D\left(P_i^{(n)}|\pi\right)$ tends to a finite limit. Hence we obtain from (4.10)

$$\sum_{j=1}^{J}\sum_{h=1}^{J} p_{i|h}(n)\cdot p_{h|j}\log\frac{p_{h|j}p_{i|h}(n)}{p_{h|j}\frac{\pi_h}{\pi_j}p_{i|j}(n+1)} \to 0,$$

as $n \to \infty$. But then the variation distance bound on the Kullback distance, see (5.10) in Chapter 2, yields

$$\sum_{j=1}^{J}\left|p_{i|h}(n)\cdot p_{h|j} - p_{h|j}\frac{\pi_h}{\pi_j}p_{i|j}(n+1)\right| \to 0 \tag{4.11}$$

as $n \to \infty$, which also implies that

$$\sum_{j=1}^{J}\pi_h p_{h|j}\left|\frac{p_{i|h}(n)}{\pi_h} - \frac{p_{i|j}(n+1)}{\pi_j}\right| \to 0 \tag{4.12}$$

as $n \to \infty$.

Since we have assumed that $p_{h|j} > 0$, the equation (4.6) gives $\pi_j > 0$ for all j (a fact which has already been used many times above). Therefore

$$\pi_h p_{h|j}\left|\frac{p_{i|h}(n)}{\pi_h} - \frac{p_{i|j}(n+1)}{\pi_j}\right| \le \sum_{j=1}^{J}\pi_h p_{h|j}\left|\frac{p_{i|h}(n)}{\pi_h} - \frac{p_{i|j}(n+1)}{\pi_j}\right|.$$

Hence (4.12) gives for all i and j that

$$\frac{p_{i|h}(n)}{\pi_h} - \frac{p_{i|j}(n+1)}{\pi_j} \to 0, \tag{4.13}$$

as $n \to \infty$. In a similar fashion we can show by studying $D\left(P_i^{(n)}|\pi\right) - D\left(P_i^{(n+m)}|\pi\right)$ that for all i and j

$$\frac{p_{i|h}(n)}{\pi_h} - \frac{p_{i|j}(n+m)}{\pi_j} \to 0 \tag{4.14}$$

as $n \to \infty$ and as $m \to \infty$. With $h = j$ in (4.14) we see that we are dealing with a Cauchy sequence (Khuri 1993, p. 137). This implies that there is a c_{ij} such that

$$\lim_{n\to\infty} \frac{p_{i|j}(n)}{\pi_j} = c_{ij}.$$

But from (4.13) we obtain

$$c_{ij} = c_{ih} \tag{4.15}$$

for all i, j and h. But since $\sum_{j=1}^{J} p_{i|j}(n) = 1$ we obtain

$$1 = \sum_{j=1}^{J} \lim_{n\to\infty} p_{i|j}(n) = \sum_{j=1}^{J} \pi_j c_{ij}.$$

But then (4.15) implies that $c_{ih} = 1$ for all i and h. Thereby we have obtained (4.9) as asserted. ∎

Proposition 7.4.3 *Consider a Markov chain with stationary transition probabilities and a finite state space. If*

$$p_{i|j} > 0, \tag{4.16}$$

for all i and j, then

$$\lim_{n\to\infty} \pi(n) = \pi \tag{4.17}$$

for **any** *initial distribution $\pi(0)$.*

Proof: Iteration of (3.3) gives

$$\pi(n) = \pi(0)P^n$$

for any initial distribution $\pi(0)$. This we can express as

$$p(X_n = j) = \sum_{k=1}^{J} p_{k|j}(n) \cdot p(X_0 = k).$$

Letting n grow to infinity we have

$$\lim_{n\to\infty} p(X_n = j) = \sum_{k=1}^{J} \lim_{n\to\infty} p_{k|j}(n) \cdot p(X_0 = k),$$

as the sum is finite. By assumption (4.16) Proposition 7.4.2 above implies that

$$\begin{aligned}\sum_{k=1}^{J} \lim_{n\to\infty} p_{k|j}(n) \cdot p(X_0 = k) &= \sum_{k=1}^{J} \pi_j \cdot p(X_0 = k) \\ &= \pi_j \cdot \sum_{k=1}^{J} p(X_0 = k) = \pi_j,\end{aligned}$$

since $p(X_0 = k)$ is a probability distribution. ∎

Hence we have found that (4.16)

$$p_{i|j} > 0$$

for all i and j is a sufficient condition for a Markov chain with a finite state space to be ergodic. There are other conditions and criteria for ergodicity. For example (Helms 1997) proves that if there is a positive integer N such that

$$p_{i|j}(N) > 0$$

for all i and j, then the chain is ergodic.

7.4.3 Classification of the State Space

Definition 7.4.2 (1) The state $j \in S$ can be **reached** from state $i \in S$, if there is a positive integer n such that

$$p_{i|j}(n) > 0.$$

(2) The transition matrix P or the chain $\{X_n\}_{n=0}^{\infty}$ is called **irreducible** if each state can be reached from each other state. ∎

It can be proved that every irreducible Markov chain with stationary transition probabilities and a finite state space has a *unique* invariant distribution. We refer to (Norris 1997) for an account of these and other additional properties of Markov chains.

7.5 Exercises

7.5.1 On the Markov Property and the Invariant Distribution

1. Let $\{X_n\}_{n=0}^{\infty} \in \text{Markov}\,(P, p_{X_0})$.

   (a) Prove that the Markov property implies that X_{n+1} and X_{n-1} are *conditionally independent* given X_n.

   (b) Show that the conditional probabilities

   $$p_{i|j}(n) = P\left(X_{m+n} = j | X_m = i\right), n \geq 1, i, j \in S \tag{5.1}$$

   are independent of m.

   (c) Show that

   $$\begin{aligned} &P\left(X_{m+1} = j_{m+1}, \ldots X_n = j_n | X_0 = j_0, X_1 = j_1, \ldots, X_m = j_m\right) \\ &= P\left(X_{m+1} = j_{m+1}, \ldots X_n = j_n | X_m = j_m\right). \end{aligned} \tag{5.2}$$

2. Assume that there is a positive integer N such that

   $$p_{i|j}\left(N\right) > 0$$

   for all i and j. Prove that if the vector π is a solution of

   $$\pi_j = \sum_{k=1}^{J} \pi_k \cdot p_{k|j}, j = 1, \ldots, J,$$

   then π is unique.

3. Let $Z_1, \ldots, Z_n, \ldots$ be the successive outcomes of rolling a die i.e. $Z_n \in \{1, 2, 3, 4, 5, 6\}$ for any n. We consider $Z_1, \ldots, Z_n, \ldots$ to be independent. Furthermore

   $$S_n = Z_1 + \ldots + Z_n$$

   and

   $$X_n = S_n \bmod(5), \qquad n = 1, \ldots$$

   which means that X_n is the remainder when S_n is divided by 5.

   (a) Show that $\{X_n\}_{n=0}^{\infty}$ is a Markov chain.

(b) Determine the probability transition matrix P for $\{X_n\}_{n=0}^{\infty}$.

(c) Determine the invariant distribution π for P.

4. The Markov chain $\{X_n\}_{n=0}^{\infty}$ with the state space $S = \{1, 2, 3\}$ has the transition matrix

$$P = \begin{pmatrix} 0 & \frac{1}{2} & \frac{1}{2} \\ 1 & 0 & 0 \\ 1 & 0 & 0 \end{pmatrix}. \tag{5.3}$$

Compute $P(2n)$ and $P(2n-1)$ for all $n \geq 1$ and draw conclusions about the ergodicity of the chain.

5. Let $P = \left(p_{i|j}\right)_{i=1,j=1}^{J,J}$ be a transition matrix with the properties

$$\sum_{j=1}^{J} p_{i|j} = 1 \qquad \text{and} \qquad \sum_{i=1}^{J} p_{i|j} = 1. \tag{5.4}$$

We call such a P a *doubly stochastic matrix.*

(a) Show that if

$$\lim_{n\to\infty} p_{i|j}(n) = \pi_j$$

holds true for all j then the invariant distribution is the uniform distribution.

(b) Show a converse, namely that if the invariant distribution for a Markov chain is the uniform distribution, then the corresponding transition matrix is doubly stochastic.

6. Let

$$P = \begin{pmatrix} 1-p & p \\ q & 1-q. \end{pmatrix} \tag{5.5}$$

be a transition matrix.

(a) Verify that

$$P^n = \frac{1}{p+q}\begin{pmatrix} q & p \\ q & p \end{pmatrix} + \frac{(1-p-q)^n}{p+q}\begin{pmatrix} p & -p \\ -q & q \end{pmatrix}. \tag{5.6}$$

(b) Find $\lim_{n\to\infty} P^n$. Compare the result to (3.7) and (3.7).

7. The random variables $\{Z_n\}_{n=-1}^{\infty}$ are independent and identically distributed with $Z_n \in \text{Be}\left(\frac{1}{2}\right)$ for all n. We set

$$X_n = [Z_n, Z_{n-1}], \qquad n = 0, 1, 2, \ldots,$$

so that $X_0 = [Z_0, Z_{-1}]$, $X_1 = [Z_1, Z_0]$ and so on.

(a) Show that $\{X_n\}_{n=0}^{\infty}$ is a Markov chain.

(b) Find the invariant distribution of the chain $\{X_n\}_{n=0}^{\infty}$.

7.5.2 Stationary Markov Processes

8. A stochastic process $\{Z_n\}_{n=0}^{\infty}$ with values in $\mathcal{Z} = \{z_1, \ldots, z_r\}$ is called **(strictly) stationary** is the finite dimensional distributions are invariant to shift, or

$$P\left(Z_{n_1} = z_{l_1}, \ldots, Z_{n_k} = z_{l_k}\right) = P\left(Z_{n_1+h} = z_{l_1}, \ldots, Z_{n_k+h} = z_{l_k}\right) \quad (5.7)$$

for any sequence $z_{l_1}, \ldots, z_{l_k}$ of symbols in $\mathcal{Z}$ and any $n_1, \ldots, n_k$ and for any integer $h > 0$. Let

$$\pi = (\pi_1, \ldots, \pi_J)$$

be the invariant distribution of a time homogeneous Markov chain $\{X_n\}_{n=0}^{\infty}$ with $\pi_j > 0$ for all j. Verify that if $\pi_j = P(X_0 = j)$ for all j then $\{X_n\}_{n=0}^{\infty}$ is stationary in the sense of (5.7).

7.5.3 Entropy Rate of a Stationary Markov Chain

The *entropy rate* of a stochastic process$\{X_n\}_{0=1}^{\infty}$ with a finite alphabet is defined as

$$H_{\infty}(X) := \lim_{n \longmapsto \infty} \frac{H(X_0, X_1 \cdots, X_n)}{n+1}. \quad (5.8)$$

9. Prove that this limit exists for stationary random processes and is given by

$$H_{\infty}(X) = \lim_{n \to \infty} H\left(X_n | X_{n-1}, \ldots, X_0\right),$$

where $H\left(X_n | X_n, \ldots, X_0\right)$ is defined as the entropy of the conditional distribution or

$$H\left(X_n | X_{n-1}, \ldots, X_0\right) = \sum_{x_0} \cdots \sum_{x_{n-1}} p_{X_{n-1},..,X_0}\left(x_{n-1}, \ldots, x_0\right) h,$$

where

$$h = -\sum_{x_n} p_{X_n|X_{n-1},..,X_0}(x_n|x_{n-1},\ldots,x_0)\log p_{X_n|X_{n-1},..,X_0}(x_n|x_{n-1},..,x_0).$$

10. Let $\{X_n\}_{0=1}^{\infty}$ be a stationary Markov chain with the transition matrix $P = (p_{i|j})_{i=1,j=1}^{J,J}$ and the invariant distribution π. Show that the entropy rate $\mathcal{H}$ of $\{X_n\}_{0=1}^{\infty}$ is equal to

$$H_\infty(X) = -\sum_{i=1}^{J}\pi_i\sum_{j=1}^{J} p_{i|j}\cdot\log p_{i|j}.$$

11. Let $\{X_n\}_{0=1}^{\infty}$ be a time invariant Markov chain with the transition matrix the transition matrix

$$P = \begin{pmatrix} 1-\alpha & \alpha \\ 0 & 1 \end{pmatrix}. \tag{5.9}$$

Determine the entropy rate of this Markov chain.

7.5.4 High-order Markov Chains

12. Suppose that $\{Z_n\}_{n=0}^{\infty}$ is a second order Markov chain with the state space

$$\mathcal{M} = \{1,2,\ldots,m\}$$

and that

$$p_{i,k|j} = P(Z_n = j|Z_{n-1} = i, Z_{n-2} = k), \qquad n \geq 1, i, j \in \mathcal{M} \tag{5.10}$$

are the time-homogeneous transition probabilities. The invariant distribution is a bivariate distribution

$$\mu_{j,k} = P(Z_{n-1} = j, Z_n = k).$$

Hence the probabilities $\mu_{j,k}$ satisfy

$$\mu_{j,k} = \sum_{i=1}^{m}\mu_{i,j}p_{i,k|j} \tag{5.11}$$

and

$$\sum_{j=1}^{m}\sum_{k=1}^{m}\mu_{j,k} = 1. \tag{5.12}$$

The process $\{Z_n\}_{n=0}^{\infty}$ can be transformed into a first order Markov chain by the definition

$$X_n = [Z_{n-1}, Z_n].$$

The effect of this is to enlarge the state space.

Take for example $\{Z_n\}_{n=0}^{\infty}$ assuming values in the binary state space $\mathcal{M} = \{1, 2\}$. The second order probability transition probabilities are fully specified by the four transition probabilities

$$p_{1,1|2} = P(Z_n = 2|Z_{n-1} = 1, Z_{n-2} = 1),$$

$$p_{2,2|1} = P(Z_n = 1|Z_{n-1} = 2, Z_{n-2} = 2),$$

$$p_{1,2|1} = P(Z_n = 1|Z_{n-1} = 2, Z_{n-2} = 1),$$

$$p_{2,1|2} = P(Z_n = 1|Z_{n-1} = 1, Z_{n-2} = 2).$$

The corresponding first order Markov chain X_n defined above then has the state space

$$\{(1,1), (1,2), 2, 1), (2,2)\}$$

and the transition matrix

$$P = \begin{pmatrix} 1 - p_{1,1|2} & p_{1,1|2} & 0 & 0 \\ 0 & 0 & p_{1,2|1} & 1 - p_{1,2|1} \\ 1 - p_{2,1|2} & p_{2,1|2} & 0 & 0 \\ 0 & 0 & p_{2,2|1} & 1 - p_{2,2|1} \end{pmatrix}, \tag{5.13}$$

Find the invariant distribution for this chain.

13. **MTD**
A high order Markov property can be modelled by **mixture transition distribution** (MTD). We let $Q = \left(q_{i|j}\right)_{i=1,j=1}^{J,J}$ be a matrix with $q_{i|j} \geq 0$, $\sum_{j=1}^{J} q_{i|j} = 1$. Actually the original construction in (Raftery 1985) deals with *columnwise* stochastic matrices. A mixing distribution is given by $\lambda_i \geq 0, i = 1, \ldots, l$, $\sum_{i=1}^{l} \lambda_i = 1$. We consider a sequence of

random variables $\{X_n\}_{n=0}^{\infty}$ assuming values in $\{1, \ldots, J\}$ such that for all n and $j_l, j_1, \ldots, j_n \in S$,

$$P\left(X_n = j_n | X_{n-l} = j_{n-l}, \ldots, X_{n-1} = j_{n-1}\right) = \sum_{i=1}^{l} \lambda_i q_{j_{n-i}|j_n} \tag{5.14}$$

for a positive integer l. The process $\{X_n\}_{n=0}^{\infty}$ is called an MTD(l) process. The integer l is called the order of the process.

(a) How can the values of an MTD(l) process be generated?

(b) Introduce the vector process

$$\left[X_n, \ldots, X_{n-l+1}\right],$$

which assumes values in the alphabet J^l. Find its transition matrix, i.e., a matrix with the arrays

$$P\left(X_n = j_n, \ldots, X_{n-l+1} = j_{n-l+1} | X_{n-1} = j_{n-1}, \ldots, X_{n-l} = j_{n-l}\right).$$

(Raftery 1985).

7.6 References and Further Reading:

T.M. Cover and J.A. Thomas (1991): *Elements of Information Theory.* John Wiley and Sons, New York.

I. Csiszár (1963): Eine informationstheoretische Ungleichung und ihre Anwendung auf den Beweis der Ergodizität von Markoffschen Ketten. *Publications of the Mathematical Institute of the Hungarian Academy of Science.* Ser. A, Fasc. 1-2, 8, pp. 85–108.

R.M. Gray (1988): *Probability, Random Processes and Ergodic Properties.*, Springer Verlag, New York, Berlin, etc..

L.L. Helms (1997): *Introduction to Probability Theory With Contemporary Applications.* W.H. Freeman and Company. New York.

O. Häggström (2001): *Finite Markov Chains and Algorithmic Applications.* `http://www.math.chalmers.se/~olleh/surveys.html`

J.G. Kemeny, J.L. Sell and A.W. Knapp (1976): *Denumerable Markov Chains.* Second Edition. Springer Verlag, New York, Berlin, etc..

A.I. Khuri (1993): *Advanced Calculus with Applications in Statistics.* John Wiley and Sons, New York.

S. Kullback (1966): An Information-Theoretic Derivation of Certain Limit Relations for A Stationary Markov Chain. *SIAM Journal of Control*, August, pp. 454–459.

T. Lindvall (1992): *Lectures on the Coupling Method.* John Wiley and Sons Inc. , New York.

J.R. Norris (1997): *Markov Chains.* Cambridge University Press, Cambridge.

A.R. Raftery (1985): A Model for High-order Markov Chains. *Journal of the Royal Statistical Society*, B, 47, pp. 528 - 539.

Chapter 8

Learning of Markov Chains

8.1 Background

8.1.1 General Summary

Now we apply the steps of Bayesian modelling to a (training) sequence using a family of Markov models. A Markov model is completely specified by a transition matrix and an initial distribution. Probabilistic learning needs:

(1) a Markovian probability distribution which specifies the probability of any sequence conditioned by the transition matrix and the initial distribution;

(2) a prior which expresses the uncertainty about the transition matrix and the initial distribution.

When (1) is combined in a known fashion with the training sequence we obtain the likelihood function of the sequence with respect to a family of Markov models. The likelihood function is combined with (2) via Bayes' rule to produce a *posterior distribution* for the parameters of the family of Markov models. Using (1) and (2) we may also compute predictive distributions and do model comparison by means of Bayes factors. Model family comparison is especially concerned with finding the *order of the Markov chain*, as defined in Chapter 7, a technique appearing in modelling DNA sequences in Chapter 9.

8.2 ML for Markov Chains

8.2.1 Preliminaries

Let $\underline{\theta}$ be the transition probability matrix

$$\underline{\theta} = \begin{pmatrix} \theta_{1|1} & \theta_{1|2} & \dots & \theta_{1|J} \\ \theta_{2|1} & \theta_{2|2} & \dots & \theta_{2|J} \\ \vdots & \vdots & \vdots & \vdots \\ \theta_{J|1} & \theta_{J|2} & \dots & \theta_{J|J} \end{pmatrix}. \tag{2.1}$$

We are concerned with estimating the model $\underline{\theta}$ in the family of probabilistic models $p(\mathbf{x}|\underline{\theta})$ for a (training) sequence $\mathbf{x}$ of $n+1$ symbols in S^{n+1}

$$\mathbf{x} = (j_0 j_1 \dots j_n) \in S^{n+1}.$$

It was shown in Chapter 7 that for Markov chains

$$p(\mathbf{x}|\underline{\theta}) = P(X_0 = j_0, X_1 = j_1, \dots, X_n = j_n|\theta) = \pi_{j_0}(0) \prod_{l=1}^{n} \theta_{j_{l-1}|j_l}. \tag{2.2}$$

Then we propose

MODEL FAMILY:

CONDITIONED ON $\pi_{j_0}(0)$ AND $\Theta = \underline{\theta}$, THE SYMBOLS IN $\mathbf{x}$ ARE AN OUTCOME OF A MARKOV CHAIN $\{X_n\}_{n \geq 0}$, WITH STATIONARY TRANSITION PROBABILITIES $\underline{\theta}$. ■

Here we have at most $J^2 - J$ transition parameters and the $J-1$ initial probabilities to estimate using the data $\mathbf{x}$. We make an approximation to the effect that we omit the initial distribution $\pi(0)$ as a part of the estimation problem. One way to think of this is that we know (or fix at will) the initial symbol in advance. Moreover, given just one training sequence we have just one single observation of the initial state. Let us first consider the maximum likelihood estimate of all of the unknown transition parameters.

As a function of $\underline{\theta}$ for fixed $\mathbf{x}$ the ensuing approximate or conditional likelihood function is

$$L(\underline{\theta}) = \prod_{l=1}^{n} \theta_{j_{l-1}|j_l}. \tag{2.3}$$

The corresponding log likelihood function is

$$\mathcal{L}(\underline{\theta}) = \sum_{l=1}^{n} \ln \theta_{j_{l-1}|j_l}. \tag{2.4}$$

8.2.2 ML of the Transition Matrix

We introduce a notation for the number of times we see a transition from i to j in $\mathbf{x} = (j_0 j_1 \dots j_n)$. Thus

$$n_{i|j} = \text{ the number of } l \text{ such that } 1 \le l \le n,\ j_{l-1} = i, j_l = j. \tag{2.5}$$

Using the frequency counts $n_{i|j}$ we can write the likelihood function as

$$L(\underline{\theta}) = \prod_{i=1}^{J} \prod_{j=1}^{J} \theta_{i|j}^{n_{i|j}}. \tag{2.6}$$

Obviously $n_{i|j}$s are the sufficient statistics for this model family. The log likelihood in (2.4) will be

$$\mathcal{L}(\underline{\theta}) = \sum_{i=1}^{J} \sum_{j=1}^{J} n_{i|j} \ln \theta_{i|j}. \tag{2.7}$$

Let also

$$n_i = \text{ the number of } l \text{ such that } 0 \le l \le n-1,\ j_l = i, \tag{2.8}$$

so that n_i is equal to the number of times the sequence $\mathbf{x}$ visits the state i, excluding the possible visit at the final time. Then we have

Proposition 8.2.1 *The maximum likelihood estimate $\widehat{\theta}_{i|j}$ of $\theta_{i|j}$ is*

$$\widehat{\theta}_{i|j} = \frac{n_{i|j}}{n_i}, \tag{2.9}$$

for all i and j.

Proof: Since the constraints

$$\theta_{i|j} \ge 0, \qquad \sum_{j=1}^{J} \theta_{i|j} = 1 \tag{2.10}$$

hold separately for each row in the transition matrix we can maximize $\mathcal{L}(\underline{\theta})$ in (2.7), which is a separable sum of the corresponding terms, by an independent maximization for each row. Thus for each row

$$\underline{\theta}_i = \left(\theta_{i|1}, \ldots, \theta_{i|J}\right)$$

we should maximize

$$\mathcal{L}(\underline{\theta}_i) = \sum_{j=1}^{J} n_{i|j} \ln \theta_{i|j} \tag{2.11}$$

as a function of $\theta_{i|1}, \ldots, \theta_{i|J}$ so that the constraints are satisfied. Therefore we may repeat the computation from Chapter 3. Let us set

$$\widehat{\theta}_i = \left(\frac{n_{i|1}}{n_i}, \ldots, \frac{n_{i|J}}{n_i}\right).$$

Since $n_{i|1} + n_{i|2} + \ldots + n_{i|J} = n_i$, as every transition from i (possibly back to i) indicates necessarily a visit to i and the final time was excluded, we see that $\widehat{\theta}_i$ satisfies the constraints. Take now an arbitrary $\underline{\theta}_i$ satisfying the constraints. Then we get as in Chapter 3

$$\mathcal{L}\left(\widehat{\theta}_i\right) - \mathcal{L}(\underline{\theta}_i) = n_i D\left(\widehat{\theta}_i | \underline{\theta}_i\right) \geq 0,$$

where $D\left(\widehat{\theta}_i | \underline{\theta}_i\right)$ is the Kullback distance, which has been proved to be non-negative. Equality holds if and only if $\widehat{\theta}_i = \theta_i$. ■

8.2.3 An Example of Full Likelihood

Suppose that the model family consists of *stationary* Markov chains with a binary state space S and with the transition probability matrices

$$A = \begin{pmatrix} 1-p & p \\ q & 1-q. \end{pmatrix}. \tag{2.12}$$

We wish now to estimate p and q using an observed sequence $\mathbf{x}$ of $n+1$ symbols in S^{n+1}

$$\mathbf{x} = (j_0 j_1 \ldots j_n) \in S^{n+1}.$$

In the preceding the initial distribution p_{X_0} was not a part of the estimation problem. If the chain is stationary then the initial distribution is an invariant distribution, which contains the unknown parameters. We can, of course, still throw away the initial distribution, but this means a loss of information, which is asymptotically insignificant.

The full likelihood function $L(p,q) = p(\mathbf{x}|A)$ for the stationary model given the sequence $\mathbf{x}$ turns out to be equal to

$$L(p,q) = \frac{p^a \cdot (1-p)^b \cdot q^c \cdot (1-q)^d}{p+q}, \tag{2.13}$$

where, using the notations for the number of state transitions in the sequence $\mathbf{x}$ introduced above,

$$a = j_0 + n_{0|1}, \qquad b = n_{0|0}, c = 1 - j_0 + n_{1|1}, \qquad d = n_{1|1}.$$

Hence there is no longer an explicit solution of the log likelihood equation obtained by setting the partial derivatives of $\ln L(p,q)$ equal to zero. In (Bisgaard and Travis 1991) it is shown that this system of equations has a unique solution which is a maximum.

8.3 The Whittle Distribution

For any given $\mathbf{x} = (j_0 j_1 \ldots j_n)$ we make the frequency counts (2.5) or

$$n_{i|j} = \text{ the number of } l \text{ such that } 1 \le l \le n,\ j_{l-1} = i, \qquad j_l = j.$$

Let us set

$$n_{i|\cdot} = \sum_{j=1}^{J} n_{i|j}, n_{\cdot|j} = \sum_{i=1}^{J} n_{i|j} \text{ for all } i \text{ and } j. \tag{3.1}$$

Thus $n_{i|\cdot}$ is the frequency count of i in the prefix $(j_0 j_1 \ldots j_{n-1})$ of $\mathbf{x}$ and $n_{\cdot|j}$ is the frequency count of j in the suffix $(j_1 \ldots j_n)$ and $\mathbf{x}$. Therefore

$$n_{i|\cdot} - n_{\cdot|i} = \delta_{ij_0} - \delta_{ij_n}, \tag{3.2}$$

where δ_{ij} is Kronecker's delta (i.e., $\delta_{ij} = 1$ if $i = j$ and $\delta_{ij} = 0$ if $i \neq j$) and

$$\sum_{i=1}^{J} n_{i|\cdot} = \sum_{j=1}^{J} n_{\cdot|j} = n. \tag{3.3}$$

Let

$$F = \left(n_{i|j}\right)_{i=1,j=1}^{J,J} \tag{3.4}$$

be any $J \times J$ matrix of non-negative integers which satisfy (3.2) and (3.3). Hence the knowledge of F and of j_0 determine j_n uniquely and F and j_n determine j_0.

Proposition 8.3.1 (Whittle's multinomial coefficient) *Let F be an $J\times J$ - matrix of non-negative integers $n_{i|j}$ such that $\sum_{i=1}^{J}\sum_{j=1}^{J} n_{i|j} = n$ and such that $n_{i|\cdot} - n_{\cdot|i} = \delta_{iu} - \delta_{iv}$ for some u and v in S. Let*

$N_{u,v}^{(n)}(F)$ = the number of sequences $\mathbf{x} = (j_0 j_1 \ldots j_n)$ having the frequency count F and satisfying $j_0 = u$, $j_n = v$.

Then

$$N_{u,v}^{(n)}(F) = \frac{\prod_{i=1}^{J} n_{i|\cdot}!}{\prod_{i=1}^{J}\prod_{j=1}^{J} n_{i|j}!} \cdot F_{vu}^{*}, \tag{3.5}$$

where F_{vu}^{*} is the (u,v)th cofactor of the matrix F^{*} with components

$$f_{ij}^{*} = \begin{cases} \delta_{ij} - \dfrac{n_{i|j}}{n_{i|\cdot}} & \text{if } n_{i|\cdot} > 0, \\ \delta_{ij} & \text{if } n_{i|\cdot} = 0. \end{cases}$$

Proof: A proof is given in (Billingsley 1962, pp. 14–15). The cofactor F_{vu}^{*} is $(-1)^{u+v}$ multiplied by the determinant of the matrix obtained obtained by deleting row u and column u, as defined in any text on matrices. ■

Then we readily obtain t the probability that $\mathbf{x} = (j_0 j_1 \ldots j_n)$ has F as its transition count and $j_0 = u$ and $j_n = v$, denoted by $P_W(F)$, is (Whittle 1955) nothing else but

$$P_W(F) = \pi_u(0) \cdot F_{vu}^{*} \cdot \frac{\prod_{i=1}^{J} n_{i|\cdot}!}{\prod_{i=1}^{J}\prod_{j=1}^{J} n_{i|j}!} \cdot \prod_{i=1}^{J}\prod_{j=1}^{J} \theta_{i|j}^{n_{i|j}}. \tag{3.6}$$

A homogeneous Markov chain is thus seen to resemble a set of independent multinomial processes.

The Whittle distribution turns out to be useful in computing statistical properties of occurrences of words, a problem of considerable biological interest as shown in Chapter 9.

8.4 Model Averaging

8.4.1 Posterior Distributions for Rows in the Transition Matrix

Let us assume that our uncertainty about *the rows of* $\underline{\theta}$ in (2.1)

$$\theta_i = \left(\theta_{i|1}, \ldots, \theta_{i|J}\right)$$

is modeled by *independent random variables* that have their respective Dirichlet densities for $i = 1, \ldots, J$ see section 3.8 in Chapter 3. These we formulate as

$$Dir\left(\theta_i; \alpha_i \cdot q_{i|1}, \ldots, \alpha_i \cdot q_{i|J}\right) = \frac{\Gamma\left(\alpha_i\right)}{\prod_{j=1}^{J} \Gamma\left(\alpha_i q_{i|j}\right)} \cdot \prod_{j=1}^{J} \theta_{i|j}^{\alpha_i q_{i|j}-1}, \tag{4.1}$$

where

$$\alpha_i > 0, \qquad q_{i|j} > 0, \qquad \sum_{j=1}^{J} q_{i|j} = 1.$$

Then we use as the simultaneous prior density the **multivariate Dirichlet density** or

$$\prod_{i=1}^{J} \text{Dir}\left(\theta_i; \alpha_i \cdot q_{i|1}, \ldots, \alpha_i \cdot q_{i|J}\right) \tag{4.2}$$

suggested by (Martin 1967, ch. 2), see also (Basawa and Rao 1980 pp. 65 - 68). Hence the posterior density is in view of (2.3) equal to

$$p\left(\underline{\theta}|\mathbf{x}\right) = \frac{\prod_{i=1}^{J} \frac{\Gamma(\alpha_i)}{\prod_{j=1}^{J} \Gamma\left(\alpha_i q_{i|j}\right)} \prod_{j=1}^{J} \theta_{i|j}^{n_{i|j}+\alpha_i q_{i|j}-1}}{p(\mathbf{x})}, \tag{4.3}$$

where $p(\mathbf{x})$ is the standardization that makes $p\left(\underline{\theta}|\mathbf{x}\right)$ a probability density in $\underline{\theta}$.

8.4.2 Predictive Probability

As an obvious extension of the predictive probabilities in Chapter 3 we might ask what is our probability

$$P\left(X_{n+1} = j | X_n = i; \mathbf{x}\right),$$

where the notation indicates that the probability is based on a given training sequence $\mathbf{x} = (j_0 j_1 \dots j_n) \in S^{n+1}$? In view of our Markov modelling of $\mathbf{x}$ in the preceding subsection one answer could be

$$\widehat{P}_{\mathrm{ML}}\left(X_{n+1} = j | X_n = i; \mathbf{x}\right) = \widehat{\theta}_{i|j}$$

plugging in the maximum likelihood estimate of the transition probability.

In a completely observabilistic sense we would consider only the single sequence $\mathbf{x}$ and ask for $P\left(X_{n+1} = j|\mathbf{x}\right)$ and provide the answer as

$$\widehat{P}_{\mathrm{ML}}\left(X_{n+1} = j|\mathbf{x}\right) = \widehat{\theta}_{j_n|j},$$

since j_n is the last symbol in the sequence.
There are other ways of addressing the stated question. Using the sequence $\mathbf{x}$ we may take some posterior density for $\underline{\theta}$ and then provide a new transition matrix by *model averaging*

$$P^*\left(X_{n+1} = j|X_n = i; \mathbf{x}\right) = \int \theta_{i|j} p\left(\underline{\theta}|\mathbf{x}\right) d\underline{\theta}. \tag{4.4}$$

Using (4.2)

$$\begin{aligned} P^*\left(X_{n+1} = r|X_n = s; \mathbf{x}\right) &= \int \theta_{r|s} p\left(\underline{\theta}|\mathbf{x}\right) d\underline{\theta} \\ &= \frac{\prod_{i=1}^{J} \frac{\Gamma(\alpha_i)}{\prod_{j=1}^{J} \Gamma\left(\alpha_j q_{i|j}\right)} I_{i|j}\left(r, s\right)}{p(\mathbf{x})}, \end{aligned}$$

where

$$I_{i|j}\left(r, s\right) = \int \prod_{j=1}^{J} \theta_{r|s} \theta_{i|j}^{n_{i|j} \alpha_i q_{i|j} - 1} d\theta_i.$$

Here the integration is with respect to all of the parameters in $\underline{\theta}$, which are, however, separated to their respective domains as rows of $\underline{\theta}$. Using the well known formulae for evaluating the various Dirichlet integrals (appendix to Chapter 3 and Chapter 6) we obtain in (4.4) the expression

$$P^*\left(X_{n+1} = r|X_n = s; \mathbf{x}\right) = \frac{n_{s|r} + \alpha_s q_{s|r}}{n_s + \alpha_s}. \tag{4.5}$$

The parameters α_s and $q_{s|r}$ play, as before, the role of pseudo counts of observations or of regularizers.

8.5 MC Order Comparison Using the Bayes Ratio

In the literature on biological sequence analysis the problem of estimating the order of a time-homogeneous Markov chain and/or testing against a null model, the multinomial process with the same state space. We now state a relevant criterion, without restricting ourselves to any specific biological situation, using Bayesian model comparison. We compute the Bayes ratio

$$B(\mathbf{x}) = \frac{q_M(\mathbf{x})}{q_{M_0}(\mathbf{x})}, \tag{5.1}$$

where under the model family M the training sequence $\mathbf{x}$ is related to the parameters in a transition matrix $\underline{\theta}$ as above and with the multivariate Dirichlet density (4.2) as prior. Under the model family M_0 the training sequence $\mathbf{x}$ is related to the parameters in a conditional independence model with a Dirichlet prior (cf. Chapter 3).

As in the chapter quoted we obtain

$$q_{M_0}(\mathbf{x}) = \frac{\Gamma(\alpha)}{\Gamma\left(\prod_{i=1}^{J} \alpha q_i\right)} \cdot \frac{\prod_{i=1}^{J} \Gamma(\alpha q_i + n_i^*)}{\Gamma(n+\alpha)}, \tag{5.2}$$

where n_i^* is equal to the number of times the symbol i appears in $\mathbf{x}$. Note that by the definitions valid here n_i^* is equal to n_i for some $J-1$ symbols and n_i+1 for the remaining of them. In view of the formulas above we have that

$$\begin{aligned} q_M(\mathbf{x}) &= \prod_{i=1}^{J} \frac{\Gamma(\alpha_i)}{\prod_{j=1}^{J} \Gamma(\alpha_i q_{i|j})} \int_{\underline{\theta}} \prod_{j=1}^{J} \theta_{i|j}^{n_{i|j}+\alpha_i q_{i|j}-1} d\underline{\theta} \\ &= \prod_{i=1}^{J} \frac{\Gamma(\alpha_i)}{\prod_{j=1}^{J} \Gamma(\alpha_i q_{i|j})} \cdot \frac{\prod_{j=1}^{J} \Gamma\left(n_{i|j} + \alpha_i q_{i|j}\right)}{\Gamma(n_i + \alpha_i)}. \end{aligned} \tag{5.3}$$

The idea in (Milosaljevic and Jurka 1993) can be recapitulated as searching a database for sequences $\mathbf{x}$ such that $-\log B(\mathbf{x})$ exceeds a threshold. The threshold can be taken as the length of the codeword for $\mathbf{x}$ compressed by a suitable algorithm, an application of the theory in Chapter 2.

8.6 Consistency of the ML

Billingsley (1961) has established a number of properties for the maximum likelihood estimator given in (2.9) under assumptions that hold if the MC is ergodic and has a finite irreducible state space.

The fundamental property, the proof of which is omitted but can be found in (Billingsley 1961), is that there is a law of large numbers for ergodic Markov chains, stated as follows:

Proposition 8.6.1 *Let $\{X_n\}_{n=0}^{\infty}$ be an ergodic Markov chain with the invariant distribution $\pi = (\pi_j)_{j=1}^{J}$ on the finite state space $S = \{1, 2, \ldots, J\}$ such that $\pi_i > 0$ for all i.*
If ϕ is any measurable and bounded real valued function on $S \times S$, then

$$\lim_{n\to\infty} \frac{1}{n} \sum_{j=1}^{n} \phi\left(X_{j-1}, X_j\right) = E_{\underline{\theta}}\left[\phi\left(X_0, X_1\right)\right]. \tag{6.1}$$

with probability one. Here

$$E_{\underline{\theta}}\left[\phi\left(X_0, X_1\right)\right] = \sum_{i=1}^{J}\sum_{j=1}^{J} \phi\left(i, j\right) \pi_i \theta_{i|j}. \tag{6.2}$$

■

A short and simple proof of a special case of this ergodic theorem using the strong Markov property is given in (Norris 1997). We can apply this law of large numbers with suitable functions ϕ to find the limits of the ML estimates in Proposition 8.2.1 of the transition probabilities as

$$\widehat{\theta}_{i|j} = \frac{n_{i|j}}{n_i} \to \theta_{i|j} \tag{6.3}$$

with probability one, as $n \to \infty$. Next we prove that log likelihood per symbol converges to the entropy rate. We divide actually the log likelihood function by $|\mathbf{x}| - 1 = n$, but then we have also omitted the initial symbol.

Proposition 8.6.2 *Let $\{X_n\}_{n=0}^{\infty}$ be an ergodic Markov chain with the invariant distribution $\pi = (\pi_i)_{i=1}^{J}$ on the finite state space $S = \{1, 2, \ldots, J\}$ such that $\pi_i > 0$ for all i. Then*

$$\lim_{n\to\infty} -\frac{1}{n}\mathcal{L}\left(\widehat{\underline{\theta}}\right) = H_{\infty}\left(X; \underline{\theta}\right), \tag{6.4}$$

with probability one, where

$$H_{\infty}(X;\underline{\theta}) = -\sum_{i=1}^{J}\pi_i\sum_{j=1}^{J}\theta_{i|j}\cdot\log\theta_{i|j}$$

is the entropy rate of the Markov chain.

Proof: From (2.7)

$$\frac{1}{n}\mathcal{L}\left(\widehat{\underline{\theta}}\right) = \sum_{i=1}^{J}\frac{n_i}{n}\sum_{j=1}^{J}\frac{n_{i|j}}{n_i}\ln\widehat{\theta}_{i|j}.$$

Then by Gibbs' inequality (5.8) in Chapter 2 we obtain for sufficiently large n

$$\frac{1}{n}\mathcal{L}\left(\widehat{\underline{\theta}}\right) \geq \sum_{i=1}^{J}\frac{n_i}{n}\sum_{j=1}^{J}\frac{n_{i|j}}{n_i}\ln\theta_{i|j} \geq \sum_{i=1}^{J}\liminf_{n\to\infty}\left[\frac{n_i}{n}\sum_{j=1}^{J}\frac{n_{i|j}}{n_i}\ln\theta_{i|j}\right].$$

Also, for sufficiently large n

$$\frac{1}{n}\mathcal{L}\left(\widehat{\underline{\theta}}\right) \leq \sum_{i=1}^{J}\limsup_{n\to\infty}\left[\frac{n_i}{n}\sum_{j=1}^{J}\frac{n_{i|j}}{n_i}\ln\widehat{\theta}_{i|j}\right].$$

Since the ergodic property in (6.1) implies that

$$\lim_{n\to\infty}\frac{n_i}{n} = \pi_i > 0 \qquad (6.5)$$

with probability one and (6.3) holds, we get the claimed result. ∎

The values of the invariant distribution π and of the transition probabilities $\theta_{i|j}$ in the entropy rate $H_{\infty}(X;\underline{\theta})$ are in this context seen as the fixed '*true values*' of the respective probabilities. This may be emphasized by writing the result above as

$$\lim_{n\to\infty} -\frac{1}{n}\mathcal{L}\left(\widehat{\underline{\theta}}\right) = -\sum_{i=1}^{J}\pi_i^{(o)}\sum_{j=1}^{J}\theta_{i|j}^{(o)}\cdot\log\theta_{i|j}^{(o)}.$$

Then (6.3) has the interpretation that the maximum likelihood estimate $\widehat{\underline{\theta}}$ converges to asymptotically to the 'true value' $\underline{\theta}_o$ with probability one or is *asymptotically consistent.*

The notion of the true value $\underline{\theta}_o$ corresponds partially to thinking that the sequence $\mathbf{x}$ used in computing the maximum likelihood estimate is actually generated by the model Markov $(\underline{\theta}_o, p_{X_0})$ in the family, where $\underline{\theta}_o$ is (initially) unknown to us.

The proof above deals with estimating a probability distribution as a table of numbers. In the parametric case the transition matrix $\underline{\theta} = \underline{\theta}(\underline{\mu})$ is a function of some parameters $\underline{\mu}$ to be estimated. An illustration of this situation is given in Exercise 7 below. Various extensions of the proof and of the consistency property above to this more complicated case are found in (Ranneby 1975, 1978 and Roussas 1965). The results in (Höglund 1974) are concerned with the *exact estimate* (see the pertinent exercise of Chapter 4) for a Markov chain written in the *canonical* form. A Markov chain is canonical if its transition probabilities are of the exponential family type.

8.7 Exercises

1. Verify (2.13) and find the maximum likelihood equations for p and q.

2. Prove (6.3) and (6.5).

3. **Whittle distribution**
   Show that the probability of obtaining $j_1 \ldots j_n$ such that $n_{\cdot|j}$ is its frequency count and that $j_0 = u$ and $j_n = v$ is

$$\pi_u(0) \cdot (n-1)! \cdot \frac{\prod_{i=1}^{J} n_{i|\cdot}!}{\prod_{i=1}^{J} \prod_{j=1}^{J} n_{i|j}!} \cdot \prod_{i=1}^{J} \prod_{j=1}^{J} \theta_{i|j}^{n_{i|j}}. \tag{7.1}$$

   using the notations in (3.6).

4. **Posterior Distributions**
   Let

$$\underline{\theta} = \begin{pmatrix} 1-p & p \\ q & 1-q. \end{pmatrix} \tag{7.2}$$

   be a transition matrix for a Markov chain with the state space $\{1, 2\}$. Let us consider the prior density for the 2×2 matrix $\underline{\theta}$ given by

$$\phi(\underline{\theta}) = f_1(p) \cdot f_2(q),$$

where

$$f_1(p) = \begin{cases} \dfrac{\Gamma(\alpha_{11}+\alpha_{12})}{\Gamma(\alpha_{11})\cdot\Gamma(\alpha_{12})} p^{\alpha_{11}-1}(1-p)^{\alpha_{12}-1} & 0 \le p \le 1, \\ 0 & \text{elsewhere,} \end{cases}$$

and

$$f_2(q) = \begin{cases} \dfrac{\Gamma(\alpha_{21}+\alpha_{22})}{\Gamma(\alpha_{21})\cdot\Gamma(\alpha_{22})} q^{\alpha_{21}-1}(1-q)^{\alpha_{22}-1} & 0 \le q \le 1, \\ 0 & \text{elsewhere.} \end{cases}$$

Thus $\phi(\underline{\theta})$ is a product of two Beta densities (also known as a matrix Beta density) $\mathcal{B}\mathrm{e}(p;\alpha_{11},\alpha_{21})$ and $\mathcal{B}\mathrm{e}(q;\alpha_{21},\alpha_{22})$. Marc Moore (1976) has shown that the posterior density for $\underline{\theta}$ is in this case asymptotically like a multivariate normal density with suitable parameters.

(a) Compute

$$P(X_n = 2 \mid X_0 = 1)$$

using model averaging. *Hint:* Use the *explicit* formula for $\underline{\theta}^n$.

(b) Give a recursive formula for $P(X_{n+1} = 2 \mid X_0 = 1)$ in terms of $P(X_n = 2 \mid X_0 = 1)$.

5. **MAP**
Find the maximum a posteriori estimate of the transition matrix of a Markov chain

$$\underline{\theta}_{\mathrm{MAP}} = \arg\max_{\underline{\theta}} p(\underline{\theta}|\mathbf{x})$$

using the posterior density in (4.3).

6. **Model Comparison**
Let $\mathbf{x}$ be a binary sequence of length n. Compute Bayes' ratio

$$B(\mathbf{x}) = \frac{q_{M_2}(\mathbf{x})}{q_{M_1}(\mathbf{x})}, \tag{7.3}$$

when M_1 is the family of Markov chains (first order MC) with a binary state space and M_2 is the family of second order Markov chains with a binary state space. Find an asymptotic expression for large n. (Katz 1981).

7. **Sufficient statistics for a binary Markov chain**
Consider a Markov chain $\{X_n\}_{n=0}^{\infty}$, with the binary digits $S = \{0, 1\}$ as the state space and specified by

$$p = P(X_n = 1) = 1 - P(X_n = 0)$$

and

$$P(X_n = 1 \mid X_{n-1} = 1) = \lambda.$$

Let $q = 1 - p$.

(a) Show that the transition matrix is

$$\underline{\theta} = \begin{pmatrix} \frac{1 - 2p + \lambda \cdot p}{q} & \frac{(1-\lambda)p}{q} \\ 1 - \lambda & \lambda \end{pmatrix}, \tag{7.4}$$

and check that the entries in this matrix are numbers between zero and one.

(b) What are the properties in this chain in the cases $\lambda = p$, $\lambda > p$ and $\lambda < p$.

(c) Show that the probability of a binary sequence $\mathbf{x} = j_0 \ldots j_{n-1}$ of length n conditioned on $\underline{\theta}$ with regard to this model family is

$$p(\mathbf{x} \mid \underline{\theta}) = C \cdot \eta_1^r \cdot \eta_2^s \cdot \eta_3^t, \tag{7.5}$$

where

$$\begin{aligned} C &= \frac{(1 - 2p + \lambda p)^{n-1}}{q^{n-2}}, \\ \eta_1 &= \frac{\lambda (1 - 2p + \lambda p)}{p(1-\lambda)}, \\ \eta_2 &= \frac{(1-\lambda)^2 pq}{(1 - 2p + \lambda p)^2}, \\ \eta_3 &= \frac{(1 - 2p + \lambda p)}{pq(1-\lambda)}, \end{aligned}$$

and

$$r = \sum_{i=1}^{n-1} j_{i-1} j_i, \qquad s = \sum_{i=0}^{n-1} j_i, t = j_0 + j_{n-1}.$$

Check what happens if $\lambda = p$.

(d) Show that the random variables

$$R = \sum_{i=1}^{n-1} X_{i-1}X_i, S = \sum_{i=0}^{n-1} X_i, T = X_0 + X_{n-1}$$

form a *sufficient statistic* for (λ, p). *Hint*: Recall the *factorization criterion* from Chapter 4.

(e) Show that the joint distribution of (R, S, T) is

$$P(R = r, S = s, T = t) = \binom{2}{t} \cdot \binom{s-1}{r} \cdot \binom{n-s-1}{s-r-1} \cdot C \cdot \eta_1^r \cdot \eta_2^s \cdot \eta_3^t.$$

(Klotz 1973).

8. Describe the ML for a sequence modelled by the family of MTD(l) discussed in Exercise 13 of Chapter 7. The parameters of the family are the matrix Q and the mixing distribution. (A.E. Raftery and S. Tavaré 1994).

8.8 References and Further Reading:

- Mathematical statistics for Markov chains:

I.V. Basawa and B.L.S. Prakasa Rao (1980): ***Statistical Inference for Stochastic Processes.*** Academic Press, London, New York, 1980, ch. 4.

P. Billingsley (1961): ***Statistical Inference for Markov Processes, Statistical Research Monographs, vol. II***, The University of Chigago Press, Chigago, Illinois, 1961.

S. Bisgaard and L.E. Travis (1991): Existence and uniqueness of the solution of the likelihood equations for binary Markov chains. ***Statistics and Probability Letters***, 12, pp. 29–35.

T. Höglund (1974): Central limit theorem and statistical inference for finite Markov chains. ***Zeitschrift für Wahrscheinlichkeitstheorie und verwandte Gebiete***, 29, pp. 123–151.

R.W. Katz (1981): On Some Criteria for Estimating the Order of a Markov Chain. ***Technometrics***, 23, pp. 243–249.

J. Klotz (1973): Statistical Inference in Bernoulli Trials with Dependence. ***Annals of Statistics***, 1, pp. 373–379.

J.R. Norris (1997): ***Markov Chains.*** Cambridge University Press, Cambridge.

B. Ranneby (1975): Asymptotic Properties of the Empirical Transition Frequencies for a Stationary Chain with Discrete Time. *Scandinavian Journal of Statistics*, 2, pp. 103–105.

B. Ranneby (1978): On Necessary and Sufficient Conditions for Consistency of MLEs in Markov Chain Models. *Scandinavian Journal of Statistics*, 2, pp. 103 – 105.

G. Roussas (1965): Extensions to Markov Processes of a Result by A. Wald about the Consistency of the Maximum Likelihood Estimate. *Zeitschrift für Wahrscheinlichkeitstheorie und Verwandte Gebiete*, 4, pp. 69–73.

- Whittle distribution and sample distributions for Markov chains

P. Billingsley (1961): Statistical Methods in Markov Chain. *Annals of Mathematical Statistics*, 32, pp. 12–40.

P. Whittle (1955): Some Distribution and Moment Formulae for the Markov Chain. *Journal of the Royal Statistical Society Ser. B*, 17, pp. 235–42.

- Bayesian probability for Markov chains

D. Freedman (1962): Invariants under mixing which generalize de Finetti's theorem. *Annals of Mathematical Statistics*, 33, pp. 916–923.

M. Moore (1976): Remarks on the asymptotic behaviour of posterior laws. *Utilitas Mathematica*, 11, pp. 237–259.

J.J. Martin (1967): *Bayesian Decision Problems and Markov Chains.* John Wiley & Sons, New York, London and Sydney.

A.E. Raftery and S. Tavaré (1994): Estimation and Modelling Repeated Patterns in High Order Markov Chains with the Mixture Distribution Model. *Applied Statistics*, 43, pp. 179–199.

- Bayesian probability for searching databases

A.Milosavljevic and J. Jurka (1993): Discovering simple DNA sequences by the algorithmic similarity method. *Computer Applications in Biological Sciences* (CABIOS) 9(4), pp. 407–411.

Chapter 9

Markovian Models for DNA sequences

9.1 DNA and Modelling

This chapter is a brief summary of some of the findings suggesting that a straightforward Markov analysis of biological sequences is not likely to be achieving all it is meant for, and also as a short overview of more elaborate constructions based on a Markovian modelling approach (MTD, *'frame dependent' Markov chains, interpolated Markov models*), excluding at this point the hidden Markov models. Another minor overview is thereafter given of the theory statistics of DNA words under a first order Markov model, for which several explicit results can be given (Cowan 1991, Prum et al. 1995, Robin and Daudin 1999).

We recapitulate first some minimum piece of terminology. A more extensive and precise, yet concise, introduction to the same definitions is (Fitch 1999). A full scale introduction to molecular biology is given in (Alberts et al. 1994).

A DNA molecule consists of a long string of linked nucleotides. Each nucleotide is composed of a nitrogenous *base*, five carbon sugar and a phosphate group. There are four different bases. The pyrimidines are thymine (T) and cytosine (C) and the purines adenine (A) and guanine (G). The DNA occurs naturally as a double helix composed of two polynucleotide strands with the bases facing inwards. The two single strands run antiparallel and are connected by hydrogen bonding between complementary bases. Because the two strands are complementary, it is sufficient for many purposes to represent a DNA molecule by a sequence of bases in a single strand. The top of the strand is known as the $5'$ end and the bottom is known as the $3'$ end. The $5'$ and $3'$ refer to the way in which the bases are attached to the sugarphosphate backbone. Sequences are always read from $5'$ to the $3'$ end.

The socalled central dogma of genetics is that DNA is transcribed into messenger ribonucleic acid (mRNA), which is translated into proteins (and that no information is transferred in the opposite direction). Triples of bases are called *codons* which code for amino acids, the building blocks of proteins. Three of the codons are 'stop' codons, which signal that the translation should stop. The remaining 61 codons code for 20 amino acids. Hence an RNA sequence can be read in three different *reading frames* thus specifying three different sequences of amino acids.

A *gene* is a specific sequence of codons that carry the information required to construct proteins. In higher order organisms portions of the gene that code for a protein are called *exons* and are separated by noncoding regions called *introns*. *Chromosomes* are large segments of DNA (actually the double helix wrapped around protein structures) which contain many genes. The size, structure and the numbers of chromosomes vary among different organisms. Taken together the organism's chromosomes make up the *genome*.

Engineering advances in biotechnology (Hunkapiller et al. 1991. Alphey 1997), more than a century of effort in genetic analysis (Balding et al. 2001) and the development of algorithms and computer hardware (parallel processing) have made it feasible, both computationally and financially, to determine the nucleotide sequences of entire genomes for an increasing number of organisms (Myers et al. 2000, IHG Sequencing Consortium 2001, Venter et al. 2001).

Sequence analysis by modelling is motivated by the following (Curnow and Kirkwood 1989):

- Sequence data offer information of precise and detailed character with which to refine and strengthen established methods of statistical genetics, for example, the analysis of variation between individuals and between populations can be brought down to the level of single nucleotide difference;

- Sequence data offer the opportunity to find genes by computational algorithms and statistical modelling. Also, the structure of genes may include regions that have physical and structural properties that may be characterized by statistical models;

- The comparison of sequences between species demands methods for determining similarities in evolution or function. Statistical methods are required both in the search for patterns in DNA sequences, which may be suggestive of a biological function and in the testing of hypotheses concerning the evolution of DNA sequences.

For statistical modelling of DNA sequence data it is not necessary to suppose that the sequence data is actually generated by a random mechanism with a simple algorithmic form. The forces of evolution which have formed the DNA sequences of living organisms may be in some aspects random, but are complex and variable and no single simple model could hope to describe the endpoint. One introduction to modelling of the forces of evolution is (W-H. Li 1997).

The fundamental reason to entertain a Markov chain model for DNA sequences is found in the very mechanism by which the sequence was being described above: nucleotides being chained linearly one by one. The information in the sequence is by this account contained in the local dependence between the bases and their neighbors. Below we shall, however, also pay attention to the notion that a DNA sequence also contains *overlapping words* (various k-tuples of bases).

As has been demonstrated in the preceding chapters, Markov chains offer computationally effective ways of expressing the various frequencies and local dependencies as well as for scoring: stating the (Bayesian) probability of a sequence given a model. The information content of the observed statistics, as measured by the entropy rate of the chain or some upper bound for it, can also be computed to test the significance of the results. In addition, it will turn out that various probability distributions of occurrences of words are computable for a stationary first order Markov chain.

9.2 Markov Chains for DNA sequences

In view of the theory of probabilistic learning the most immediate operation to perform on a sequence is to determine the relative frequencies of its component elements. Initial statistical analyses of DNA sequences were concerned with frequencies, *'base composition'*, on the alphabet of bases

$$S = \{A, T, C, G\}$$

and it became clear that the four bases are not uniformly distributed in any sequence and that the compositions vary within and between sequences (or between fragments). It was also found that the coding sequences do not use synonymous codons with equal frequency. Also various k-tuples of bases are non-uniformly distributed. Using the frequencies of doublets, also called dinucleotide frequencies, Smith et al. (1983) observed that exons and introns can be separated on the basis of these statistics.

It was also established that distinct correlation patterns exist between adjacent bases and that these patterns are partly common for all DNA sequences and partly different for prokaryotes and eukaryotes. In eukaryotic DNA there is the presences of periodic repeats and base sequence repeats. Surveys of these and other empirical statistical findings are found, e.g., in (Avery and Henderson 1999, W.Li 1997, Stückle et al. 1990, Tavaré and Giddings 1989, Tavaré and Song 1989, Weir 1985).

The work in, e.g., (Almagor 1983, Blaisdell 1985, Garden 1980, Phillips et al. 1989) presents Markov chain models learned from respective sequences using maximum likelihood and some criterion for choice of the order of the Markov property, as defined in Definition 7.1.2 in Chapter 7. In these and other studies it was found that the probability of finding a particular base at one position can depend not only on the immediate adjacent bases but also on several more distant bases upstream or downstream and that this dependency can vary. In addition, the base composition may vary from one segment of the sequence to another, i.e., be *heterogeneous.* The segmentation techniques for decomposing DNA sequences into homogeneous regions include the hidden Markov models, to be presented in Chapter 11.

In Scherer et al. (1994) a seventh order Markov model was imposed to exploit the frequencies of octamers (words with eight letters) to find segments with distinctive patterns.

Several methods of gene finding use Markov models of coding and noncoding regions in order to classify sequence segments as either exons or introns. The finding in (Fickett and Tung 1992) was that methods based on reading frame specific hexamer composition should be best. An MC of order $k = 5$ reflects dependences in hexamers (words with six letters).

Example 9.2.1 *[Linkage Analysis]* A *marker* in genetics is, e.g., a segment of DNA the pattern of inheritance of which can be determined. We are given a map of M ordered marker loci, these are, e.g., DNA positions. The loci are designated by i. The *inheritance pattern* at each locus can be described by a vector $\mathbf{v}_i = (v_1(i), \ldots, v_l(i))$ of l binary digits. Here $v_r(i)$ is 0 if the paternally derived allele is transmitted and 1 if the maternally derived allele is transmitted. The inheritance vector $\mathbf{v}_{i+1}$ differs from $\mathbf{v}_i$ in each component with probability θ_i. The probability θ_i is known as the recombination factor. We model the sequence of inheritance vectors as a Markov chain $\{X_i\}_{i=1}^{M}$. The state space of the chain has clearly $J = 2^n$ different states. The transition matrix is given by

$$P_{i,i+1} = \left(p_{r|s}(i, i+1)\right)_{r=1,s=1}^{J,J}, \qquad i = 1, \ldots, M-1, \tag{2.1}$$

where

$$p_{r|s}(i, i+1) = \theta_i^k (1 - \theta_i)^{n-k},$$

and k is equal to the Hamming metric, see Example 5.3.2 in Chapter 5 between $\mathbf{v}_i$ and $\mathbf{v}_{i+1}$. This is the Markov model used in linkage analysis introduced by Lander and Green (1987). The model is a *non-homogeneous* MC, since the transition matrix depends on the locus i. In genetics the inheritance vector is, however, not directly observable. ∎

9.3 Frame Dependent Markov Chains

Borodovsky and his coworkers, see, e.g., (Borodovsky and McIninch 1993), introduced a general model of a *non-homogeneous (frame dependent) kth order 3 phase MC.* A coding region is read as successive non-overlapping codons, which are instances of 3-symbol *words.* Since several different codons can code for the amino acid, the symbols may have different importance depending on their position with respect to the codon partition. Therefore Borodovsky et al. consider a Markov chain of order k with three stationary transition probability matrices, P_1, P_2, P_3. Here P_m has for $m = 1, 2, 3$ the entries

$$P\left(X_{3t+m} = j_{3t} | X_{3t-k+m} = j_{3t-k+m}, \ldots, X_{3t-1+m} = j_{3t-1+m}\right). \tag{3.2}$$

The index m represents the position of the symbol inside the codon. This is illustrated for $k = 5$ in Figure 9.1. The circles represent consecutive DNA bases, the numbers indicate the codon position. The three arrows carry with them the different transition probabilities in (3.2) for each m. For each m the next base is generated by a distribution conditionally on the five previous bases and on the codon position m. The kth order 3-phase MC

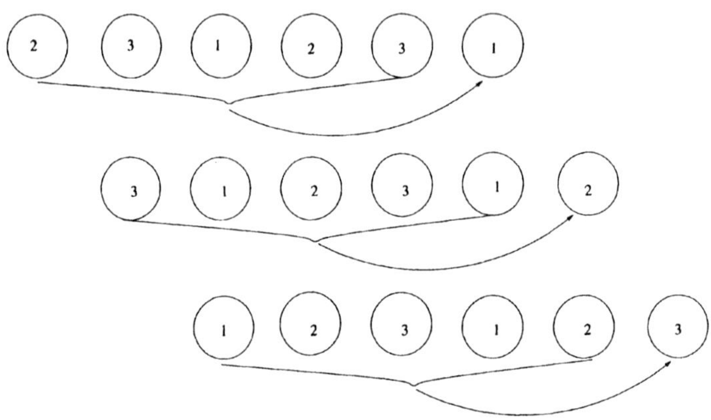

Figure 9.1: Frame dependent Markov chains, $m = 1, 2, 3, \qquad k = 5$

model is augmented by a homogeneous MC for non-coding regions and is also estimated together with a kth order 3-phase MC model for the other strand of the DNA sequence. Hence the task of learning this type of model from data requires the estimation of seven transition matrices with orders that are equal to $k = 5$ for coding regions of prokaryotic DNA. The probability of an observed fragment is calculated and the posterior probability of each hypothesis about the coding region is computed. The relevant algorithms and their implementation in software are known as GeneMark[1]. The frame dependent Markov Chain incorporated in GeneMark have been also invoked by this team of researchers in a gene finder using hidden Markov models discussed later in Chapter 11.

[1]`http://genemark.biology.gatech.edu/GeneMark/gm_info.html`

9.4 MTD and Interpolated Markov Models

If Markov chains of high order are applied straightforwardly the number of parameters in the transition matrix becomes discouraging. For example, a 5th order Markov chain from DNA sequence data involves a transition matrix with $3 \cdot 4^5 = 3072$ parameters. Sequence data is sparse for learning of this model family, since there must be many occurrences of all possible 6-tuples and some of these words may be rare, even, e.g., for a bacterial genome of length 1.8 million bases.

MTD(l) models, cf. Exercise 13 of Chapter 7, have $l-1$ parameters in the mixture distribution and $J \cdot (J-1)$ parameters in the matrix Q. Then if $J = 4$ and $l = 5$ there are 16 parameters to be estimated. In (Raftery and Tavaré 1994) an MTD(2) model provides a good description of the intron sequences in two special parts of Mouse genome when they are coded in the $\{A/G, C, T\}$ alphabet. In this case MTD(2) is, in fact, used as an approximation of a function of a first order Markov chain. Some properties of functions or *lumpings* of Markov chains are presented in Section 13.4 below.

There are different ideas about defining parsimoniously parametrized Markov chains with a 'local Markov property' or with a memory of a size that lies in some sense 'in-between' the integer sizes of the memory. One interesting approach is the *variable length* MC studied in (Bühlmann and Wyner 1999, Ron et al. 1996). The details of variable length MC are excluded here. An application of VLMC to splice site detection is given in (Cawley 2000).

Salzberg et al. (1998,1999) have introduced another interesting (and practically successful, it seems) computational probabilistic model based on an *interpolation of Markov models* (IMM) of orders $k = 0, 1, \ldots, 8$. The corresponding software system is called **GLIMMER** and is designed to identify coding regions for bacteria. The rationale for interpolation of Markov models is according to (Salzberg et al. 1998, Delcher et al. 1999) that there are some words that occur frequently enough so that they can be useful in assigning probabilities to sequences.

An IMM uses thus a combination of all the k-order conditional probabilities

$$P\left(X_t = j_t | X_{t-k} = j_{t-k}, \ldots, X_{t-1} = j_{t-1}\right), \tag{4.3}$$

where $j_{t-l} \in S = \{A, T, C, G\}$ for $l = 0, 1, \ldots, 8$, estimated from data. The idea is that for oligomers[2] that occur frequently, the IMM can use an 8th

[2] *An oligomer is a molecule consisting of just a few monomers*, monomers are single

order model, while it can use a 5th order or even lower order model for rare oligomers. The probabilities computed are smoothed by the lower order models, for which much more data is available.

Formally the ideas are represented by the following expressions. The probability of a sequence $\mathbf{x}= j_0 j_1 \ldots j_n$ with regard to the IMM is computed as

$$P(\mathbf{x}|\mathrm{IMM}) = \sum_{l=0}^{n} IMM_8(\mathbf{w}_l). \tag{4.4}$$

Here $\mathbf{w}_l$ is the oligomer ending at position l of $\mathbf{x}$ and $IMM_8(\mathbf{w}_l)$ is the interpolated Markov model score, computed recursively as

$$IMM_k(\mathbf{w}_l) = \lambda_k(\mathbf{w}_{l-1}) \cdot P_k(\mathbf{w}_l) + (1 - \lambda_k(\mathbf{w}_{l-1})) \cdot P_{k-1}(\mathbf{w}_l). \tag{4.5}$$

In the formula (4.5) the symbol $\lambda_k(\mathbf{w}_{l-1})$ is a numeric weight (measure of plausibility) associated with the $k$mer ending at position $l-1$ in the sequence $\mathbf{x}$ and the $P_k(\mathbf{w}_l)$ is the estimate of the probability of j_l in the kth order model. The 0th order model is nothing else but the prior probability distribution on S. In this manner the IMM exploits flexibly the statistics of words of lower order, not giving weight to the rare words.

One question is, of course, how the probabilities in the IMM can be learned (from microbial genomes). The up to date details (involving the construction of a 'context tree' and the pruning of it by an estimate of a certain mutual informations) are somewhat beyond the scope of these notes and the interested reader is referred to the original contributions quoted above. Additionally, the GLIMMER[3] software uses the idea of frame-dependent Markov chains of the eighth order. A set of known microbial coding sequences must be available in the training set.

One possible procedure for setting the weights $\lambda_k(\mathbf{w}_{l-1})$ in (4.5) that is tangential to the ideas implemented by Salzberg and his coworkers is Turing's estimate of word frequencies, discussed in the Exercise 2 at the end of this chapter.

9.5 Probability Computations for Words

Some part of the information in a DNA sequence can be sought in patterns or *overlapping words* (k-tuples of bases, e.g., ATAT). In particular, one might

molecules forming a polymer.

[3] `http://www.cs.jhu.edu/labs/compbio/glimmer.html`

wish to find words that are rare or frequent. This gives also information on sequence heterogeneity. Words with non-random internal statistics and an unexpectedly high frequency of occurrence are candidates for *motifs* in a DNA sequence. Hence a check on a probabilistic model is that it can correctly reflect the observed statistics of words in databases of long strings.

The extensive literature on patterns in random sequences and on the applications in biological sequence analysis has been recently surveyed in (Reinert et.al 2000). A concise introduction is found in (Blom et al. 1993, Chapter 14). Two of the results in this area are recapitulated below, one on the probability of the first occurrence and the other on the expected frequency of a word, when modelling a sequence by a stationary first order MC. These results are exact and explicit and require only the elementary Markov chain ideas discussed in the preceding chapters.

9.5.1 First Occurrence of a Word

Probability distributions for, e.g., distances between occurrences of words in a random sequence can be useful in analysis of DNA sequences. A recursive formula, owed to Robin and Daudin (1999), for computing the probability of the first occurrence of a word under a first order Markov chain model is proven next. The result and its proof in (Robin and Daudin 1999) generalize (Blom and Thorburn 1982). Similar results for semi-Markov chains are given in (Rudander 1996).

A key tool in this and other analyses of the statistics of words (or patterns in random sequences) is the *autocorrelation of a word*, as introduced by Guibas and Odlyzko (1981), see also (Blom and Thorburn 1982). In the literature on computational biology the autocorrelation is treated, e.g., in (Gentleman and Mullin 1989, Pevzner 1992, Kleffe and Borodovsky 1992). Actually Kleffe and Borodovsky loc. cit. propose an alternative to autocorrelation.

The autocorrelation is defined as follows. Let $\mathbf{w} = w_1 \dots w_h$ designate a word of length h with symbols in S, e.g., $\{A, T, C, G\}$. We say that there is an *overlap of length* r if the first r symbols in $\mathbf{w}$ equal the r last symbols, in the same order, or $w_j = w_{h-r+j}$, $j = 1, \dots, r$. The autocorrelation of $\mathbf{w}$ is a binary vector $\epsilon_1\epsilon_2\ldots\epsilon_h$ defined by

$$\epsilon_r = \begin{cases} 1 & \text{if there is an overlap of length } r, \\ 0 & \text{otherwise.} \end{cases}$$

Clearly $\epsilon_h = 1$ for every word. There are words that cannot have any other overlaps. The word ATAT has the autocorrelation 0101.

We consider a stationary Markov chain $(X_t)_{t=0}^{\infty} \in \text{Markov}(P, \pi)$ with the state space S and let

$$\tau(u,v) = \prod_{i=u}^{v} p_{w_{i-1}|w_i} \tag{5.6}$$

designate the probability that the chain should emit $w_u \ldots w_v$ after w_{u-1}. We denote by $P_\pi(\mathbf{w})$ is the stationary probability of observing $\mathbf{w}$ so that

$$P_\pi(\mathbf{w}) = \pi(w_1) \cdot \prod_{i=2}^{h} p_{w_{i-1}|w_i}. \tag{5.7}$$

Next we state a recursion for the probability of the first occurrence of the word $\mathbf{w}$.

Proposition 9.5.1 *If $(X_t)_{t=0}^{\infty} \in \text{Markov}(P, \pi)$, then $p(l)$, the probability of the first occurrence of the word $\mathbf{w} = w_1 \ldots w_h$ at time $t = l$ in a sample path of $(X_t)_{t=0}^{\infty}$, is for $l > h$*

$$p(l) = P_\pi(\mathbf{w}) - L_2 - L_3, \tag{5.8}$$

where

$$L_2 = \sum_{z=h}^{l-h} p(z) \cdot p_{w_k|w_1}(l - z - h + 1) \cdot \tau(2, h),$$

where $p_{w_k|w_1}(n)$ is the probability of transition from w_k to w_1 in n steps, and

$$L_3 = \sum_{z=l-h+1}^{l-1} p(z) \cdot \epsilon_{h-l+z} \cdot \tau(h - l + z + 1, h).$$

Furthermore, $p(l) = 0$ for $l < h$ and $p(h) = P_\pi(\mathbf{w})$.

Proof: It is clear by definition that $p(l) = 0$ for $l < h$. Also, $p(h) = P_\pi(\mathbf{w})$ given in (5.7), since the MC is stationary. We consider next for $l > h$ the event that $\mathbf{w}$ occurs at l but not necessarily for the first time. This event has, by stationarity, the probability $P_\pi(\mathbf{w})$ and can be decomposed into three mutually disjoint cases: **(1)** $\mathbf{w}$ occurs for the first time at l; **(2)** $\mathbf{w}$ occurs for the first time at z, with probability $p(z)$, and then at l, the distance between these two being at least h, $h \leq l - z$; and **(3)** $\mathbf{w}$ occurs for the first time at z and also at l, the distance between the two being less than h (which can happen due to overlaps).

Then the rules of probability calculus in (2.6) and (2.7) in Chapter 1 entail that the sought probability, which is the probability for **(1)**, is computed by subtracting from $P_\pi(\mathbf{w})$ the probabilities for **(2)** and **(3)**, which is the statement in (5.8). We find next the detailed expressions for the two probabilities in the right hand side of (5.8) using the Markov property.

In the case **(2)** $\mathbf{w}$ occurs for the first time at z, which has the probability $p(z)$, and then w_h is found at $z + h - 1$. As $\mathbf{w}$ occurs then at l, the Markov chain will have to move from w_h to w_1 in exactly $l - z - h + 1$ steps, which has the probability $p_{w_k|w_1}(l - z - h + 1)$, where by the Markov property we need only to condition on $X_{z+h-1} = w_h$. Thereafter Markov chain has to emit successively $w_2 \ldots w_h$, for which the probability is written as $\tau(2, k)$ by definition in (5.6). Multiplying the conditional probabilities and summing over all z with the required properties gives the asserted formula for L_2 as the probability of the events **(2)**.

In **(3)** we compute the sum of probabilities of first occurrence of $\mathbf{w}$ at time z and then at l for $l - h < z \leq l$. This can happen only because of an overlap. Assume thus that $\epsilon_{h-l+z} = 1$. We can write $z = l - h + j$ for some integer j. Then $\epsilon_{h-l+z} = \epsilon_j = 1$ and therefore the last j symbols in $\mathbf{w}$ are equal to the first j, in the same order. This means by definition of overlap that the first $h - l + z = j$ symbols of $\mathbf{w}$ are found starting at time l, so that the chain has to generate the sequence $w_{h-l+z+1}w_{h-l+z+2}\ldots w_h$ after these and the probability of this is by (5.6) equal to $\tau(h - l + z + 1, h)$. The Markov property and factorization of probabilities by conditioning yield L_3, as claimed. ∎

The paper by Robin and Daudin (1999) contains also several additional explicit results about the exact distributions of word occurrences. The algorithmic implementations of (5.8) are also of interest.

9.5.2 Conditional Expected Frequency of a Word

Expectation of the frequency of a word does not involve the autocorrelation, whereas already for the variance of a word one needs to compute the autocorrelation. We consider again a stationary Markov chain $(X_t)_{t=0}^{\infty}$ $\in$ Markov (P, π) and recapitulate a result due to Cowan (1991) as proved in (Prum et al. 1995). Whittle's multinomial coefficient appears in the result and hence we recall the setting for it.

For any given $\mathbf{x} = (j_0 j_1 \dots j_n)$ with $j_l \in S = \{A, T, C, G\}$ (represented for convenience of writing indices in summations as $S = \{1, 2, 3, 4\}$) we have the frequency counts

$$n_{i|j} = \text{ the number of } l \text{ such that } 1 \leq l \leq n, \qquad j_{l-1} = i, \qquad j_l = j.$$

These two-word frequency counts are the entries in

$$F = \left(n_{i|j}\right)_{i=1,j=1}^{4,4} \tag{5.9}$$

so that knowledge of F and of $j_0 = u$ determine the final symbol j_n uniquely. Therefore we can introduce the set

$$\mathbf{S} = \text{ all } n\text{-sequences with the first symbol } u \text{ and the same } F\ .$$

Then the number of elements in $\mathbf{S}$ is given by Whittle's multinomial coefficient, $N_{u,v}^{(n)}(F)$ in formula (3.5) in Proposition 8.3.1 in Chapter 8.

Consider again a word of length h, $\mathbf{w} = w_1 \dots w_h$, where $w_l \in \{A, T, C, G\}$. The random number of occurrences of $\mathbf{w}$ in $(X_t)_{1 \leq t \leq n}$ is given by

$$N_{\mathbf{w}} = \sum_{t=1}^{n-h+1} I_{\{X_t = w_1, \dots, X_{t+h-1} = w_h\}}, \tag{5.10}$$

where $I_{\{X_t = w_1, \dots, X_{t+h-1} = w_h\}}$ is the indicator function of the event that the word starts at position t. Our aim is now to compute the expectation $E[N_{\mathbf{w}} | \mathbf{S}]$ of $N_{\mathbf{w}}$ conditioned on $\mathbf{S}$ assuming that $\mathbf{x}$ is sampled from $\mathbf{S}$ with uniform distribution, which can be done without restriction, see Exercise 3 below.

For this let us next y and z designate the first and last symbols in the word, i.e., $\mathbf{w} = y w_2 \dots w_{h-1} z$. Then $n_{ij}(\mathbf{w})$ are the numbers of two-words in $\mathbf{w}$ and constitute the entries in the matrix $F(\mathbf{w})$.

We consider an arbitrary sequence $\mathbf{s}$ in $\mathbf{S}$, with the word $\mathbf{w}$ appearing at position i

$$\mathbf{s} = (j_0 j_1 \dots j_{i-1} y w_2 \dots w_{h-1} z j_{i+h} \dots j_n)\ .$$

The sequence $\mathbf{s} \setminus \mathbf{w}$ with the length $n-h-2$ is obtained by removing $\mathbf{w}$ from $\mathbf{s}$ except for y and z

$$\mathbf{s} \setminus \mathbf{w} = (j_0 j_1 \ldots j_{i-1} y z j_{i+h} \ldots j_n) .$$

Then the corresponding matrix of frequency counts $F(\mathbf{s} \setminus \mathbf{w})$ has the entries

$$n_{ij}(\mathbf{s} \setminus \mathbf{w}) = n_{ij} - w_{ij}$$

where, using a self-explanatory notation,

$$w_{ij} = n_{ij}(\mathbf{w}) \text{ if } (i,j) \neq (y,z)$$

and otherwise

$$w_{yz} = n_{yz}(\mathbf{w}) - 1.$$

We denote by $\mathbf{S} \setminus \mathbf{w}$ the set of all sequences of length $n-h+1$ and such that the two-word beginning at i is yz, thus with the frequency matrix $F(\mathbf{s} \setminus \mathbf{w})$. The construction establishes a one-to-one correspondence between $\mathbf{S} \setminus \mathbf{w}$ and the subset of sequences in $\mathbf{S}$ such that the h-word beginning at i is $\mathbf{w}$. The numbers of elements $\beta_i(\mathbf{w}, \mathbf{S})$ and $\beta_i(yz, \mathbf{S} \setminus \mathbf{w})$ in these sets, respectively, are thus equal

$$\beta_i(\mathbf{w}, \mathbf{S}) = \beta_i(yz, \mathbf{S} \setminus \mathbf{w}) . \tag{5.11}$$

If $\mathbf{x}$ is sampled at random from $\mathbf{S}$ we have

$$E[N_{\mathbf{w}} | \mathbf{S}] = \frac{1}{N_{u,v}^{(n)}(F)} \sum_{i=1}^{n-h+1} \beta_i(\mathbf{w}, \mathbf{S}) .$$

The same rule can be used for N_{yz}, the number of occurrences of the two-word yz in $\mathbf{S} \setminus \mathbf{w}$, so that:

$$E[N_{yz} | \mathbf{S} \setminus \mathbf{w}] = \frac{1}{N_{u,v}^{(n-h-2)}(F(\mathbf{s} \setminus \mathbf{w}))} \sum_{i=1}^{n-h+1} \beta_i(yz, \mathbf{S} \setminus \mathbf{w}) .$$

But conditioned on $\mathbf{S} \setminus \mathbf{w}$ the quantity N_{yz} is equal to $n_{yz}(\mathbf{s} \setminus \mathbf{w})$. Therefore we have in view of (5.11) proved the following result.

Proposition 9.5.2 *Let* $(X_t)_{t=0}^{\infty} \in \in \text{Markov}(P, \pi)$ *with a state space of four symbols. Then the conditional expectation of the number occurrences of* $\mathbf{w}$ *in* $(X_t)_{1 \le t \le n}$ *given the observed frequencies of all two-words is given by*

$$E[N_{\mathbf{w}} | \mathbf{S}] = n_{yz}(\mathbf{s} \setminus \mathbf{w}) \frac{N_{u,v}^{(n)}(F(\mathbf{s} \setminus \mathbf{w}))}{N_{u,v}^{(n-h-2)}(F)}, \tag{5.12}$$

where $N_{u,v}^{(n)}$ *and* $N_{u,v}^{(n-h-2)}$ *are Whittle's multinomial coefficients.* ∎

The result is owed to (Cowan 1991) and the proof is from (Prum et al. 1995).

9.6 Exercises

1. **Frame dependent 1st order 3 phase MC**
   Show by artificially renaming the alphabet $S = \{A, T, C, G\}$ in the second coding position as well as in the third coding position that the frame-dependent 1st order, 3 phase MC in the sense of Borodovsky et al. has a transition probability matrix given by the 12×12 matrix

$$\begin{pmatrix} 0 & P_1 & 0 \\ 0 & 0 & P_2 \\ P_3 & 0 & 0 \end{pmatrix}, \tag{6.1}$$

   where P_m has for $m = 1, 2, 3$ the entries depicted in (3.2) and O:s are matrices with conforming dimensions. Compute thereafter (a generic formula for) the invariant distribution of the frame dependent chain. (Kleffe and Borodovsky 1993).

2. **Size distribution and learning probabilities from scarce data**
   Let us consider a sequence of $\mathbf{x} = x_{l_1} x_{l_2} \ldots x_{l_n}$ from an alphabet $\mathcal{X}$ with L symbols. We wish to estimate the probabilities

$$\theta_j, j = 1, \ldots, L$$

   in a situation, where many symbols have been observed only a few times. Actually the original setting of this exercise deals with non-overlapping words drawn from a finite alphabet, a vocabulary. The statistics to be used in the estimates are the components of the size distribution to be defined below.

   The indicator function of any $x \in \mathcal{X}$ is $I_{\{x_{l_i}=x\}} = 1$, if $x_{l_i} = x$ and 0 otherwise. Then

$$n_j = \sum_{i=1}^{n} I_{\{x_{l_i}=x_j\}}$$

   is the number of times the symbol $x_j \in \mathcal{X}$ appears in $\mathbf{x}$. Next let

$$N_r = \sum_{j=1}^{L} I_{\{n_j=r\}}, \qquad r = 1, \ldots, L$$

be the number of symbols that occurred exactly r times in $\mathbf{x}$. The vector with the elements N_r is known as the *size distribution.* Note that

$$n = \sum_{r=1}^{\infty} r N_r.$$

Then *Turing's estimate* of θ_j is

$$\widehat{\theta}_j = (r+1)\frac{N_{r+1}}{N_r}, \tag{6.2}$$

if the symbol x_j appeared r times in $\mathbf{x}$. The point of this exercise is to investigate the rationale for Turing's estimate established by I.J. Good, as expounded in (Nádas 1991).

(a) Argue that

$$E[N_r] = \binom{n}{r} \sum_{j=1}^{L} \theta_j^r (1-\theta_j)^{n-r}.$$

We designate $E[N_r]$ also as $E_n[N_r]$.

(b) Suppose that $T \in \mathrm{U}(1,\ldots,n)$ independently of all other random variables needed. Then consider the chance θ_{l_T} of the symbol found in the T:th position in $\mathbf{x}$ and obtain

$$P(n_{l_T} = r | \theta_{l_T}) = \binom{n}{r} \theta_{l_T}^r (1-\theta_{l_T})^{n-r}.$$

Compute next

$$E[\theta_{l_T} | n_{l_T} = r] = \sum_{i=1}^{n} \theta_i P(\theta_i | n_i = r)$$

using Bayes formula for $P(\theta_i | n_i = r)$ and show with the notations from (a) that

$$E[\theta_{l_T} | n_{l_T} = r] = \frac{r+1}{n+1} \frac{E_{n+1}[N_{r+1}]}{E_n[N_r]}. \tag{6.3}$$

Next $E[\theta_{l_T} | n_{l_T} = r]$ is regarded as the true *chance* of an arbitrary symbol which occurred in the sample exactly r times. Replacing in the right hand side (6.3) the expectations in by the respective observed values we obtain (up to a factor) Turing's estimate $\widehat{\theta}_j$ in (6.2).

In Turing's estimate the size distribution appears as the proper statistic for estimating the probabilities of words. A biological application is the estimation of the probabilities of species.

The size distribution $(N_r)_{r \geq 1}$ is in mathematical population genetics also known as the *allelic partition* corresponding to $(n_1, \dots, n_L)$.

3. $(X_t)_{t=0}^{\infty} \in \text{Markov}(P, \pi)$. Verify that all the sequences in the set **S** defined in Section 9.5.2 above have the same probability.

4. Consider a sequence with the symbols chosen uniformly and independently of each other from a finite alphabet with L elements. Show that the occurrence of words $\mathbf{w} = w_1 \dots w_h$ of length h in this random sequence can be regarded as a Markov chain of order $h - 1$ and with L^h states and find its transition probability matrix using the autocorrelations of the words. This is from (Gentleman and Mullin 1989) and extends an earlier observation by William Feller.

5. Let $(X_n)_{n=0}^{\infty} \in \text{Markov}(P, \pi)$ on the state space $S = \{0, 1\}$ with the transition matrix P defined by

$$P := \begin{pmatrix} 1 - \beta & \beta \\ 1 - \alpha & \alpha \end{pmatrix}, \tag{6.4}$$

where $0 < \alpha < 1$ and $0 < \beta < 1$. We define

$$M(n, r) := \sum_{i=r}^{n} I\{X_{i-r+1} = X_{i-r+2} = \dots = X_i = 1\},$$

i.e., the number of overlapping runs of 1s of length r occurring up to time n.

Define an embedded Markov chain $\{Y_n; n = 0, 1, 2, \dots,\}$ with state space $S = \{0, \dots, r\}$, in the following way:

$$m = \min\{i \in Z_+; X_{n-i} = 0\},$$

where Z_+ is the set of non-negative integers, so that

$$Y_n := \min[m, r].$$

(a) Show that the Markov chain $(Y_n)_{n=0}^{\infty}$ is stationary and has the transition matrix P^Y, defined by

$$P^Y := \begin{pmatrix} 1-\beta & \beta & 0 & \dots & 0 & 0 & 0 \\ 1-\alpha & 0 & \alpha & \dots & 0 & 0 & 0 \\ 1-\alpha & 0 & 0 & \dots & 0 & 0 & 0 \\ \vdots & \vdots & \vdots & \ddots & \vdots & \vdots & \vdots \\ 1-\alpha & 0 & 0 & \dots & 0 & \alpha & 0 \\ 1-\alpha & 0 & 0 & \dots & 0 & 0 & \alpha \\ 1-\alpha & 0 & 0 & \dots & 0 & 0 & \alpha \end{pmatrix}.$$

(b) Show that $M(n,r) = \sum_{i=r}^{n} I\{Y_i \in S_1\}$, where $S_1 = \{r\}$ and $I\{Y_i \in S_1\}$ is the indicator function. (Erhardsson 1999).

6. Show that the matrix $P_{i,i+1}$ in (2.1) of Example 9.2.1 is, in fact, a transition matrix.

9.7 References and Further Reading:

- Technologies for DNA sequence analysis

L. Alphey (1997): *DNA Sequencing. From experimental methods to bioinformatics.* Springer Verlag, New York.

T. Hunkapiller, R.J. Kaiser, B.F. Koop and L. Hood (1991): Large-Scale and Automated DNA Sequence Determination. *Science*, 254, pp. 59–67.

International Human Genome Sequencing Consortium (2001): Initial sequencing and analysis of the human genome. *Nature*, 409, pp. 860–921.

E.W. Myers et al. (2000): A Whole-Genome Assembly of *Drosophilia. Science*, Vol. 287, 24, March, pp. 2196–2204.

J.G. Venter, M.D. Adams, E.W. Myers et al. (2001): The Sequençe of the Human Genome. *Science*, vol. 291, 16 February, pp. 1304–1351.

- Molecular biology, genetics and evolution

B. Alberts, D. Bray, J. Lewis, M. Raff, K. Roberts and J.D. Watson (1994): *Molecular Bioloy of* **The Cell**, *Third Edition*, Garland Publishing, New York.

D.J. Balding, M. Bishop and C.Cannings (2001): *Handbook of Statistical Genetics*, John Wiley and Sons, Chichester, New York.

W.M. Fitch (1999): An Introduction to Molecular Biology for Mathematicians and Computer Programmers. *Mathematical Support for Molecular Biology*, DIMACS Series in Discrete Mathematics and Theoretical Computer Science, Vol. 47, pp. 1–31.

W-H. Li (1997): *Molecular Evolution*. Sinauer Associates, Inc., Publishers, Sunderland Massachusetts.

- Statistics of DNA sequences :

R.N. Curnow, T.B.L. Kirkwood (1989): Statistical Analysis of Deoxyribonucleic Sequence Data–a Review. *Journal of the Royal Statistical Society*, **A**, 152, Part 2, pp. 199–220.

J.W. Fickett and C-S. Tung (1992): Assessment of protein coding measures. *Nucleid Acids Research*, 20, pp. 6441–6450.

W. Li (1997): The study of correlation structures of DNA sequences: critical review. *Computers and Chemistry*, Vol. 21, pp. 257–271.

T.F. Smith, M. Waterman and R. Sadler (1983): Statistical characterization of nucleic acid sequence functional domains. *Nucleic Acids Research*, 11, pp. 2205–2220.

E.E Stückle, C. Emmrich, U. Grob U, and P.J. Nielsen (1990): Statistical analysis of nucleotide sequences. *Nucleic Acids Research*, 18, pp. 6641–6647.

B.S. Weir (1985): Statistical analysis of molecular genetic data. *IMA Journal of Mathematics Applied in Medicine and Biology*, 2, pp. 1–39.

- Markov chains and linkage analysis

E.S. Lander and P. Green (1987): Construction of multilocus genetic linkage maps in humans. *Proceedings of the National Academy of Sciences USA*, 84, pp. 2363–2367.

- Markov modelling of DNA sequences for various organisms:

H. Almagor (1983): A Markov Analysis of DNA Sequences. *Journal of Theoretical Biology*, 104, pp. 633–645.

P.J. Avery and D.A. Henderson (1999): Fitting Markov chain models to discrete state series such as DNA sequences. *Applied Statistics*, 48, part 1, pp. 53–61.

B.E. Blaisdell (1985): Markov Chain Analysis Finds a Significant Influence of Neighboring Bases on the Occurrence of of a Base in Eukaryotic Nuclear DNA Sequenes both Protein-Coding and Non-Coding. *Journal of Molecular Evolution*, 21, pp. 278–288.

M. Borodovsky and J.D. McIninch (1993): GENMARK: Parallell gene recognition for both DNA strands. *Computers and Chemistry*, 17, pp 123–133.

P.W. Garden (1980): Markovian Analysis of Viral DNA/RNA Sequences. *Journal of Theoretical Biology*, 82, pp. 674–684.

G. J. Phillips, J. Arnold and R. Ivarie (1987): Mono- through hexanucleotide composition of the *Escherichia coli* genome: a Markov chain analysis. *Nuclear Acid Research*, 15. pp. 2611–2626.

S. Scherer, M.S. McPeek and T.P. Speed (1994): Atypical regions in large genomic DNA sequences. *Proceedings of the National Academy of Sciences USA*, 91, pp. 7134–7138.

S. Tavaré and B. Giddings (1989): Some statistical aspects of the primary structure of nucleotide sequences. M.S. Waterman (ed.): *Mathematical methods for DNA sequences.* CRC Press, Boca Raton, Florida.

S. Tavaré and B. Song (1989): Codon preferences and primary sequence structure in protein-coding region. *Bulletin of Mathematical Biology*, 51, pp. 95–115.

- MTD, Variable Length, and Interpolated Markov models:

P. Bühlmann and A.J. Wyner (1998): Variable length Markov chains. *Annals of Statistics*, 27, pp. 480–513.

S.E. Cawley (2000): *Statistical models for DNA sequencing and analysis.* Ph.D. Thesis, Department of Statistics, University of California, Berkeley. `http://stat-www.berkeley.edu/users/scawley/Thesis/index.html`

A.L. Delcher, D. Harmon, S. Kasif, O. White and S.L. Salzberg (1999): Improved microbial gene identification with **GLIMMER**. *Nucleic Acid Research*, 27, pp. 4636–4641.

A.E. Raftery and S. Tavaré (1994): Estimation and Modelling Repeated Patterns in High Order Markov Chains with the Mixture Distribution Model. *Applied Statistics*, 43, pp. 179–199.

D. Ron, Y. Singer and N. Tishby (1996): The power of amnesia: Learning probabilistic automata with variable memory length. *Machine Learning*, 25, 117–149.

S.L. Salzberg, A.L. Delcher, S. Kasif and O. White (1998): Microbial gene identification using interpolated Markov models. *Nucleic Acid Research*, 26, pp. 544–548.

- Words in DNA Sequences

G. Blom and D. Thorburn (1982): How many random digits are required until a given sequence is obtained ? *Journal of Applied Probability*, 19, pp. 518–531.

G. Blom, L. Holst and D. Sandell (1994): *Problems and Snapshots from the World of Probability.* Springer Verlag, New York.

R. Cowan (1991): Expected frequency of DNA patterns using Whittle's formula. *Journal of Applied Probability*, 28, pp. 886–892.

T. Erhardsson (1999): Compound Poisson Approximation for Markov Chains using Stein's method. *Annals of Probability*, 27, pp. 565–596.

J.F. Gentleman and R.C. Mullin (1989): The Distribution of Occurrence of Nucleotide Subsequences, Based on Their Overlap Capability. *Biometrics*, 45, pp. 35–52.

L.J. Guibas and O.M. Odlyzko (1981): String Overlaps, Pattern Matching and Nontransitive Games. *Journal of Combinatorial Theory, Ser. A*, pp. 183–208.

J. Kleffe and M. Borodovsky (1992): First and second moment counts of words in random texts generated by Markov chains. *Computer Applications in Biological Sciences* (CABIOS), 8, pp. 433–441.

P.A. Pevzner (1992): Nucleotide sequences versus Markov models. *Computers and Chemistry*, 16, pp. 103–106.

B. Prum, F. Rodolphe and E. de Turckheim (1995): Finding Words with Unexpected Frequencies in Deoxyribonucleic Acid Sequences. *Journal of the Royal Statistical Society*, **B**, 57, pp. 205–220.

G. Reinert, S. Schbath and M. Waterman (2000): Probabilistic and Statistical Properties of Words: An Overview. *Journal of Computational Biology*, vol. 7, pp. 1–46.

S. Robin and J.J. Daudin (1999): Exact Distribution of Word Occurrences in a Random Sequence of Letters. *Journal of Applied Probability*, 36, pp. 179–193.

J. Rudander (1996): *On the first occurrence of a given pattern in a semi-Markov process*. Uppsala Dissertations in Mathematics 2. Department of mathematics, Uppsala university.

• Estimating probabilities from scarce data

A. Nádas (1991): Good, Jelinek, Mercer, and Robbins on Turing's Estimate of Probabilities. *American Journal of Mathematical and Management Sciences*, 11, pp. 299–308.

Chapter 10

Hidden Markov Models: an Overview

10.1 Introduction

10.1.1 The HMM

A *hidden Markov model* (HMM) (earlier also known as a *probabilistic function of a Markov chain*, or as a *Markov source*, or as a *Markov regime model*) is a stochastic process generated by two interrelated probabilistic mechanisms. These are an underlying Markov chain with a finite number of states, and a set of random functions, each associated with its respective state. At discrete instants of time the process is assumed to be in some state and an observation is generated by the random function corresponding to the current state. The underlying Markov chain then changes its state according to its transition matrix.

The observer sees only the output of the random functions associated with each state and cannot directly observe the states of the underlying Markov chain. Hence the Markov chain is hidden but the name for the model family is *hidden Markov model.*

In principle, the outputs from the states of the hidden Markov chain may be multivariate random processes having some continuous joint probability density function. In this text we shall, however, limit the treatment to output processes with a discrete finite alphabet.

10.1.2 The Thumb Tack Again

Let us recall modelling of a sequence of flips of a thumb tack proposed and discussed by D. Heckerman (see Chapter 3). The model entertained there was a sequence of Bernoulli random variables that were conditionally independent given the chance of success. Next we make a variation of the theme as in (Rabiner 1989).

Let us now imagine a sequence of flips performed by a person who stands behind a barrier (like a curtain) through which we cannot see what is happening. The individual behind the curtain does not tell us what is exactly going on there but reports only the results of each flip of the thumb tack.

Thus a sequence of *hidden* flips of the thumb tack is being performed with the observation sequence consisting of a sequence of reported heads (H) and tails (T). Let us suppose that a series reported to us would be

$$HTHHHTTTTTTTTTTTHHHTTTTTTTTTTHH.$$

We might pay attention to the long uninterrupted runs of tails. That might be modeled by means of a scenario, where we think of two different techniques of tossing a thumb tack, the first suitably adjusted to produce mostly T and the other one more or less a random flip of thumb tack. The choice between the techniques might be done by some method of randomization before every flip.

10.2 Standard HMM

10.2.1 Definition and notations

An HMM can be seen as a model family for a sequence of symbols from an alphabet $\mathcal{O} = \{o_1, o_2, \ldots o_K\}$. The model is based on the idea of a hidden sequence of Markovian state transitions. More formally an HMM is characterized by (I)−(III) in the following:

(I) **Hidden Markov Chain** A Markov chain $\{X_n\}_{n=0}^{\infty}$ assuming values in a finite state space $S = \{1, 2, \ldots, J\}$ with J states. The conditional probabilities

$$a_{i|j} = P\left(X_n = j | X_{n-1} = i\right), \qquad n \geq 1, \qquad i, j \in S \qquad (2.1)$$

are assumed to be time-homogeneous. The transition matrix is designated by

$$A = \left(a_{i|j}\right)_{i=1,j=1}^{J,J} \tag{2.2}$$

with the familiar constraints

$$a_{i|j} \geq 0, \sum_{j=1}^{J} a_{i|j} = 1.$$

At time $n = 0$ the state X_0 is specified by the probability distribution $\pi_j(0) = P(X_0 = j)$ with

$$\pi(0) = (\pi_1(0), \ldots, \pi_J(0)).$$

(II) **Observable Random Process** A random process $\{Y_n\}_{n=0}^{\infty}$ with a finite state space $\mathcal{O} = \{o_1, o_2, \ldots o_K\}$, where K need not equal J. The processes $\{Y_n\}_{n=0}^{\infty}$ and $\{X_n\}_{n=0}^{\infty}$ are for any fixed n related by the conditional probability distributions

$$b_j(k) = P(Y_n = o_k | X_n = j).$$

We set

$$B = \{b_j(k)\}_{j=1,k=1}^{J,K}.$$

We shall call this the *emission probability matrix.* This is another stochastic matrix in the sense that

$$b_j(k) \geq 0, \qquad \sum_{k=1}^{K} b_j(k) = 1.$$

(III) **Conditional Independence**

Assumption 10.2.1 For any sequence of states $j_0 j_1 \ldots j_n$ the probability of the sequence $o_0 o_1 \ldots o_n$ is

$$P(Y_0 = o_0, \ldots, Y_n = o_n \mid X_0 = j_0, \ldots, X_n = j_n, B) = \prod_{l=0}^{n} b_{j_l}(l). \tag{2.3}$$

In other words
The emitted symbols are conditionally independent given the state sequence. ∎

The postulate (III) is crucial for all the mathematical developments that are to follow. The detailed consequences of combining (III) and the Markov property will be established in Chapter 13. The assumption (III) is sometimes made fairly tacitly in tutorials and applied literature on HMM. The process $\{Y_n\}_{n=0}^{\infty}$ can always be represented as function of a Markov chain (see exercise) and is in general not a Markov chain.

By virtue of these assumptions we may write the *joint probability* of $o_0 \ldots o_n$ and $j_0 \ldots j_n$ as

$$\begin{aligned} P(Y_0 &= o_0, \ldots, Y_n = o_n, X_0 = j_0, \ldots, X_n = j_n; A, B, \pi(0)) \\ &= P(Y_0, \ldots, Y_n \mid X_0, \ldots, X_n, B) \cdot P(X_0, \ldots, X_n, A, \pi(0)) \\ &= \pi_{j_0}(0) \cdot \prod_{l=0}^{n} b_{j_l}(l) \cdot \prod_{l=1}^{n} a_{j_{l-1}|j_l}, \end{aligned}$$

where the last equality follows by (2.3) above and by (1.9) from Chapter 7. Rearranging the last expression gives

$$\begin{aligned} P(Y_0 &= o_0, \ldots, Y_n = o_n, X_0 = j_0, \ldots, X_n = j_n; A, B, \pi(0)) \\ &= \pi_{j_0}(0)\, b_{j_0}(0) \cdot \prod_{l=1}^{n} a_{j_{l-1}|j_l} b_{j_l}(l). \end{aligned}$$

Then we obtain the joint probability of $o_0 \ldots o_n$ as a marginal distribution by summing over all possible paths of the state sequence. This is written as:

$$P(Y_0, \ldots, Y_n; A, B, \pi(0)) = \sum_{j_o=1}^{J} \cdots \sum_{j_n=1}^{J} \pi_{j_0}(0)\, b_{j_0}(0) \prod_{l=1}^{n} a_{j_{l-1}|j_l} b_{j_l}(l). \quad (2.4)$$

Hence the finite dimensional distributions of $\{Y_n\}_{n=1}^{\infty}$ are **fully specified** by our choice of the stochastic matrices A, B and of the initial distribution $\pi(0)$. Therefore we may use the compact notation for the model

$$\lambda = (A, B, \pi(0)).$$

Then we have

THE MODEL FAMILY:

CONDITIONED ON $\lambda = (A, B, \pi(0))$,

THE STRING $\mathbf{o} = o_0 \ldots o_n$ HAS THE PROBABILITY

$$\begin{aligned} P(\mathbf{o}) &= P(Y_0 = o_0, \ldots, Y_n = o_n; \lambda) \\ &= \sum_{j_o=1}^{J} \cdots \sum_{j_n=1}^{J} P(Y_0 = o_0 \ldots, Y_n = o_n, X_0 = j_0, \ldots, X_n = j_n; \lambda), \end{aligned}$$

where

$$\begin{aligned} P(Y_0 &= o_0, \ldots, Y_n = o_n, X_0 = j_0, \ldots, X_n = j_n; \lambda) \\ &= \pi_{j_0}(0) \cdot \prod_{l=0}^{n} b_{j_l}(l) \cdot \prod_{l=1}^{n} a_{j_{l-1}|j_l}. \end{aligned}$$

■

Remark 10.2.1 (Initial Emission) The various contributions and tutorials on HMM contain one minor discrepancy in detail with respect to the the formulation of $P(\mathbf{o})$ in (2.5). Levinson et al. (1983) and Baum's and Petrie's papers in the 1960's write

$$P(Y_1 \ldots, Y_n, X_0, \ldots, X_n; \lambda) = \pi_{j_0}(0) \cdot \prod_{l=1}^{n} b_{j_l}(l) \prod_{l=1}^{n} a_{j_{l-1}|j_l}, \tag{2.5}$$

which means that there is no emission in the initial state or that the initial state is a *silent state*. Another way to interpret this is that the initial symbol is always known to be the same. Our formula (2.5) follows the presentation in Rabiner's tutorial (1989). Thus care is required in reading and in detailed comparison of various formulas in the sequel and in the references. ■

Remark 10.2.2 Jelinek (1999) expresses the hidden Markov models in terms of an emission distribution that is a function of the state transition. Formally this means that $b_j(k) = P(Y_n = o_k | X_n = j)$ is replaced by

$$b_{i|j}(k) = P(Y_n = o_k | X_{n-1} = i, X_n = j).$$

This constitutes, however, a completely equivalent formulation to that used above, as is easily seen (Jelinek 1999, p. 36). ■

10.2.2 Generative Modelling of Sequences

Phrases like 'probabilistic mechanisms' and similar used above can be seen as evocative of the following generator of sequences in $\mathcal{O}^{N+1}$:

1) Choose an initial state $X_0 = j_0$ according to the distribution $\pi(0)$;

2) Set $n = 0$.

3) Choose $Y_n = o_k$ according to the symbol probability distribution in the state j_n which is $b_{j_n}(k)$;

4) Transit to a new state $X_{n+1} = j_{n+1}$ according to the state probability distribution $a_{j_n|j_{n+1}}$;

5) Set $n \to n+1$ and return to step 3) if $n \leq N$, otherwise terminate.

The above procedure can be used as both generator of observations and as a model for how a given sequence $\mathbf{o} = o_0 \ldots o_n \in \mathcal{O}^{N+1}$ was generated by an appropriate HMM.

10.3 Examples and Applications

Example 10.3.1 (Finite Mixtures) The marginal distribution for any Y_n is by the preceding definition of HMM

$$P(Y_n = o_k) = \sum_{j=1}^{J} P(Y_n = o_k | X_n = j) \cdot \pi_j(n).$$

Assume that all the rows of A are identical, that is

$$a_{i|j} = w_j \text{ for } j = 1, 2, \ldots, J, \text{ for all } i.$$

Then $\pi(n) = (w_1, w_2, \ldots, w_J)$ for all $n \geq 1$ and for any $\pi(0)$, as is immediately seen, recalling $\pi(n) = \pi(0) \cdot A^n$ from Chapter 7. In other words, for $n \geq 1$

$$P(Y_n = o_k) = \sum_{j=1}^{J} P(Y_n = o_k | X_n = j) \cdot w_j,$$

where $P(Y_n = o_k | X_n = j)$ does not depend on n. The assumption of identical rows of A means that X_n are independent random variables and the assumption of conditional independence in (2.3) is the assumption of pairwise independence for (X_n, Y_n). But this is nothing else than the family of finite mixture of distribution models presented in Chapter 4. ∎

Example 10.3.2 (Burst Noise) Consider a Markov chain $\{X_n\}_{n=0}^{\infty}$ with the binary state space $S = \{b, g\}$ and the transition matrix

$$A = \begin{pmatrix} 1-p & p \\ q & 1-q \end{pmatrix}.$$

This Markov chain is hidden in another process $\{Y_n\}_{n=0}^{\infty}$ with the state space $\mathcal{O} = \{0, 1\}$ and such that the assumption (2.3) is valid. The process $\{Y_n\}_{n=0}^{\infty}$ has the emission probability matrix

$$B = \begin{pmatrix} \rho & 1-\rho \\ 1 & 0 \end{pmatrix},$$

where $0 \leq \rho \leq 1$. This example is owed to Gilbert (1960). Gilbert considered this scheme for modelling burst noise in binary communications channels. In that context we think of Z_n as a received bit, S_n as a transmitted bit and Y_n as a noise bit with

$$Z_n = S_n +_2 Y_n,$$

where $+_2$ is addition modulo two, ($1 +_2 1 = 0 +_2 0 = 0$ and $1 +_2 0 = 0 +_2 1 = 1$). If the Markov chain assumes the state g (for 'good'), then the emission probabilities are such that only digital zeroes will be emitted and there will be no transmission errors. In the state b (bad) the errors will be interspersed with error-free transmissions. A graphical recapitulation of the model generating Y_n is Figure 10.1. ∎

Example 10.3.3 (Noisy Markov Chains) A Markov chain $\{X_n\}_{n=0}^{\infty}$, with the binary digits $S = \{0, 1\}$ as the state space, has the transition matrix

$$A = \begin{pmatrix} 1-p & p \\ p & 1-p \end{pmatrix}. \tag{3.1}$$

We observe another process $\{Y_n\}_{n=0}^{\infty}$ with the state space $\mathcal{O} = \{0, 1\}$ such that

$$\begin{aligned} P(Y_0 &= o_0, \ldots, Y_n = o_n | X_0 = j_0, \ldots, X_n = j_n; B) \\ &= \prod_{l=0}^{n} \epsilon^{|o_l - j_l|} \cdot (1-\epsilon)^{1-|o_l - j_l|}. \end{aligned}$$

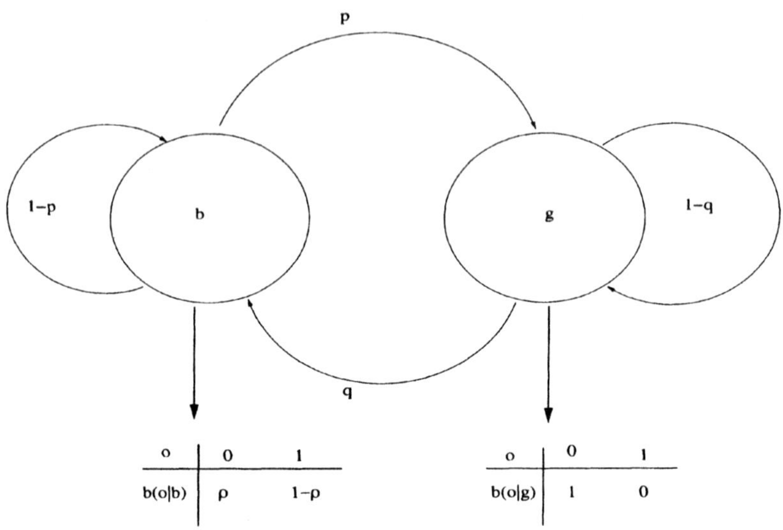

Figure 10.1: Burst noise

Clearly the assumption 10.2.1 is satisfied and when this setting is completed by some initial distribution $\pi(0)$, we have specified a hidden Markov model. The HMM has the emission probability matrix

$$B = \begin{pmatrix} 1-\epsilon & \epsilon \\ \epsilon & 1-\epsilon \end{pmatrix}.$$

This is a Markov chain observed in noise, where the probability of error in observation is ϵ. A system of random variables (X_n, Y_n) having these distributions is

$$X_{n+1} = X_n +_2 W_n$$
$$Y_n = X_n +_2 V_n,$$

where $+_2$ is addition modulo two and $\{W_n\}_{n=0}^{\infty}$ is a sequence of independent Bernoulli random variables with probability of success $= p$, and $\{V_n\}_{n=0}^{\infty}$ is a sequence of independent Bernoulli random variables with probability of success $= \epsilon$. We assume here that $\{W_n\}_{n=0}^{\infty}$ and $\{V_n\}_{n=0}^{\infty}$ are independent of each other. ■

Example 10.3.4 (A Null Model in Sequence Analysis) In biological sequence analysis the following model turns out to play a certain role to be clarified later. A Markov chain $\{X_l\}_{l=0}^{\infty}$ has the state space $\{G, F\}$ and the following transition matrix

$$A = \begin{pmatrix} 1-\alpha & \alpha \\ 0 & 1 \end{pmatrix}.$$

The chain starts always in state G and ends (is absorbed) in the state F. The state F is *a silent state*, which means that no symbol is emitted there. For all l with $X_l = G$, a symbol o_l from a finite alphabet $\mathcal{O}$ is emitted. ∎

10.4 Influence Diagrams, Nonstandard HMM

A standard hidden Markov model can be regarded as the graph in Figure 10.2. This kind of graph is known as an *influence diagram* or a *belief network* (Smith 1989). Each node in the graph represents a random variable describing the state X_n or the observation Y_n at some time n. This is quite different from the state diagrams exemplified by, e.g., Figure 10.1. In Figure 10.2 the edges (arrows) represent direct influences. The graph expresses the model assumptions. The Markov assumption states that if we know which state j is visited at time n, no other information from the past is relevant about the future. In the Figure the variable X_n separates the previous variables from the future: by removing X_n from the graph, the variables X_m $(m > n)$ become disconnected from X_m $(m < n)$.

In terms of influence diagrams one can easily grasp and/or develop other than standard hidden Markov models. Examples of non-standard models are given in Figure 10.3, which is an autoregressive HMM, Figure 10.4, a coupled HMM, and Figure 10.5, a factor HMM. The key to reading the Figures is to see circles as containing states of a hidden chain and rectangles as containing the emitted symbols at respective times. A detailed study of factor HMMs as in Figure in 10.5 is given in (Ghahramani and Jordan 1997). Boys et al. (2000) apply autoregressive HMM in biological sequence analysis.

Given the standard hidden Markov model family above, there are evidently three basic problems of interest that must be solved for the model to be useful in applications to sequence data. These problems are the following (Rabiner 1989, Juang & Rabiner 1991):

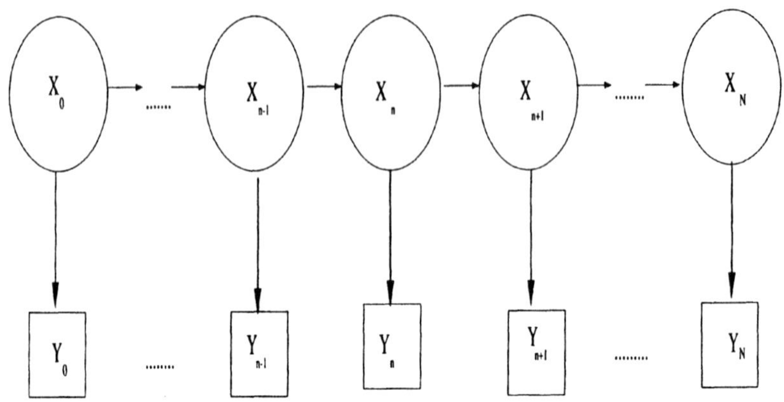

Figure 10.2: Standard HMM: Influence diagram

(1) **The Evaluation or Scoring Problem**
The real concern in the evaluation problem is computational efficiency. Without complexity constraints one can straightforwardly evaluate

$$P\left(Y_0 = o_0, \ldots, Y_n = o_n; \lambda\right)$$

directly as this probability is defined in (2.5). Since the summation involves J^{n+1} possible sequences, the total computational requirements are of the order $2(n+1) \cdot J^{n+1}$ operations. The need to compute without the exponential growth in computation is the first basic problem of HMM. The solution is known as the *forward–backward procedure* derived in Chapter 14 using the results in Chapter 13.

(2) **The Decoding or Alignment Problem**
We are often interested in uncovering the most likely state sequence that led to the observed sequence $(o_0 \ldots o_n)$. In the context of hidden Markov models for biological sequence families, see Chapter 12, this is the *alignment problem.* There are, of course, several ways to define the decoding criterion. Finding the sequence $j_0^* \ldots j_n^*$ that maximizes

$$P\left(X_0 = j_0, \ldots, X_n = j_n, Y_0 = o_0, \ldots, Y_n = o_n; \lambda\right)$$

for a fixed observed sequence $o_0 \ldots o_n$ is in practice the most frequently used criterion, since it can be implemented by the *Viterbi algorithm,*

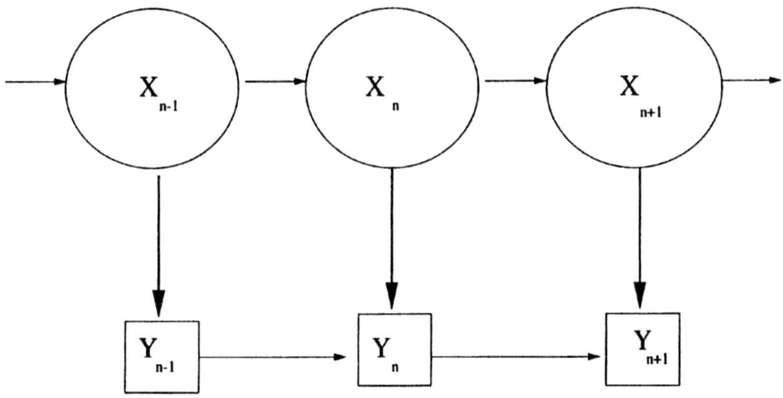

Figure 10.3: Autoregressive HMM: Influence diagram

actually already present the computation of the evolutionary distance in Chapter 5 and derived independently from scratch again in Chapter 14.

(3) **The Estimation or Training Problem**
Given an observed sequence $\mathbf{o} = o_0 \ldots o_n$, the estimation problem involves finding the model

$$\lambda = (A, B, \pi(0))$$

that specify the model most likely to generate the given sequence $o_0 \ldots o_n$ now also called a *training sequence.* In solving the estimation problem we shall analyse:

1) Maximum likelihood $\widehat{\lambda}_{\mathrm{ML}}$ is defined by.

$$\widehat{\lambda}_{\mathrm{ML}} = \arg\max_{\lambda} P\left(Y_0 = o_0, \ldots, Y_n = o_n; \lambda\right).$$

The probability in the right hand side is given in (2.5). This expression can have various degrees of complexity depending on the actual expressions for the conditional distributions in B. $\widehat{\lambda}_{\mathrm{ML}}$ is for standard HMM computed by using the *Baum–Welch algorithm,*

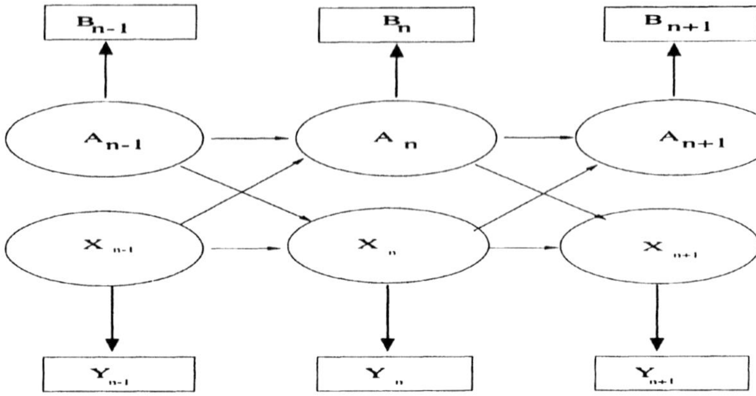

Figure 10.4: Coupled HMM: Influence diagram

which is based on the same procedures as the EM algorithm for finite mixtures in Chapter 4.

The properties of the Baum–Welch algorithm for computing $\widehat{\lambda}_{\text{ML}}$ are treated in extensive detail in Chapters 15 and 16. The property of *asymptotic consistency*, i.e., that of proving the convergence of $\widehat{\lambda}_{\text{ML}}$ as $n \to \infty$ to the postulated 'true model' within the family, is proved for ergodic HMM in Chapter 17.

The quantities computed in the Baum–Welch learning are fundamental even for the other techniques (i.e., **2)**-**4)** below) for estimating λ.

2) MAP learning

$$\widehat{\lambda}_{\text{MAP}} = \arg\max_{\lambda} P(Y_0 = o_0, \ldots, Y_n = o_n; \lambda)\, \phi(\lambda)$$

where $\phi(\lambda)$ is a prior density. The version of MAP using the Dirichlet priors is given in Chapter 16.

3) Minimum discrimination information is owed to (Ephraim et al. 1989) and is based on minimization of a suitable Kullback distance.

4) Smooth learning algorithms A smooth reparametrization of λ is made to use the method of gradient descent. The technique

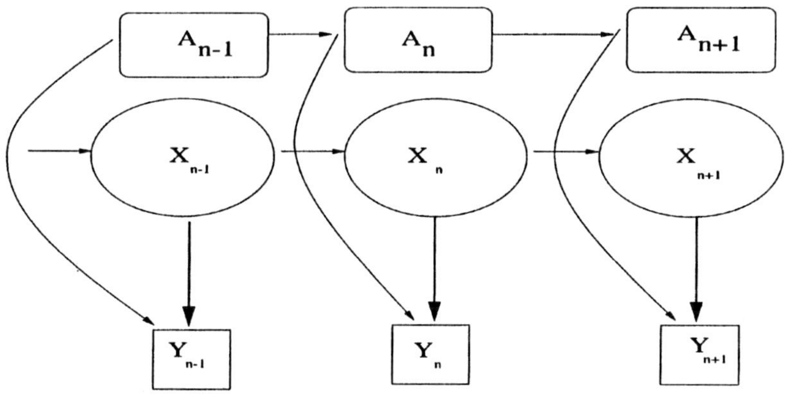

Figure 10.5: Factor HMM: Influence diagram

is developed in (Baldi et al. 1994) and will be presented in an exercise to Chapter 16.

5) Viterbi training If we set

$$V^*(\lambda) = \max_{j_0 \cdots j_n} P\left(X_0 = j_0, \ldots, X_n = j_n; Y_0 = o_0 \ldots, Y_n = o_n; \lambda\right)$$

and try to find λ so as to maximize $V^*(\lambda)$, then we may talk about Viterbi training. The method has been developed and analyzed under a different name in (Juang and Rabiner 1990) and will be treated by exercises in Chapters 15 and 16.

Against this background the hidden Markov models in Figures 10.3–10.5 are of real interest only if there exist effective algorithmic solutions for the three problems stated above. A condition for computable evaluation and learning in standard and nonstandard HMM is found in (Lucke 1996).

Smyth et al. (1997) introduce a general framework of graphical models, 'probabilistic independence networks', by means of which one can derive the computational algorithms for scoring and alignment problems in a quite different manner than done in this text. The technique of probabilistic independence networks can also be used to treat scoring and alignment for a number of non-standard HMM's.

In (Kung 1993) the *pattern completion problem* is also introduced and discussed. By this is meant the estimation of a randomly missing symbol in $\mathbf{o}$. The solution can be found by means of the Viterbi algorithm.

10.4.1 A Fourth Problem

If the states of the Markov chain in HMM have no physical or biological interpretation, there would seem to exist a fourth problem. This is that HMM's with different 'topologies' or architectures might be used to model the same sequence of data. How an architecture could be learned from data does not seem to have been widely studied and will not be discussed in this text. However, in principle, the method of Bayesian model selection should be applicable up to a certain degree. There are heuristics for trimming of architecture in HMM implemented in some of the software used in bioinformatics.

10.5 Finite State Stochastic Machines

10.5.1 Notations and Terminology

The definitions of probabilistic automata or finite state stochastic machines are variants or generalizations of HMM. References for basic properties of stochastic machines are (Ericson 1972) or (Paz 1971). The data compression model of (Ott 1967) is based on finite state stochastic machines. In fact, the models in (Churchill 1989) are stochastic sequential machines.

An (autonomous) *stochastic Mealy machine* $\mathbf{M_e}$ is the triple

$$\mathbf{M_e} = \{\mathcal{S}, \mathcal{O}, f_{Y_n,X_n|X_{n-1}}(o, s^{'}|s)\}, \tag{5.2}$$

where

$$s, s^{'} \in \mathcal{S} = \{1, 2, \cdots, J\}$$

is the *state alphabet* of $\mathbf{M_e}$ and

$$z \in \mathcal{O} = \{z_1, z_2, \cdots, z_K\}$$

is the *output alphabet* of $\mathbf{M_e}$. The probability

$$f_{Y_n,X_n|X_{n-1}}(o, s^{'}|s) = P(Y_n = o, X_n = s^{'}|X_{n-1} = s), \tag{5.3}$$

is in this context called the 'state $\mapsto$ output and state $\mapsto$ next state' map of the machine, where Y_n is the *output* of $\mathbf{M_e}$, X_n is the *state* of $\mathbf{M_e}$.

A stochastic machine can in this way be thought of being 'realized' by a factorization of the map (5.3) rewritten by the chain rule of conditional probability (2.16) in Chapter 1 as

$$f_{Y_n,X_n|X_{n-1}}(o,s'|s) = f_{Y_n|X_{n-1}}(o|s) \cdot f_{X_n|Y_n,X_{n-1}}(s'|o,s). \tag{5.4}$$

A *stochastic machine with probabilistic output* is obtained assuming in (5.4) that the current output symbol does not influence the next state, or

$$f_{Y_n,X_n|X_{n-1}}(o,s'|s) = f_{Y_n|X_{n-1}}(o|s) \cdot f_{X_n|X_{n-1}}(s'|s). \tag{5.5}$$

The state process X_n in a stochastic machine with probabilistic output is a Markov chain. Note that this differs formally from our definition of an HMM in the appearance of the one step delay in the emission probability distributions. The formula for the joint probability of the output of a stochastic machine is

$$P(Y_1 = o_{i_1}, \cdots, Y_m = o_{i_m}) = \sum_{i_0 \in S} \sum_{j_1 \in S} \cdots \sum_{j_m \in S} \pi_{i_0} \cdot \mathbf{\Phi}, \tag{5.6}$$

where

$$\mathbf{\Phi} := \prod_{k=1}^{m} f_{Y_n,X_{k+1}|X_k}(o_{i_k}, j_k|j_{k-1}). \tag{5.7}$$

10.5.2 Carlyle's Matrix Formula

We introduce, for any finite state stochastic machine with a probabilistic output, cf. (5.5), the matrix $\mathbf{P}_Y(o_k)$ with the elements

$$(\mathbf{P}_Y(o_k))_{ij} := f_{Y_n|X_{n-1}}(o_k|j) \cdot f_{X_n|X_{n-1}}(j|i). \tag{5.8}$$

Hence, as $a_{i|j} = f_{X_n|X_{n-1}}(j|i)$,

$$\mathbf{P}_Y(o_k) = A\mathbf{F}(o_k), \tag{5.9}$$

where $\mathbf{F}(z_k)$ is the diagonal matrix

$$\mathbf{F}(o_k) := \begin{pmatrix} f_{Y_n|X_{n-1}}(o_k|1) & 0 & \dots & 0 \\ 0 & f_{Y_n|X_{n-1}}(o_k|2) & \dots & 0 \\ \dots & \dots & \dots & \dots \\ 0 & 0 & \dots & f_{Y_n|X_{n-1}}(o_k|J) \end{pmatrix}. \tag{5.10}$$

Thus (5.6) becomes the Carlyle matrix formula

$$P(Y_1 = o_{i_1}, \cdots, Y_m =_{i_m}) = \pi(0) \prod_{l=1}^{m} \mathbf{P}_Y(o_{i_l})\mathbf{1} = \pi(0) \prod_{l=1}^{m} A\mathbf{F}(o_{i_l})\mathbf{1}, \quad (5.11)$$

where $\mathbf{1}$ is the $J|1$ column vector of ones, see (Carlyle 1963). We shall in the sequel prove a version of this formula for HMM by means of the forward and backward variables.

10.6 Exercises

1. Consider Gilbert's channel, Example 10.3.2 above. Assume that the Markov chain $\{X_n\}_{n=0}^{\infty}$ is *stationary*. Show that the probability of error $P(e)$ in transmission, i.e.,

$$P(e) = P(Z_n \neq S_n)$$

is

$$P(e) = (1 - \rho) \cdot \frac{q}{q + p}.$$

2. The process $\{Y_n\}_{n=0}^{\infty}$ is a function of a Markov chain $\{X_n\}_{n=0}^{\infty}$

$$Y_n = g(X_n), \quad (6.1)$$

where $g(\cdot)$ is a map from S to $\mathcal{O}$, where the number of symbols in $\mathcal{O}$ is smaller than the number of symbols in S. Show that $\{X_n, Y_n\}_{n=0}^{\infty}$ is an HMM in the sense of the definition above. What is the emission probability matrix B in this case ?

3. **HMM as a Function of a Markov Chain**
Let $\{X_n\}_{n=0}^{\infty}$ and $\{Y_n\}_{n=0}^{\infty}$ be an HMM and let $\{X_n\}_{n=0}^{\infty}$ be stationary. We define a state space Z using the Cartesian product

$$Z = S \times \mathcal{O} = \{1, \ldots, J\} \times \{o_1, o_2, \ldots o_K\}.$$

Let us also define the new Markov process $\{Z_n\}_{n=0}^{\infty}$ with values in Z by means of the transition probability matrix (a $J \cdot K \times J \cdot K$ matrix)

$$a_{(i,o_k)|(j,o_l)} = a_{i|j} \cdot b_j(o_l) = P(Z_{n+1} = (j, o_l)|Z_n = (i, o_k)).$$

Let us next define a map g from Z onto $\mathcal{O}$ as

$$g((i, o_k)) = o_k$$

and let us define a new random process $\left\{Y_n'\right\}_{n=0}^{\infty}$, which is a function of a Markov chain, by

$$Y_n' = g(Z_n), \qquad n = 0, \ldots, .$$

Show that $\left\{Y_n'\right\}_{n=0}^{\infty}$ and $\{Y_n\}_{n=0}^{\infty}$ are probabilistically equivalent in the sense that

$$P(Y_0 = o_0, Y_1 = o_1, \ldots, Y_n = o_n) = P\left(Y_0' = o_0, Y_1' = o_1, \ldots, Y_n' = o_n\right)$$

for any sequence $o_0 o_1 \ldots o_n$.

4. **Naive Rule of Decoding**
Markov chain $\{X_n\}_{n=0}^{\infty}$, with the binary digits $S = \{0, 1\}$ as the state space, has the transition matrix

$$A = \begin{pmatrix} \frac{1+p}{2} & \frac{1-p}{2} \\ \frac{1-p}{2} & \frac{1+p}{2} \end{pmatrix}, \tag{6.2}$$

where $0 \leq p < 1$. The initial distribution is the distribution

$$\pi = \left(\frac{1}{2}, \frac{1}{2}\right).$$

We observe another process $\{Y_n\}_{n=0}^{\infty}$ with the state space $\mathcal{O} = \{0, 1\}$ and the emission probability matrix

$$B = \begin{pmatrix} 1-\epsilon & \epsilon \\ \epsilon & 1-\epsilon \end{pmatrix}$$

and assuming conditional independence:

$$\begin{aligned} P(Y_0 &= o_{j_0}, \ldots, Y_n = o_{j_n} | X_0 = j_0, \ldots, X_n = j_n; B) \\ &= \prod_{l=0}^{n} \epsilon^{|o_{j_l} - j_l|} (1-\epsilon)^{1-|o_{j_l} - j_l|}. \end{aligned}$$

With these quantities we set $\lambda = (A, B, \pi)$. Clearly we have

$$P(Y_0 = o_{j_0}, \ldots, Y_n = o_{j_n} | X_0 = j_0, \ldots, X_n = j_n; B) = \epsilon^k (1-\epsilon)^{n+1-k},$$

where k is the number of mismatches or the Hamming metric, see Example 5.3.2 in Chapter 5, between the binary strings $o_{j_0} \ldots o_{j_n}$ and $j_0 \ldots j_n$. It is assumed that $0 \leq \epsilon < \frac{1}{2}$.

(a) Show that

$$P(X_0 = j_0, \ldots, X_n = j_n; A) = \left(\frac{1}{2}\right)^{n+1} \left[\frac{(1-p)}{(1+p)}\right]^m \cdot (1+p)^n,$$

where m is the number of transitions $0 \mapsto 1$, $1 \mapsto 0$ in $j_0 \ldots j_n$.

(b) Show that

$$\begin{aligned} P(X_0 &= j_0, \ldots, X_n = j_n | Y_0 = o_{j_0}, \ldots, Y_n = o_{j_n}; \lambda) \\ &= \psi(o_{j_0}, \ldots, o_{j_n}) \left[\frac{(1-p)}{(1+p)}\right]^m \cdot \left[\frac{\epsilon}{1-\epsilon}\right]^k, \end{aligned}$$

where k is defined above in terms of the Hamming metric and $\psi(o_{j_0}, \ldots, o_{j_n})$ does not depend on the state sequence $j_0 \ldots j_n$.

(c) Explain using (6.3) that if

$$(1-p)^2(1-\epsilon) \geq (1+p)^2\epsilon,$$

then the optimal decoding $(j_0^* .. j_n^*)$ defined as

$$(j_0^* .. j_n^*) = \arg\max_{j_0 \ldots j_n} P(X_0 = j_0, .., X_n = j_n | Y_0 = o_{j_0}, .., Y_n = o_{j_n}; \lambda)$$

coincides with the observed sequence ('naive rule of decoding a noisy MC')

$$(j_0^* \ldots j_n^*) = (o_{j_0} \ldots o_{j_n}).$$

Hint: Think of what happens to m and k if we replace $(j_0 \ldots j_n)$ by $(o_{j_0} \ldots o_{j_n})$ and what then happens to the posterior probability under study. (Rudemo 1974)

5. Consider the hidden Markov chain $\lambda = (A, B)$ in Example 10.3.4 above. Give the probability distribution

$$P\left(Y_0 = o_0, \ldots, Y_n = o_n; \lambda\right)$$

for $n = 0, 1, \ldots$.

6. Prove the matrix formula (5.11).

10.7 References and Further Reading:

P. Baldi and Y. Chauvin (1994): Smooth On-Line Learning Algorithms for Hidden Markov Models. *Neural Computation*, 6, pp. 307–318.

R.J. Boys, D. Henderson and D.J. Wilkinson (2000): Detecting homogeneous segments in DNA sequences using hidden Markov models. *Applied Statistics*, 49, Part 2, pp. 269–285.

J.W. Carlyle (1963): Reduced Forms for Stochastic Sequential Machines. *Journal of Mathematical Analysis and its Applications*, Vol. 7, pp. 167–175.

Y. Ephraim, A. Dembo and L.R. Rabiner (1989): Minimum Discrimination Information Approach for Hidden Markov Modelling. *IEEE Transactions on Information Theory*, 35, pp. 1000–1013.

T. Ericson (1972): Unifilar Sources and the Source Approximation Problem. *Ericsson Technics*, Vol. 28, pp. 175–290.

Z. Ghahramani and M. Jordan (1997): Factorial hidden Markov models. *Machine Learning*, 29, pp. 245–273.

E.N. Gilbert (1960): Capacity of a Burst-Noise Channel. *The Bell System Technical Journal*, Vol. 39, pp. 1253–1265.

F. Jelinek (1999): *Statistical methods for speech recognition.* MIT Press, Cambridge Massachusetts, London, England.

B.H. Juang and L.R. Rabiner (1990): The segmental K-means Algorithm for Estimating Parameters of Hidden Markov Models. *IEEE Transactions on Acoustics, Speech and Signal Processing*, 38, pp. 1639–1641.

B.H. Juang and L.R. Rabiner (1991): Hidden Markov Models for Speech Recognition. *Technometrics*, 33, pp. 251–272.

S.Y. Kung (1993): *Digital Neural Networks.* Chapter 7, PTR Prentice Hall, Englewood Cliffs, New Jersey, 07632.

S.E. Levinson, L.R. Rabiner and M.M. Sondhi (1983): An Introduction to the Applications of the Theory of Probabilistic Functions of a Markov Chain to Automatic Speech Recognition. *The Bell System Technical Journal*, 62, pp. 1053–1074.

H. Lucke (1996): Which Stochastic Models Allow Baum-Welch Training ? *IEEE Transactions on Signal Processing*, 44, pp. 2746–2756.

G. Ott (1967): Compact Encoding of Stationary Markov Sources. *IEEE Transactions on Information Theory*, IT - 13, pp. 82–86.

A. Paz (1971): *Introduction to Probabilistic Automata.* Academic Press, New York.

L.R. Rabiner (1989): A Tutorial on Hidden Markov Models and Selected Applications in Speech Recognition. *Proceedings of the IEEE*, 77, pp. 257–286.

M. Rudemo (1974): *Prediction and Filtering for Markov Processes.* TRITA-MAT-1974-6, Department of mathematics, Royal Institute of Technology, Stockholm.

J.Q. Smith (1989): Influence diagrams for statistical modelling. *Annals of Statistics*, 17, pp. 654–672.

P. Smyth, D. Heckerman and M.L. Jordan (1997): Probabilistic Independence Networks for hidden Markov probability models. *Neural Computation*, 9, pp. 227–269.

Chapter 11

HMM for DNA Sequences

11.1 Introduction

HMM were first applied in speech recognition and are now widely used in computational biology and bioinformatics. The applications include (see the list of references at the end of this chapter) linkage analysis, radiation hybrid mapping, multiple alignment and functional classification of proteins (outlined in the next chapter), prediction of protein folding, recognition of genes in bacterial and human genomes, predicting trans-membrane topology, analysis, horizontal gene transfer and prediction of DNA functional sites and identification of DNA periodical patterns and other. The present chapter contains a concise survey of some of the formal features of applications to segmentation, gene finding and radiation hybrid mapping.

Modelling by HMM in biological sequence analysis is an expression of **modularity**: a complex system is built by assembling simpler parts. Probability theory serves as the foundation whereby the parts are combined, ensuring that the system as a whole is coherent and providing ways to interface models to data. The important feature of hidden Markov modelling is flexibility and adaptability. One can choose any structure for the states, transitions and emission distributions that is appropriate for the problem at hand.

Modularity in the preceding paragraph is a notion coined in computer science but seems to have a counterpart in molecular biology which sees many proteins as consisting of a number of separately evolved, independently structured domains or modules (Patthy 1991, Sonnhammer and Kahn 1994). The

HMMs to be presented in this chapter do not, however, encompass shuffling and recombination of modules as described in (Patthy 1991, Sonnhammer and Kahn 1994).

11.2 Heterogeneity and Segmentation of Natural DNA sequences

One of the first applications of HMMs to the analysis of biological sequence data was to compositional heterogeneity of natural DNA sequences (Churchill 1989, 1992). The effort in splitting genomes into compositionally homogeneous regions is of biological interest for a number of reasons (Peshkin and Gelfand 1999):

First, the segments may have meaningful base compositions. In the human genome coding sequences tend to be CG-rich, a topic discussed in (Venter et al. 2001). On the other hand AT-rich regions in yeast genome usually occur in coding regions (Peshkin and Gelfand, 1999).

Second, the performance of many algorithms for functional mapping of genomic DNA depends on the homogeneity of the sequences. The compositional segmentation of genomes allows for use of different sets of parameters adapted to each composition.

A biologist may even be interested in describing the structure of the sequence or in detecting segments which may be atypical or in comparing structures between sequences. In addition, any good model of segmentation says something about the statistical properties of natural DNA. Discussion and applications of the segmentation models are given in (Boys et al. 2000, Muri 1997). The work in (Schmidler et al. 2000) contains a segmentation model closely related to the hidden semi-Markov models (see below in the present chapter).

The DNA segmentation problem can be studied by means of several different statistical models surveyed in (Braun and Müller 1998, Karlin and Macken 1991, Schmidler et al. 2000). In modelling with hidden Markov chains we postulate that there are in DNA sequences more or less distinct segments belonging to a small number of distinct compositional classes within which the observations follow the same distribution, and between which the sequence has different distributions. The probability of observing a particular base at a given site on the DNA would depend on the type of the segment it is in.

Example 11.2.1 (Segmentation HMM) The state and output alphabets in Churchill (1989, 1992) are taken as $S = \{1, 2\}$ and $\mathcal{O} = \{o_1, o_2\} = \{AT, GC\}$, respectively. The hidden Markov chain has the transition matrix

$$A = \begin{pmatrix} 1-p & p \\ q & 1-q \end{pmatrix}, \tag{2.1}$$

where $0 \leq p \leq 1$ and $0 \leq q \leq 1$. Connect this to the emission probability matrix

$$B = \begin{pmatrix} 1-w & w \\ r & 1-r \end{pmatrix}, \tag{2.2}$$

where $0 \leq w \leq 1$ and $0 \leq r \leq 1$. The value of r is chosen close to 0 and q is taken close to 0. Hence the state 1 of the hidden Markov chain persists, once the chain has entered it, and GC has a high probability of being emitted. Thus the model is capable of generating DNA sequences with GC-rich segments (and AT-rich segments). There is a similar model for base composition known as the *walking Markov model*, (Fickett et al. 1992), presented in Exercise 1 at the end of this chapter. ∎

Peshkin and Gelfand (1999) extend and apply Churchill's segmentation HMM for analysis of the yeast genome. In (Peshkin and Gelfand 1999) the number of segment states (states of the underlying Markov chain) is determined from yeast genome using the Bayesian model selection criterion, cf. Chapter 3 above.

11.3 Gene Finding and HMM

There are, according to (Fickett 1996, Searls 2000), two main approaches to gene finding: first the methods identifying a gene by finding a similar known object in available databases; second, model based methods that compose prototype objects, and then identify genes by matching to such prototypes. The prototype descriptions are derived *ab initio* from sequence data and provide models (of sequence families). The completed sequence models are stored in what are known as *secondary databases.*

Gene finding by HMM is a model based approach pioneered in (Krogh et al. 1994), where a hidden Markov model for finding genes in *E.Coli* was constructed. This HMM combined features of many known computational *ab initio* methods of gene finding. The details of the model are similar to the HMMs for multiple alignment of proteins to be presented in more detail in the next chapter. Surveys of gene finding by HMM are (Henderson et al. 1997, Krogh 1999).

The gene finding capability of HMM is based on its architecture of hidden states arranged in modules designed for the specific regions of the DNA: the exons, introns, the patterns found in the intragenic region and the cyclic transitions between exons and introns and on modelling the emission properties within each such state to take into account the profile characteristics of biological signals in each of the regions.

Both of the recent publications reporting the sequencing of the draft human genome (International Human Genome Sequencing Consortium 2001 p.899, Venter et al. 2001) refer to GENSCAN, software based on the HMM developed in (Burge and Karlin 1997), as one of the tools employed in gene finding. To be quite accurate, (Venter et al. 2001) deem the evidence based genefinder (called **otto**) developed by Celera as outperforming GENSCAN. A critical analysis of gene finding by statistical or homology methods is given in (Karlin et al. 2001)

The HMM in GENSCAN is an extension, sometimes known as the HMM with duration, of the standard HMM. Another name would be a **hidden semi-Markov model** (see also (Kulp et al. 1996)).

Example 11.3.1 (HMM with duration) The standard HMM in section 10.2 in Chapter 10 is here changed by imposing a suitable probability distribution for the *duration* of stay in a hidden state (for duration of stay in a standard MC see Exercise 3 at the end of this chapter).

(I) **The state space of the Hidden Markov Chain** The hidden states of the model correspond to fundamental functional units of a eukaryotic gene depicted in Figure 11.1. The circles and boxes represent the functional units of a gene or a genomic region. In the forward strand we follow the graph in the direction of the arrows starting from the intergenic noncoding region N. Then we have promoter Pr, $5'+$ untranslated region, connected to translation intiation exon E and intron state I_5, a single exon gene E_0 (translation start $\rightarrow$ stop codon), or the first exon E_F (translation start $\rightarrow$ stop codon). This is followed by introns $I_k, k = 0, 1, 2$, in three possible reading frames. The internal exons $E_k, k = 0, 1, 2$ are also divided according to the reading frame. The terminal or last exon site is E_L (acceptor splice site $\rightarrow$ stop codon), translation termination exon E, and intron state I_3, after which there is an $3'$ untranslated region and a return to N. A model for the reverse strand can be read as a mirror symmetrical image relative to the vertical axis cutting the noncoding intergenic region N. In (Burge

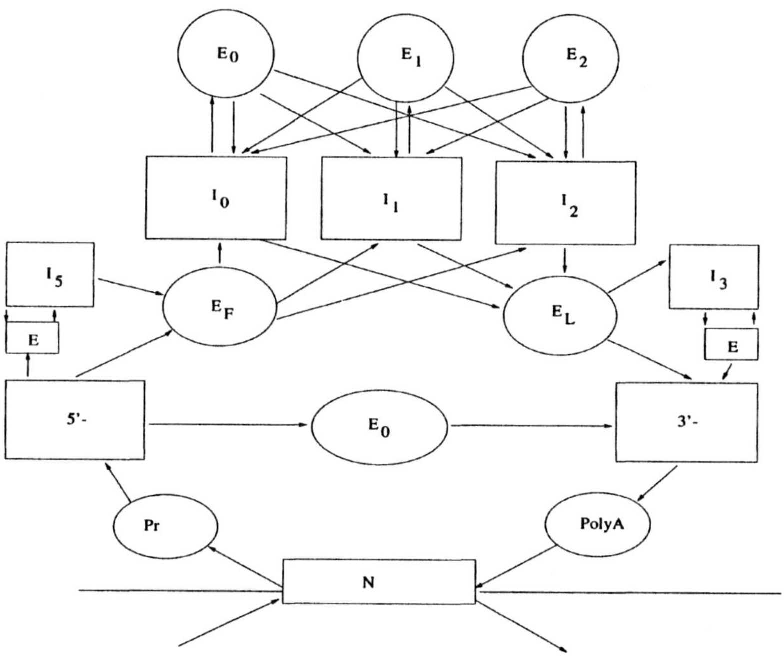

Figure 11.1: The forward strand in a gene location (Burge and Karlin 1997)

and Karlin 1997) the coding portions of the exons are described by the framedependent MC of Borodovsky and McIninch outlined in Chapter 9 above. The non-coding states are modeled using a homogeneous fifth-order MC.

(II) **Duration** Given that the MC is in state j, $X_n = j$, the discrete duration d_n is chosen from the length distribution ϕ_j. We set

$$\underline{\phi} = (\phi_1, \phi_2, \dots \phi_K),$$

where each ϕ_j is a probability distribution on the positive integers. One set of rules for constructing ϕ_j is given in (Levinson 1986). In (Burge and Karlin 1997) introns and the intergenic segments have a duration of stay given by a geometric distribution.

(III) **Observable Segments** A random process $\{Y_n\}_{n=0}^{D}$, where D is given in advance, with the finite state space $\mathcal{O} = \{o_1, o_2, \ldots o_K\}$ (which is equal to $\{A, T, C, G\}$ in gene finding) is generated by as a segment $s_n = o_{k_1} \ldots o_{k_{d_n}}$ with d_n symbols of $\mathcal{O}$ with the conditional distribution

$$\prod_{r=1}^{d_n} b_j(k_r),$$

where the emission probability is as before

$$b_j(k_r) = P\left(Y_{d_0+\ldots d_{n-1}+r} = o_{k_r} | X_n = j\right).$$

This assumption of mutual independence of the emitted symbols in a segment conditioned by a state sequence and a sequence of durations can be relaxed.

(IV) The MC moves to a new state j_{n+1} and a new duration is drawn from the distribution $\phi_{j_{n+1}}$. The complete emitted sequence is obtained by a concatenation of the segments of lengths d_n generated during the stays in the states.

By these assumptions we write the *joint probability* of $o_{l_0} \ldots o_{l_D}$ and $j_0 \ldots j_n$ as

$$P\left(Y_0 = o_{l_0}, \ldots, Y_D = o_{l_D}, X_0 = j_0, \ldots, X_n = j_n; A, B, \underline{\phi}, \pi(0)\right) \tag{3.3}$$

$$= \pi_{j_0}(0)\, \phi_{j_0}(d_0) \prod_{r=1}^{d_0} b_{j_0}(k_r) \prod_{l=1}^{n} a_{j_{l-1}|j_l} \phi_{j_l}(d_l) \prod_{r=1}^{d_l} b_{j_l}(k_r),$$

where $d_0 + d_1 + \ldots + d_n = D$ with d_n possibly truncated to make the sum equal to the pre-selected value D. The compact notation for the model is then

$$\lambda = \left(A, B, \underline{\phi}, \pi(0)\right).$$

It turns out that the problems of scoring, learning and alignment can be solved for the duration hidden Markov model by modifications of the respective standard algorithms. These modifications are treated in Exercise 9 in Chapter 14 and in Exercises 12 and 13 in Chapter 15. ■

In (Lukashin and M. Borodovsky 1998) an HMM with duration is designed for prokaryotic gene finding. The number of hidden states is taken as nine, four states in both two strands (direct and reverse) of the DNA sequence and one state as the noncoding state. The four coding states are for the start codon, a typical gene, atypical gene and the stop codon. The resulting algorithms and software are called GeneMark:hmm [1].

11.4 Radiation Hybrid Mapping

A *physical map* of a chromosome shows the relative locations and estimated distances between known *markers* along the chromosome. The markers may be genes or simply arbitrary DNA words along the chromosome. This is helpful for the sequencing projects ('map first, sequence later', International Human Genome Sequencing Consortium (2001)) in the sense that fragments can be assembled to a final sequence using a physical map. One experimental method of deriving a physical map is the radiation hybrid mapping.

Radiation hybrid mapping has been used for small scale physical mapping of the human genome. The experimental method involves exposing human cells to gamma radiation, which breaks the chromosome into random fragments. The DNA fragments are then rescued by fusion with hamster cells that retain a random subset of the human DNA fragments. Each resulting hybrid cell can be cloned to form a hybrid cell line of cells all containing the same random subset of the human genome. A radiation hybrid panel consists of a number of different hybrid cell lines.

If there is no break between a pair of markers, they are coretained, that is, the hybrid contains either both or neither of the markers. Each hybrid panel is screened against each marker. A positive result ('+') indicates that the hybrid in question retained the marker, while a negative result ('−') indicates that the marker in question was not retained. The data in the form of these binary results is put into a matrix. The task is to reconstruct the markers' positions on the original chromosome given only their retention patterns in

[1] http://genemark.biology.gatech.edu/GeneMark/whmm_info.html

the matrix outlined. Stringent mathematical studies of the properties of the computational algorithms for these problems are found in (Ben-Dor and Chor 1997) and (Ivansson 2000).

Example 11.4.1 (HMM for radiation hybrid mapping) A number of Hidden Markov models for the computational reconstruction of the markers' positions have been devised in (Boehnke et al. 1991, Lange et al. 1995, Slonim 1996). A separate HMM as in Figure 11.2 is assigned for each hybrid. For each marker i, $i = A, B, C, \ldots$ the HMM has the hidden state space $S_i = \{0, 1\}$. The state of the model at a given marker corresponds to the number of copies of that marker actually retained by the given hybrid, i.e., none (0) or one (1) in the haploid case. The observed data is either + or $-$. In addition there is a probability of screening error, so that a retained marker can be missed and erroneously detected. Let θ_{i-1} be the probability that there is at

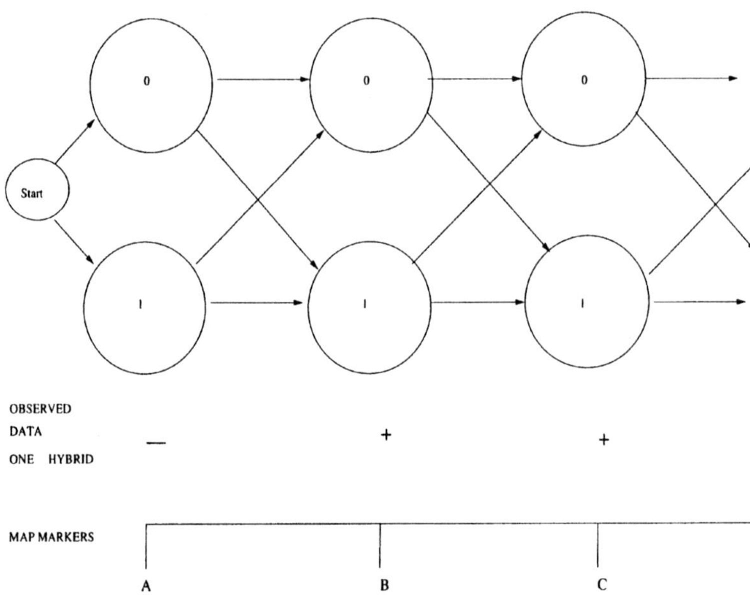

Figure 11.2: The HMM for the radiation hybrid mapping in the haploid case

least one break between $i - 1$ and i and let r_i be the probability of retention of the marker i. Retention and breaks are supposed to occur independently of each other.

Then the transition probabilities between the states at markers $i-1$ and i and with the state space $S_i = \{0, 1\}$ is

$$A_{i-1,i} = \begin{pmatrix} 1-\theta_{i-1}r_i & \theta_{i-1}r_i \\ \theta_{i-1}(1-r_i) & 1-\theta_{i-1}(1-r_i) \end{pmatrix}. \tag{4.4}$$

Here $\theta_i r_i = P_{i-1,i}(X_i = 1|X_{i-1} = 0)$ and $\theta_i(1-r_i) = P_{i-1,i}(X_i = 0|X_{i-1} = 1)$. This is notably an example of a *non-homogeneous* MC. The emission probability B matrix is now given by

$$B = \begin{pmatrix} 1-\alpha & \alpha \\ 1-\beta & \beta \end{pmatrix}, \tag{4.5}$$

where $\alpha = P(Y_i =' +'|X_i = 0)$, the probability of false positives, and β is equal to $P(Y_i =' -'|X_i = 1)$, the probability of false negatives. ■

11.5 Exercises

1. **Walking Markov models**

   Fickett et al. (1992) introduced the walking Markov modelling (WMM) of base composition of DNA sequences. A WMM is a modification of Churchill's segmentation HMM in Example 11.2.1. The transition from AT -rich segments to CG -rich segments (and vice versa) takes place in a smooth manner in a WMM, as will be seen in this exercise.

   For a more detailed account of the biological rationale for WMM we refer the interested reader to the original paper. For the present purpose only the formal definition of WMM is of interest.

   We let $\{W_n\}_{n=0}^{\infty}$ be a random walk with reflecting boundaries at the points c and d, where $0 < c < d < 1$. In other words $c \leq W_n \leq d$ for all n. The reflecting random walk we can define by

$$W_{n+1} = \text{sat}(W_n + \xi_n; c, d)$$

   where

$$\text{sat}(x; c, d) = \begin{cases} c & \text{if } x \leq c, \\ x & \text{if } c < x < d, \\ d & \text{if } d \leq x, \end{cases}$$

and $\{\xi_n\}_{n=0}^{\infty}$ is a sequence of independent random variables with, e.g.,

$$\xi_n = \begin{cases} -\delta & \text{with probability } p, \\ \delta & \text{with probability } 1-p, \end{cases}$$

where δ is a small number in comparison with $d-c$. Then, conditionally on $W_n = w$, the emitted symbol Y_n is selected from the alphabet $S = \{A, T, C, G\}$ with the probabilities

$$b_w = \begin{pmatrix} \frac{w}{2} & \frac{w}{2} & \frac{1-w}{2} & \frac{1-w}{2} \end{pmatrix}$$

independently of other emitted symbols.

(a) Is WMM an instance of HMM in the sense defined in the preceding? If not, what are the differences between standard HMM and WMM?

(b) Assuming that $\delta = \dfrac{d-c}{l}$ for some integer l and that $W_0 \equiv c + p\delta$ for some (non-random) integer p, compute the invariant distribution of the MC $\{W_n\}_{n=0}^{\infty}$.

2. **HMM for radiation hybrid mapping**
Explain how the entries in the transition probability matrix (4.4) are obtained.

3. **HMM for locating regulatory regions**
HMM for locating regulatory genes developed in (Crowley et al. 1997) the transition matrix of the underlying Markov chain is the same as in Churchill's segmentation HMM

$$A = \begin{pmatrix} 1-p & p \\ q & 1-q \end{pmatrix}. \tag{5.6}$$

Here the two states of the chain are 'regulatory region' (of DNA) or 'not a regulatory region'. A regulatory region is a part of chromosomal DNA that serve as signals for turning the gene expression off or on. The emission alphabet is a catalogue of words that act as these signals. In (Crowley et al. 1997) the number of regulatory regions expected in a sequence is reflected by selection of prior densities on p and q as the beta priors (see appendix 3.8.3 in Chapter 3 above). How should α_1 and α_2 in $\mathcal{B}e\,(p; \alpha_1, \alpha_2)$ and $\mathcal{B}e\,(q; \alpha_1, \alpha_2)$ be chosen to reflect the prior guess that regulatory regions of length 200 to 600 base pairs occur zero to four times every 5000 base pairs in a region of genomic DNA?

3. **HMM with duration**
   Let $\{X_n\}_{n=0}^{\infty}$ be a Markov chain with a finite state space and the transition matrix
   $$A = \left(a_{i|j}\right)_{i=1,j=1}^{J,J}.$$
   (a) Compute probability of the time of *duration of stay* in state i denoted by $\phi\,(n)$ for the event that the chain stays in the state i for exactly n steps after having started in state i. Formally the sought probability is
   $$\phi\,(n) = P\,(X_1 = i, X_2 = i, \ldots, X_n = i, X_{n+1} \neq i | X_0 = i)\,.$$
   (b) What is the rationale for introducing HMM with duration?

4. **HMM with duration**
   Which numerical value should be assigned to $a_{i|i}$ in an HMM with duration? (Levinson 1986).

11.6 References and Further Reading:

- Modularity of proteins

L.Patthy (1991): Modular exchange principles in proteins. *Current Opinion in Structural Biology*, 1, pp. 351–361.

E.L.L. Sonnhammer and D. Kahn (1994): Modular arrangement of proteins as inferred from analysis of homology. *Protein Science*, 3, pp. 482–492.

- HMM and the statistics of DNA sequence segmentation:

R.J. Boys, D. Henderson and D.J. Wilkinson (2000): Detecting homogeneous segments in DNA sequences using hidden Markov models. *Applied Statistics*, 49, Part 2, pp. 269–285.

J.V. Braun and H-G. Müller (1998): Statistical methods for DNA Sequence Segmentation. *Statistical Science*, 13, pp. 142–162.

G.A. Churchill (1989): Stochastic Models for Heterogeneous DNA Sequences. *Bulletin of Mathematical Biology*, 51, pp. 79–94.

G.A. Churchill (1992): Hidden Markov Models and the Analysis of Genome Structure. *Computers and Chemistry*, 16, pp. 107–115.

J.W. Fickett, D.C. Torney and D.R. Wolf (1992): Base Compositional Structure of Genomes. *Genomics*, 13, pp. 1056–1064.

S. Karlin and C. Macken (1991): Some Statistical Problems in the Assessment of Inhomogeneities of DNA Sequence Data. *Journal of the American Statistical Association*, 86, pp. 27–35.

F. Muri (1998): Modelling bacterial genomes using hidden Markov models. *Proceedings in computational statistics: 13th symposium in Bristol*, pp. 89–100.

L. Peshkin and M.S. Gelfand (1999): Segmentation of yeast DNA using hidden Markov models. *Bioinformatics*, 15, pp. 980–986.

- Surveys of gene finding:

J.W. Fickett (1996): The Gene Identification problem: an overview for developers. *Computers in Chemistry*, 20, pp. 103–118.

D.B. Searls (2000): Bioinformatics tools for whole genomes. *Annual Reviews of Genomics and Human Genetics*, 01, pp. 251–279.

- Hidden Markov models in gene finding:

C. Burge and S. Karlin (1997): Prediction of Complete Gene Structures in Human Genomic DNA. *Journal of Molecular Biology*, 268, pp. 78–94.

C.B. Burge and S. Karlin (1998): Finding genes in genomic DNA. *Current Opinion in Structural Biology*, 8, pp. 346–354.

J. Henderson, S. Salzberg and K.H. Fasman (1997): Finding Genes in DNA with Hidden Markov Model. *Journal of Computational Biology*, 4, 1997, pp. 127–141.

S. Karlin, A. Bergman and A.J. Gentles (2001): Annotation of the *Drosophila* genome. *Nature*, 411, 17 May, pp. 259–260.

A. Krogh, I.S. Mian and D. Haussler (1994): A hidden Markov model that finds genes in *E. Coli* DNA. *Nucleic Acids Research*, 22, pp. 4768–4778.

A. Krogh (1999): An introduction to hidden Markov models for biological sequences. In S.L. Salzberg, D.B. Searls and S. Kasif: *Computational Methods in Molecular Biology*, Elsevier, Amsterdam etc. ,pp. 45–63.

D. Kulp, D. Haussler, M.G. Reese, F.H. Beckman (1996): A generalized hidden Markov model for the recognition of genes in DNA. *Intelligent systems in molecular biology*, 4, pp. 134–142.

A.V. Lukashin and M. Borodovsky (1998): GeneMark.hmm: new solutions for gene finding. *Nucleic Acids Research*, 26, pp. 1107–1115.

A. Salamov and V. Solovyev (2000): Ab initio gene finding in *Drosophilia* genomic DNA. *Genome Research*, 10, pp. 516–522.

T. Yada, M. Nakao, Y. Totoki and K. Nakai (1999): Modelling and predicting transcriptional units of *Escherichia coli* genes using hidden Markov models. *Bioinformatics*, 15, pp. 987–993.

- Human Genome Sequencing:

IHG – International Human Genome Sequencing Consortium (2001): Initial sequencing and analysis of the human genome. *Nature*, 409, pp. 860–921.

J.G. Venter, M.D. Adams, E.W. Myers et al. (2001): The Sequence of the Human Genome. *Science*, vol. 291, 16 February, pp. 1304–1351.

- Radiation Hybrid Mapping:

A. Ben-Dor and B. Chor (1997): On Constructing Radiation Hybrid Maps. *Journal of Computational Biology*, 4, pp. 517–533.

M. Boehnke, K. Lange and D.R. Cox (1991): Statistical Methods for Multipoint Radiation Hybrid Mapping. *American Journal of Human Genetics*, 49, pp. 1174–1188.

L. Ivansson (2000): *Computational Aspects of Radiation Hybrid Mapping.* Doctoral Dissertation, Royal Institute of Technology (KTH), Stockholm.

K. Lange, M. Boehnke, D.R. Cox and K.L. Lunetta (1995): Statistical Methods for Polyploid Radiation Hybrid Mapping. *Genome Research*,5, pp. 136–150.

D.K. Slonim (1996): Building Human Genome Maps with Radiation Hybrids. Chapter 4 in *Learning from Imperfect Data in Theory and Practice*, Ph.D. Thesis, Department of EECS, MIT, Cambridge, Massachusetts.

- Linkage analysis:

E.S. Lander and P. Green (1987): Construction of multilocus genetic linkage maps in humans. *Proceedings of the National Academy of Sciences USA*, 84, pp. 2363–2367.

- More Hidden Markov models in computational biology and bioinformatics:

L. Bize, F. Muri, F. Samson, F. Rodolphe, S. Dusko Enrlich, B. Prum and P. Bessires: Searching gene transfers on Bacillus subtilis using hidden Markov models. *Proceedings of the third annual conference on computational molecular Biology.* 1999. pp. 43–49.

G.A. Churchill (1995): Accurate Restoration of DNA Sequences. *Case Studies in Bayesian Statistics II; Lecture Notes in Statistics*, LNS 105, Springer Verlag 1995, pp. 90–125.

E.M. Crowley, K. Roeder, M. Bina (1997): A Statistical Model for Locating Regulatory Regions in Genomic DNA. *Journal of Molecular Biology*, pp. 8–14.

A. Krogh, B. Larsson, G.v. Heijne and E.L.L. Sonnhammer (1998): Predicting Transmembrane Protein Topology with a Hidden Markov Model: Application to Complete Genomes. *Journal of Molecular Biology*, 305, pp. 567–580.

- Hidden semi-Markov models:

S.E. Levinson (1986): Continuously variable duration hidden Markov models for automatic speech recognition. *Computer, Speech and Language*, 1, pp. 29 - 45.

Chapter 12

Left to Right HMM for Sequences

12.1 Introduction

12.1.1 Contents

This chapter presents in mathematical terms some of the main features of the hidden Markov models invoked in modelling (protein) sequence families and in multiple alignment. The techniques are introduced in (Krogh et al. 1994, Baldi et al. 1994, Baldi and Chauvin 1994, White et al. 1994). These constitute a generalization of a method known as *profile analysis* (Gribskov et al. 1987) and are therefore often called *profile* HMM's. The books (Baldi and Brunak 1994) and (Durbin et al. 1998) contain specialized treatments of profile hidden Markov models by some of the original contributors. Brief surveys are found in (Eddy 1996, Krogh and Hughey 1996, Krogh 1999).

In the present chapter profile HMM is introduced in terms of a simple case of *motif-based hidden Markov models* (Grundy et al. 1997)[1] offered as a solution to finding approximate common substrings in a set of strings. This problem can be formulated in a completely non-biological manner as that of finding one or several (or none) approximate shared substrings in a number of sequences. The biological interpretation of the problem is that the approximate common substrings are conserved motifs common to a family

[1] `http://metameme.sdsc.edu/`

We introduce a 'terminating random walk' or an absorbing Markov chain $\{X_n\}$ that has a special configuration of $\square_i$:s and $\diamond_i$:s as its state space. The graph in Figure 12.1 shows a hidden Markov model for the **ACS** problem:

- A state shown as a rectangular box $\square$ in Figure 12.1 is a **match** state. These states correspond to the symbols in the substring $\mathbf{s}$.
- A state shown by a diamond shaped box $\diamondsuit$ in Figure 12.1 models **insertions** of random symbols between two occurrences of the common substring. These states have self loops (transitions from the state back to itself).

The arrows show the moves in the graph corresponding to non-zero transition probabilities. The chain moves linearly between two match states with probability one. In the case depicted in Figure 12.1 the common substring has the length $W = 3$ and will occur twice. We introduce the symbol

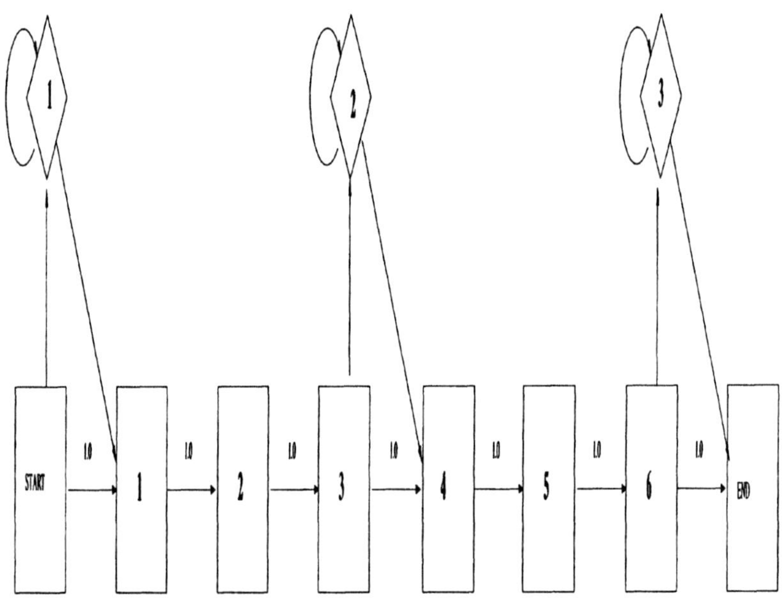

Figure 12.1: An HMM for ACS

derlying idea is already treated in the section on probabilistic modelling of motifs in Chapter 4 above.

The **ACS** problem is formulated as follows: we are given a set of sequences or fragments $\mathbf{o}^{(l)}$ of symbols from $\mathcal{O}$. The sequences are essentially random except that in each of them is imbedded one possibly different approximate copy of some *unknown string* $\mathbf{s}$ of length W (or perhaps none).

The copies of $\mathbf{s}$ are approximate in the sense that none or some symbols have been changed to some other symbol by a random process which allows for mutations but does not include *insertions* or *deletions* of symbols inside $\mathbf{s}$. The length W of the common substring is known in advance and the same for all sequences $\mathbf{o}^{(l)}$.

The biological motivation of the **ACS** problem by the following considerations (Tateno et al. 1997). An eukaryotic gene is in general divided into subregions, introns, and exons and a protein is composed of domains and modules. Those subregions rather than the entire region are known to essentially determine the biological functions of the gene or the protein. In particular, motifs have been regarded as playing a significant role in the functioning of the protein in which they reside. By **motif** is meant in particular the a specific arrangement of evolutionarily conserved amino acids in a protein. Motifs are expected to be related to biological functions. *Kimura's neutral theory of molecular evolution* (Li 1997) holds that as the biological importance of a region in a gene or a protein increases in evolution, the evolutionary pressure on the region becomes higher, making it more invariable or conserved.

We model the approximate substring $\mathbf{s}$ and the strings in which this appears as generated by a hidden Markov model. We shall eventually find the sought sequence $\mathbf{s}$ by estimating a model (or a probability distribution) for it using the sequences $\mathbf{o}^{l}$. The model is an HMM that can generate sequences with the required properties.

First we establish two different types (I)$-$(II) of probability distributions on an alphabet $\{o_1, \ldots, o_K\}$ conditional on two sets of abstract states $\Box_i$ and $\diamond_i$:

(I) $P(o_l|\Box_i)$ having low entropy.

(II) $P(o_l|\diamond_i)$ with high entropy.

$\Box_i$ and $\diamond_i$ are the positions in those segments of the sequences $\mathbf{o}$ that contain the symbols of the approximate word and the random symbols, respectively.

of biological sequences (for the notion of family see the quotation in 6.1.1 in Chapter 6) and are thought of as (short) stretches of molecules with highly similar structure interspersed in relatively unstructured sequence. The profile HMM for a sequence family is then obtained as a generalization of the motif-based hidden Markov model.

A characteristic feature of the profile models for sequence families is the socalled *left to right architecture* imposed on the transition matrix of the hidden Markov chain.

12.1.2 Left to Right HMM

Consider the Markov chain with J states and the transition matrix

$$A = \begin{pmatrix} a_{1|1} & a_{1|2} & \dots & 0 \\ 0 & a_{2|2} & \dots & 0 \\ \vdots & \vdots & \vdots & \vdots \\ 0 & 0 & \dots & 1 \end{pmatrix} \tag{1.1}$$

with the properties $a_{i|j} = 0,\ i < j,\ a_{i|j} = 0,\ j > i + \triangle$ for some positive integer $\triangle$, where $\triangle$ gives the maximum number of states that the process can jump in one step. In the profile models for sequence families $\pi_j(0) = \delta_{1,j}$, known as Kronecker's delta, so that the chain is always started in the state 1, which is called the *START state.*

The state J is called the *END state*, since $a_{J|J} = 1$ and $a_{J|i} = 0,\ i < J$. In probability terminology the state J is called an **absorbing state** . Thus some of the properties of *transient Markov chains* (Kemeny et al. 1976, ch. 5) or *A chains*, where states are either absorbing or transient, are of relevance here.

Linked to a process which emits symbols that are conditionally independent given the state sequence, this gives an HMM with a left to right architecture.

12.2 Motif-Based Hidden Markov Models

12.2.1 The Approximate Common Substring Problem

The **approximate common substring(ACS)** problem is recapitulated here according to (Bailey 1995, Bailey and Elkan 1995). In fact the un-

$$T = \min\{n \geq 1 | X_n = END\}, \quad \text{i.e., the time to hit END}$$

and from Figure 12.1 the state space

$$\mathcal{S} = \{START, \diamond_1, \Box_1, \Box_2, \Box_3, \Box_2, \Box_4, \Box_5, \Box_6, \diamond_3, END\}.$$

The chain starts always in $START$ and moves probabilistically from left to right. The $START$ state and the END state emit no symbols.

We link the Markov chain to the distributions $P(o_l|\Box_i)$ and $P(o_l|\diamond_i)$ by conditional independence.

The transitions in the state graph are not accessible to us, but we observe a sequence $\{o_{l_n}\}_{n=0}^{T}$. The approximate common substrings are generated by emissions during the sojourns in the match states $\Box_i$. This platform for solution of **ACS** is in other words a sequence generator.

Once methods for learning, scoring and aligning are known, we can find the common substring in the ACS problem in the sense that we have estimates of $P(o_l|\Box_i)$, and for each $\mathbf{o}^{(l)}$ we have the best state path telling us where we should locate the approximate string in each $\mathbf{o}^{(l)}$.

One can build a model for several different approximate common substrings or motifs by inserting a new blocks of modules of given in Figure 12.1. The HMM for the **ACS** above is a special case of the full profile HMM.

12.2.2 Profile HMM

By design the profile HMM should detect distant similarities between sequences invoking an imbedded model of the processes in molecular evolution. The sequences forming a family (defined by some criteria, see Section 6.1.1 in Chapter 6) can be seen as being related to an original sequence that has undergone random insertion, deletion and mutation at certain positions in the course of the evolution (Li 1997).

To describe the profile HMMindex profile HMM let us imagine consensus sequences or common substrings, generically denoted by $\mathbf{s}$. We can think of reading the string $\mathbf{s}$ from left to right and at each position of $\mathbf{s}$ deciding whether there is an insertion of a new symbol, a deletion of the present symbol or a mutation/retention of the symbol in $\mathbf{s}$. A similar idea is implemented in (Allison et.al 1992).

This is formalized in a state space graph arranged linearly according to the positions. Each position has a **module** with three states, each with its own emission properties over the alphabet $\mathcal{O}$. A module in any position is

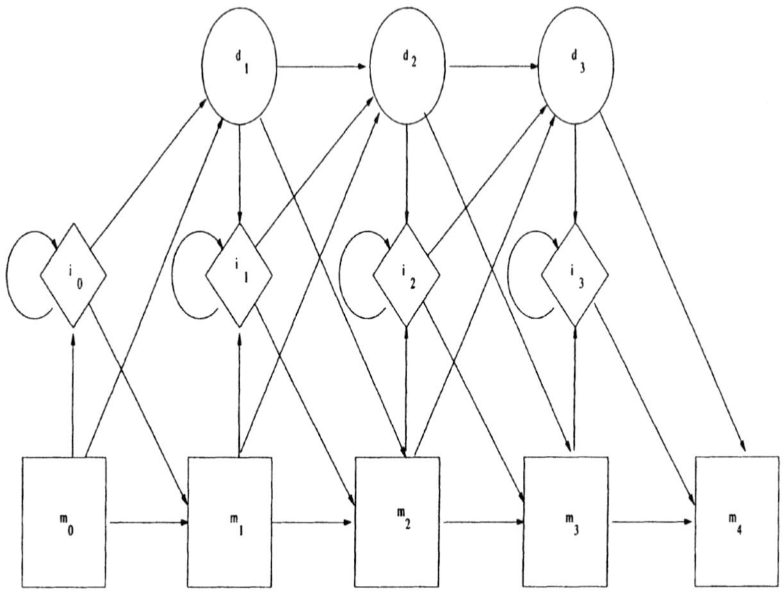

Figure 12.2: The architecture of a profile HMM

given as follows:

- A state shown as a rectangular box □ in Figure 12.2 is a **match** state. These states correspond to the symbols in the consensus sequence **s** in the sense of emitting with a high probability the respective symbol in **s**. The emission probabilities for these states are the counterparts of the profile probability distributions in Chapter 6.

- A state shown by a diamond shaped box ◇ in Figure 12.2 models **insertions** of random symbols between two positions, i.e., two symbols in the consensus sequence. These states have self loops. Each insertion state has a distinct emission probability distribution.

- A state shown by a circle ○ in Figure 12.2 models a **deletion** (relative to the consensus sequence), corresponding to skipping a symbol in **s**. No symbol is emitted from these states.

We can build a model for several different (conserved) motifs or approximate common substrings by inserting new modules of the kind shown in Figure 12.2. This architecture has been incorporated in gene finding in (Krogh et al. 1994).

States of neighboring positions (modules) in Figure 12.2 are connected by arrows, which depict the permitted (non-zero) transitions in the state graph. Transitions into match or delete states always move forward in the model, whereas transitions into insert states do not. Multiple insertions between every one of the match states can occur, unlike in the ACS architecture, since the self loop on the insert state allows transition from the insert state to itself.

A hidden Markov chain $\{X_n\}$ moves from left to right in the graph of Figure 12.2 through the modules $[\Box_i, \bigcirc_i, \diamond_i]$ according to the transitions shown.

12.2.3 Plan 7 in HMMER

A variation of the profile HMM discussed in the preceding subsection is found in (Eddy 1998) and depicted in Figure 12.3. In this architecture there are no transitions from delete states to insert states or from insert states to delete states. Thus the number of non-zero transition probabilities in a module, corresponding to the arrows actually depicted, is reduced to seven, which explains the attribute 7 in the name of the architecture. Plan 7 architecture in (Eddy 1998) contains other novelties. Perhaps the most important of these from a probability calculus point of view is that there is a state, a small circle $\circ$ after the main body of modules that permits also a transition back to the first of the match states.

12.3 Multiple Alignment by profile HMM

Let us suppose that we have a set with r strings or sequences $\mathbf{o}^{(l)}$ of symbols from the alphabet $\mathcal{O}$. In case a profile HMM with an architecture built by means of blocks of modules as defined in Figure 12.2 has been already been estimated or will be trained using the r sequences $\mathbf{o}^{(l)}$ using a learning algorithm, we can solve the decoding problem, i.e, we can find for each sequence $\mathbf{o}^{(l)}$ the best path from START to END, in which each symbol of a sequence is related to a successive match or insertion state along the path.

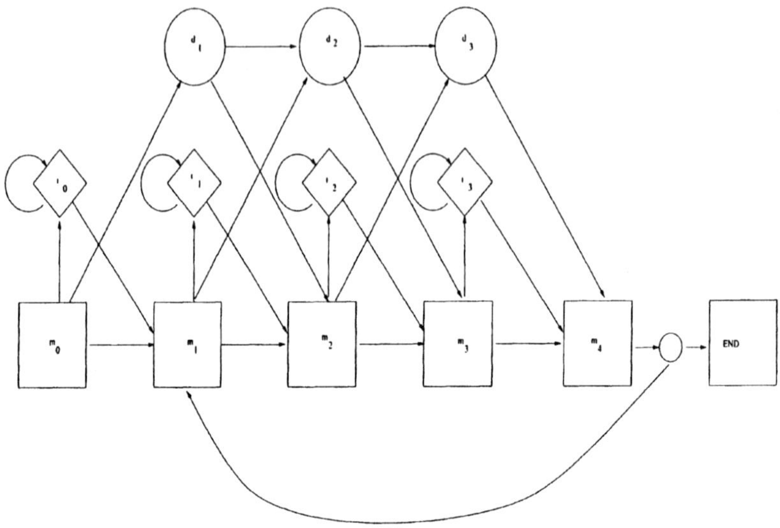

Figure 12.3: Plan 7

The transitions between the states are constrained by the connecting lines in the architecture. These best paths can be arrayed as a multiple alignment, in the sense of (2.2) in Chapter 6.

The best path is in this context called **alignment of a sequence to the HMM**. A multiple alignment is obtained with the format $\mathcal{A}$ in (2.2) in Chapter 6 by arranging the sequences $\mathbf{o}^{(l)}$, $l = 1, \ldots, r$, so that the line of match states defines the basic columns and distributing $'-'$ on those positions in the sequences that do not use the match states in the best path (cf. Chapter 5). This makes no distinction between insertion and deletion. If needed, the positions of insertions can be marked, too, e.g., by lowercase symbols like a, t, c, g. Then $'-'$ is added to those sequences that make no insertions on those positions. For results of applying these methods to biological sequence families we refer to (Baldi et al. 1994, Grundy et al. 1997, Hughey and Krogh 1996, Krogh et al. 1994).

12.4 Scoring with HMM

12.4.1 HMM Alignment to a Profile

We look at

$$\begin{aligned}\min_{j_0 \cdots j_n} - \log P\left(Y_0\right. &= \left. o_0 \ldots, Y_n = o_n, X_0 = j_0, \ldots, X_n = j_n; \lambda\right) \\ &= \min_{j_0 \cdots j_n} \sum_{l=1}^{n} \left[- \log b_{j_l}\left(l\right) - \log a_{j_{l-1}|j_l} \right].\end{aligned}$$

This is similar to aligning a sequence to a profile as discussed in Chapter 6. The term $-\log b_{j_l}(l)$ represents the penalty for aligning the letter o_{j_l} to the position represented by the state j_l in the model. The term $-\log a_{j_{l-1}|j_l}$ represents the penalty for using the transition from j_{l-1} to j_l. If $j_{l-1} \mapsto j_l$ is a transition from a match state to a delete state or from a match state to an insert state, then $-\log a_{j_{l-1}|j_l}$ represents a **gap open** penalty. If it is from delete state to a delete state or an insertion state to an insertion state then it is **gap extend** penalty (cf. Chapter 5).

Then it can be seen that the probabilistic score for a gap of length k is with the subindices from Figure 12.2

$$\log a_{m_j|i_j} + (k-1) \log a_{i_j|i_j} + \log a_{i_j|m_{j+1}},$$

which can be regarded as an affine gap penalty (cf. Chapter 5).

Hence

$$\min_{j_0 \cdots j_n} - \log P\left(Y_0 = o_0 \ldots, Y_n = o_n, X_0 = j_0, \ldots, X_n = j_n; \lambda\right)$$

can be used as a score for identifying a sequence to a family. An advantage of the profile HMM is that all these penalties depend on the position in the model and can be learned from the available sequences.

12.4.2 Log Odds Score

In any statistical method of discriminating between a family of sequences and those sequences that are not in the family the key quantity is the probability that a given sequence was generated by a model for the family. This can also be done for HMM via the forward-backward procedure.

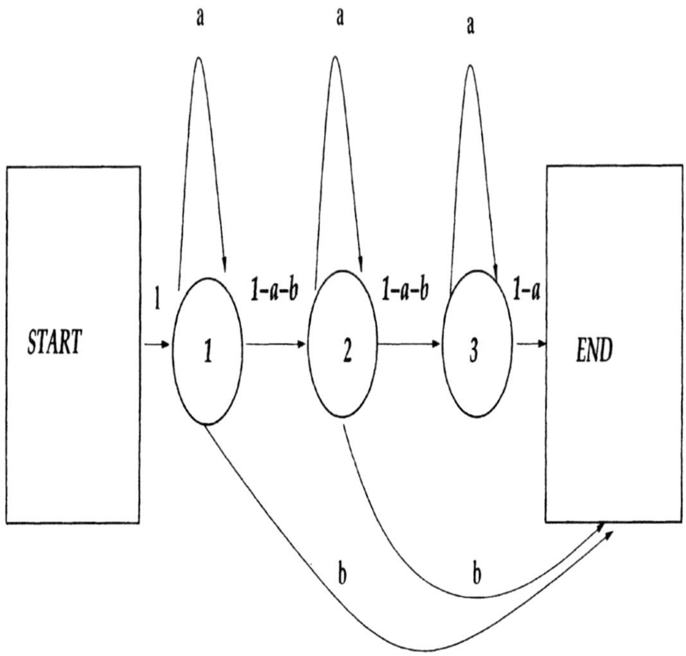

Figure 12.4: A Markov chain for β-strand

The chain will eventually leave the transient state without ever returning to it. Examples of this are all other but the END states in the left to right architectures.

Definition 12.5.2 Let the Markov chain $\{X_n\}_{n=0}^{\infty}$ have the transition matrix P and a finite state space. Then

$$T_j = \min\{n > 0 : X_n = j\}$$

is the *first return time* to the state j, if $X_0 = j$. ■

Let

$$f_{j|j}(n) = P\left(X_1 \neq j, X_2 \neq j, \ldots, X_{n-1} \neq j, X_n = j | X_0 = j\right)$$

Hence $\lambda_\emptyset$ is a very simple special case of profile HMM architecture.

12.5 Fragment Lengths

12.5.1 Transience

We discuss next left to right Markov chains without focusing exclusively on the models described above. Let A be the stationary transition matrix of a Markov chain with a finite state space $S = \{1, \ldots, J\}$. To give a motivation for the focus of this section we start with an example found in the bioinformatics literature.

Example 12.5.1 Lengths of Protein Segments
In (White et al. 1994) an HMM is constructed for predicting the structural classes of proteins from their amino acid sequences. This involves the taxonomy of the protein tertiary structures that are often defined according to the occurrence of segments that are called α-helices, α-loops and β-strands. An important empirical parameter is the length of the various segments of the protein.

In (White et al. 1994) a β-strand is given a hidden state space of the form given in Figure 12.4. The domains of the protein superclasses (see Section 6.1.1 in Chapter 6) are then modeled by combining modules of β-strands with modules of the other segments (not displayed here). The state space in Figure 12.4 has clearly a left to right architecture, where the means and variances of the length of the generated sequences can be computed by the techniques to be introduced in this section. By tuning in the parameters in an appropriate manner desired means and standard deviations for β-strand lengths are obtained. ∎

Definition 12.5.1 If there is a state i such that $a_{i|i} = 1$, then we say that the state is **absorbing**. ∎

If the chain enters an absorbing state, then it cannot leave it (is absorbed there) and we regard the Markov chain as terminated. The *END* states of the left to right architectures are examples of termination.

If there is at least one absorbing state, then some states must be **transient**. Such is a state j with the property that

$$P(X_n = j \text{ for some } n > 0 | X_0 = j) < 1.$$

For a profile HMM it is known that the log likelihood scores $-\log P(\mathbf{o}; \lambda)$ are dependent on the length of $\mathbf{o}$ and on the number of modules (match states) in the profile HMM. Therefore the the log likelihood scores cannot be directly used to discriminate a sequence.

In *log odds scoring* the discrimination is based on a normalization of the negative log likelihood score, which means a comparison to a null hypothesis, usually a simpler statistical model, which is intended to represent a whole set of sequences without reference to the subset comprised of a specific family of biological sequences.

Thus if $\mathbf{o} = o_0 \ldots o_n$ is a sequence of symbols in $\mathcal{O}$ and λ is an HMM, the *log odds score* is

$$\text{score}(\mathbf{o}) = \log \frac{P(\mathbf{o}; \lambda)}{P(\mathbf{o}; \lambda_\emptyset)}, \tag{4.1}$$

where

$$P(\mathbf{o}; \lambda_\emptyset) = P(Y_0 = o_0, \ldots, Y_n = o_n; \lambda_\emptyset)$$

is the probability of $\mathbf{o}$ with respect to a null model (and analogously for $P(\mathbf{o}; \lambda)$). The question is the appropriate choice of a null model $\lambda_\emptyset$. A discussion of this issue is found in (Barrett et al. 1997).

12.4.3 A Null Model for Scoring

In HMMER User's Guide (Eddy 1998) S. Eddy proposes the following null model $\lambda_\emptyset$. A Markov chain $\{X_l\}_{l=0}^{\infty}$ has the state space $\{G, F\}$ and the following transition matrix

$$A_\emptyset = \begin{pmatrix} 1-\alpha & \alpha \\ 0 & 1 \end{pmatrix}. \tag{4.2}$$

The chain starts always in state G and the state F is the END state. The state F is silent (no symbols are emitted). For all l with $X_l = G$, a symbol o_l from a finite alphabet $\mathcal{O}$ is emitted. The emission probability distribution is denoted by

$$B_\emptyset = \{b(o_k)\}_{k=1}^{K}$$

so that $P(Y_l = o_k) = b(o_k)$. For example we can have $K = 4$ and $b(o_k) = \frac{1}{4}$, or in the case of amino acids, we may choose the emission distribution from some database for proteins. We assume that the emission variables $\{Y_l\}_{l=0}^{\infty}$ are conditionally independent given a state sequence. Let us now set

$$\lambda_\emptyset = (A_\emptyset, B_\emptyset).$$

be the probability for that a chain starting at j will return to j for the first time at time n. Then we have clearly that the expected time of return is

$$E\left[T_j | X_0 = j\right] = \sum_{n=1}^{\infty} n \cdot f_{j|j}(n). \tag{5.1}$$

Also

$$P\left(X_n = j \text{ for some } n > 0 | X_0 = j\right) = \sum_{n=1}^{\infty} f_{j|j}(n). \tag{5.2}$$

If a Markov chain is irreducible (as defined in section 7.4.3 of Chapter 7) and has a finite state space, then

$$P\left(X_n = j \text{ for some } n > 0 | X_0 = j\right) = 1$$

for all j and the states are called *recurrent.*

12.5.2 The Probability Distribution for The Length of Generated Sequence

Let $\{X_n\}_{n=0}^{\infty}$ be a Markov chain with a finite state space which has only one absorbing state and the rest of states are transient. Such chains are sometimes called A chains. The results to be presented here constitute a simplified version of Chapter 5 in (Kemeny et al. 1976).

We renumber, if necessary, S so that the absorbing state is state J. These chains are ergodic in a degenerate sense, since

$$\lim_{n \to \infty} P\left(X_n = J\right) = 1. \tag{5.3}$$

The proof for this is omitted. We define *the time to absorption* as

$$T = \min\{n : X_n = J\}.$$

We drop the subscript from the notation, since in this section there is no risk of confusion. The distribution of T is obviously also the distribution for the length of a sequence (fragment) generated by a left to right hidden Markov chain (that has no other silent states than the START state). Then the state space can be written as

$$S = S_* \cup \{J\},$$

where $S_* = \{1, 2, \ldots, J-1\}$ is the set of transient states. Let us consider the following decomposition of the transition matrix

$$A = \begin{pmatrix} P_0 & \mathbf{P}_1 \\ \mathbf{0} & 1 \end{pmatrix}, \tag{5.4}$$

where P_0 is the restriction of A to $S_* \times S_*$ and $\mathbf{P}_1$ is column vector containing the transition probabilities to the absorbing state. In Example 12.5.1 we obtain, omitting the start state, from Figure 12.4 the matrix

$$P_0 = \begin{pmatrix} a & 1-a-b & 0 \\ 0 & a & 1-a-b \\ 0 & 0 & a \end{pmatrix}$$

and

$$\mathbf{0} = \begin{pmatrix} 0 & 0 & 0 \end{pmatrix}, \mathbf{P}_1 = \begin{pmatrix} b & b & 1-a \end{pmatrix}^T,$$

where ^{T} denotes transposition of a row to a column.

Proposition 12.5.2 *Let $\{X_n\}_{n=0}^{\infty}$ be a Markov chain with a finite state space which has only one absorbing state J and the rest being transient. Then the time to absorption T has the distribution*

$$P(T = k | X_0 = i) = \left[P_0^{k-1}\mathbf{P}_1\right]_i, \tag{5.5}$$

where $i \in S_*$ and $[x]_i$ denotes the ith component in the column vector x.

Proof: Let us straightforwardly compute the desired distribution by marginalization and by the chain rule (2.16) in Chapter 1:

$$\begin{aligned} P(T = k | X_0 = i) &= \sum_{j=1}^{J} P(T = k | X_{k-1} = j, X_0 = i) P(X_{k-1} = j \mid X_0 = i) \\ &= \sum_{j=1}^{J} P(T = k | X_{k-1} = j, X_0 = i) \left[A^{k-1}\right]_{i|j} \\ &= \sum_{j=1}^{J} P(X_k = J | X_{k-1} = j, X_0 = i) \left[A^{k-1}\right]_{i|j}, \end{aligned}$$

where $\left[A^{k-1}\right]_{i|j}$ is the element on position i,j in A^{k-1}. The Markov property gives

$$\begin{aligned}\sum_{j=1}^{J} P(X_k &= J|X_{k-1}=j, X_0=i)\left[A^{k-1}\right]_{i|j} \\ &= \sum_{j=1}^{J} P(X_k=J|X_{k-1}=j)\left[A^{k-1}\right]_{i|j} = \sum_{j=1}^{J-1} a_{j|J}\left[A^{k-1}\right]_{i|j},\end{aligned}$$

since $P(X_k=J|X_{k-1}=J)=0$, as the chain is in this case terminated already at time $k-1$. But for $j=1,\ldots,J-1$ we have $a_{j|J}=[\mathbf{P}_1]_j$. It can be verified by an induction that

$$A^k = \begin{pmatrix} P_0^k & Q_k \\ \mathbf{0} & 1 \end{pmatrix},$$

where $Q_k = \left(\sum_{j=0}^{k-1} P_0^j\right)\mathbf{P}_1$. Hence we have for $i,j \in S_*$

$$\left[A^{k-1}\right]_{i|j} = \left[P_0^{k-1}\right]_{i|j}.$$

Thus

$$P(T=k|X_0=i) = \sum_{j=1}^{J-1}\left[P_0^{k-1}\right]_{i|j}[\mathbf{P}_1]_j,$$

as claimed. ∎

12.5.3 Expected Length of a Generated Sequence

We deal with the same A chain as in the preceding subsection. T is the time to absorption $T=\min\{n : X_n = J\}$. Let B_j be

B_j = the number of times the chain visits the state j before being absorbed.

Then, since $T = B_1 + B_2 + \ldots + B_{J-1}$,

$$m_{ij} = E\left[B_j|X_0=i\right]$$

where j and i are in S_*. Then

$$E\left[T|X_0=i\right] = m_{i1} + m_{i2} + \ldots + m_{iJ-1}.$$

We define the $J-1 \times J-1$ matrix M by

$$M = (m_{ij})_{i,j=1}^{J,J}.$$

Proposition 12.5.3

$$M = (I - P_0)^{-1}, \tag{5.6}$$

where I the $J-1 \times J-1$ identity matrix.

Proof: Fix i and j in S_*. Let us define the indicator function

$$I_{\{X_n=j\}} = \begin{cases} 1 & \text{if } X_n = j, \\ 0 & \text{if } X_n \neq j. \end{cases} \tag{5.7}$$

Then we see that

$$B_j = \sum_{n=0}^{\infty} I_{\{X_n=j\}}$$

and by homogeneity in time

$$\begin{aligned} m_{ij} = E\left[B_j|X_0 = i\right] &= \sum_{k=1}^{J} E\left[B_j|X_1 = k, X_0 = i\right] \cdot P\left(X_1 = k|X_0 = i\right) \\ &= \sum_{k=1}^{J} E\left[B_j|X_1 = k, X_0 = i\right] \cdot a_{i|k}. \end{aligned}$$

But we have that

$$\begin{aligned} E\left[B_j|X_1 = k, X_0 = i\right] &= E\left[\sum_{n=0}^{\infty} I_{\{X_n=j\}}|X_1 = k, X_0 = i\right] \\ &= E\left[I_{\{X_0=j\}}|X_1 = k, X_0 = i\right] \\ &\quad + E\left[\sum_{n=1}^{\infty} I_{\{X_n=j\}}|X_1 = k, X_0 = i\right] \\ &= E\left[I_{\{X_0=j\}}|X_1 = k, X_0 = i\right] \\ &\quad + E\left[\sum_{n=1}^{\infty} I_{\{X_n=j\}}|X_1 = k\right], \end{aligned}$$

where we have used the Markov property in the second term. By computing expectations of indicator functions we have

$$E\left[I_{\{X_0=j\}}|X_1=k,X_0=i\right]=P\left[X_0=j|X_1=k,X_0=i\right].$$

Here $P\left[X_0=j|X_1=k,X_0=i\right]$ is equal to 1, if $i=j$ and is equal to zero, if $i\neq j$, which we write as

$$P\left[X_0=j|X_1=k,X_0=i\right]=\delta_{ij}\quad \text{Kronecker's delta.}$$

Furthermore we have

$$E\left[\sum_{n=1}^{\infty} I_{\{X_n=j\}}|X_1=k\right]=m_{kj}.$$

(Note that this follows since the chain has time homogeneous transitions). In summary we have obtained for $i,j\in S_*$ that

$$m_{ij}=\delta_{ij}+\sum_{k=1}^{J-1} m_{kj}\cdot a_{i|k},$$

since $m_{kj}=0$ if $k=J$, since if we start in the absorbing state we stay there. In matrix form this is

$$M=I+P_0M,$$

and the solution is at least formally $M=\left(I-P_0\right)^{-1}$. ∎

12.5.4 Quasi stationarity?

An A chain is ergodic in the degenerate sense in (5.3). However, for many A chains there exists a *quasi stationary distribution* that models these chains as stationary over any reasonable time scale.

We are still concerned with the same generic A chain as in the preceding sections. We let T be the time to absorption $T=\min\{n:X_n=J\}$.

Then a **normalized quasi stationary distribution** for the A chain $\{X_n\}_{n=0}^{\infty}$ is a probability distribution $h(i)$ on S_* with the property that

$$P\left(X_n=i\mid T>n\right)=h(i), i\in S_*, \tag{5.8}$$

when $P\left(X_0=i\right)=h(i), i\in S_*$. This implies that the chain with the initial distribution h has the following stationarity property: the conditional distribution of $X_n\ldots,X_{n+r}$ given that $T>n+m+r$ is the same for all $n>0$.

The impossibility of deriving the existence of a quasi stationary distribution for the left to right architecture simply by referring to the known necessary and sufficient conditions (Kesten 1995, p. 656) is because in a left to right architecture P_0 is not irreducible (except in a trivial case), as the states are not communicating.

There is a close connection between quasi stationarity and the existence of the **Yaglom limit**, which is the limit of the ratio

$$\frac{P(X_k = j \mid X_0 = i)}{P(T > k \mid X_0 = i)}$$

with $i, j \in S_*$ as k increases to ∞. In the HMM architecture we have $X_0 = 1$ with probability one. The study of the Yaglom limit is given as a project in the exercises.

12.6 Exercises

12.6.1 ACS

1. Draw the *trellis* corresponding to the hidden Markov model for an approximate common substring problem given in Figure 12.1.

12.6.2 Duration of Stay and Absorption

2. Let $\{X_n\}_{n=0}^{\infty}$ be a Markov chain with a finite state space and the transition matrix

$$A = \left(a_{i|j}\right)_{i=1,j=1}^{J,J}.$$

   (a) Compute probability of the time of *duration of stay* in state i i.e. the probability $\phi(n)$ for the event that the chain stays in the state i for exactly n steps after having started in state i. Formally we thus have

$$\phi(n) = P(X_1 = i, X_2 = i, \ldots, X_n = i, X_{n+1} \neq i | X_0 = i).$$

   (b) Compute the expected duration of stay in state i given that the chain started in that state.

3. A Markov chain with the state space $\{0, 1\}$ has the transition matrix

$$A = \begin{pmatrix} 1-\alpha & \alpha \\ 0 & 1 \end{pmatrix}. \tag{6.9}$$

   (a) Let T be *the time to absorption* for this chain. Compute the distribution

$$P(T = k | X_0 = 0), \qquad k \geq 1.$$

   (b) Compute

$$E[T | X_0 = 0].$$

   using *two different methods.*

4. A Markov chain with the 5 states $S = \{1, 2, 3, 4, 5\}$ has the transition matrix

$$A = \begin{pmatrix} p & q & 0 & 0 & 0 \\ 0 & p & q & 0 & 0 \\ 0 & 0 & p & q & 0 \\ 0 & 0 & 0 & p & q \\ 0 & 0 & 0 & 0 & 1 \end{pmatrix}, \tag{6.10}$$

where $p + q = 1$. Compute the expected time to absorption given that $X_0 = 1$, i.e.,

$$E[T | X_0 = 1].$$

5. Consider a Markov chain with J states and the transition matrix

$$A = \begin{pmatrix} p & q & 0 & \dots & 0 \\ 0 & p & q & \dots & 0 \\ 0 & \vdots & \dots & \vdots & 0 \\ 0 & \vdots & \dots & p & q \\ 0 & \dots & \dots & 0 & 1 \end{pmatrix}, \tag{6.11}$$

where $p + q = 1$. Compute the expected time to absorption given that $X_0 = 1$ i.e.

$$E[T | X_0 = 1].$$

6. Verify by straightforward summation that

$$\sum_{k=1}^{\infty} P(T = k \mid X_0 = i) = 1$$

for $i \in S_*$.

12.6.3 A Project on the Yaglom Limit

The state space is written as

$$S = S_* \cup \{J\},$$

where S_* is the set of transient states. From the proof of Proposition 12.5.2 we note that

$$A^k = \begin{pmatrix} P_0^k & Q_k \\ \mathbf{0} & 1 \end{pmatrix}.$$

Let $\mathbf{e}_1$ be the following $1 \times J - 1$ vector

$$\mathbf{e}_1 = (1, 0, \ldots, 0),$$

let $\mathbf{e}$ be the following $J - 1 \times 1$ vector of ones

$$\mathbf{e} = (1, 1, 1, \ldots, 1)^T$$

and let

$$\mathbf{f}_j = (0, \ldots, 1, \ldots 0)^T$$

be a $J - 1 \times 1$ vector with one on position j and zeros in the rest of the positions. Then we note that

$$\mathbf{P}_1 = (I - P_0)\,\mathbf{e},$$

where I is the $J - 1 \times J - 1$ identity matrix, since all row sums are equal to one. Thus we write

$$\mathbf{e}_1 P_0^k \mathbf{e} = \mathbf{e}_1 (I - Q_k)\,\mathbf{e} \tag{6.12}$$

as the probability of the event that the chain has not been terminated at step k after having started in state 1. In addition we obtain

$$\mathbf{e}_1 P_0^k \mathbf{f}_j \tag{6.13}$$

as the probability of hitting the state $j \in S_*$ at step k having started in state 1. Thus it holds that the ratio in the Yaglom limit is written as

$$\frac{P(X_k = j \mid X_0 = 1)}{P(T > k \mid X_0 = 1)} = \frac{\mathbf{e}_1 P_0^k \mathbf{f}_j}{\mathbf{e}_1 P_0^k \mathbf{e}} \tag{6.14}$$

Then, if P_0 is *irreducible and non-cyclic*, it is easily seen using matrix calculus that

$$P_0^k = \rho_1^k \mathbf{w}\mathbf{v}^T + O\left(k^{m-1} \mid \rho_2 \mid^k\right), \tag{6.15}$$

and

$$\mathbf{e}_1 P_0^k \mathbf{e} = O\left(\rho_1^k\right), \tag{6.16}$$

where ρ_1 and ρ_2 are the largest and second largest left eigenvalues of P_0, respectively, and m is the multiplicity of ρ_2. Also, $\mathbf{v}$ is the left eigenvector corresponding to ρ_1 and $\mathbf{w}$ is the right eigenvector corresponding to the largest right eigenvalue of P_0. Then the vector

$$\mathbf{w}\mathbf{v}^T$$

whose elements are of the form $w_i v_i$, $i = 1, \ldots, J-1$ is the quasi stationary distribution.

The details of proofs of these assertions are found in (Bhat 1962, Darroch and Seneta 1965, Matthews 1970).

7. (a) Find the Yaglom limit in Exercise 3 above. Give the quasi stationary distribution in (5.8), too.

   (b) In (Grundy 1998) one finds a completely connected Meta-MEME model here depicted in a reduced form in Figure 12.5. Here the motifs are not connected in a particular order, instead there is a complete set of connections and self loops between the motifs. Within each rectangle there is a left to right model. The edges between the motifs are taken as insert states. This kind of architecture can be used to represent shuffling of modules. Can the properties (6.15) and (6.16) be made to hold true in in this kind of completely connected network of motifs?

8. A transition probability matrix A is of the form

$$A = \begin{pmatrix} P_0 & \mathbf{P}_1 \\ \mathbf{0} & 1 \end{pmatrix}, \tag{6.17}$$

where

$$P_0 = \begin{pmatrix} p_1 & q_1 & 0 \\ 0 & p_2 & q_2 \\ p_3 & 0 & 0 \end{pmatrix}, \tag{6.18}$$

$p_i + q_i = 1$, $i = 1, 2, 3$, $\mathbf{P}_1^T = (0, 0, q_3)$ and $\mathbf{0}$ is a row vector with three zeros as elements. Here the left to right architecture has been modified as in Plan 7 in (Eddy 1998), see Figure 12.3.

Investigate P_0^k to see if there is a quasi stationary distribution as found in (6.15). If this fails, explain the reason.

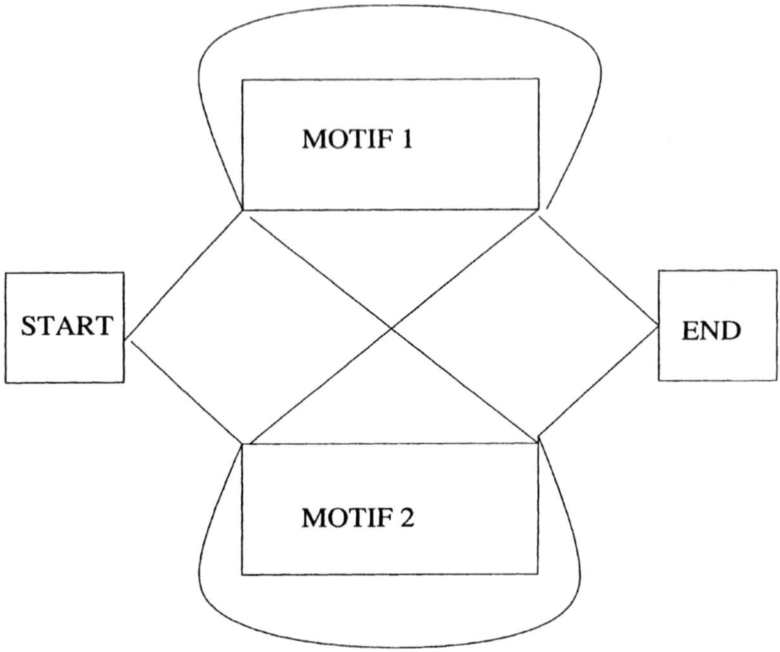

Figure 12.5: A completely connected architecture

12.6.4 Expected length

9. Compute the expected length of the β-strands generated by the architecture in example 12.5.1 above.

10. **Null Model for Scoring**
We consider the model in section 12.4.3 above. Let $T_\emptyset$ be the time to absorption of the chain $\{X_l\}_{l=0}^{\infty}$. In other words, the random variable $T_\emptyset$ counts *the number of transitions in* G and the transition to the state F. Then if $\mathbf{o} = o_0 \ldots o_n$ is given sequence of $n+1$ symbols from $\mathcal{O} = \{o_1, \ldots, o_K\}$, then the probability for this string to be generated by $\lambda_\emptyset$ is given by

$$P(\mathbf{o}; \lambda_\emptyset) = P(Y_0 = o_0, \ldots, Y_n = o_n | T_\emptyset = n+1; \lambda_\emptyset) \cdot P(T_\emptyset = n+1; \lambda_\emptyset)$$

(a) Compute

$$\log P(\mathbf{o}; \lambda_{\emptyset}).$$

(b) Consider

$$\mathcal{L} = -\log P(Y_0 = o_0, \ldots, Y_n = o_n | T_{\emptyset} = n + 1; \lambda_{\emptyset}).$$

Compute the expectation and variance

$$E_{\emptyset}[\mathcal{L}] \text{ and } \mathrm{Var}_{\emptyset}[\mathcal{L}],$$

where $E_{\emptyset}$ and $\mathrm{Var}_{\emptyset}$ denote the expectation and the variance, respectively, with respect to the distribution $\lambda_{\emptyset}$. Verify that these are effectively proportional to the length of the string $\mathbf{o}$. The final answer may contain non-explicit quantities such as the Shannon entropy of $B_{\emptyset}$.

12.7 References and Further Reading:

- Profile Hidden Markov models:

P. Baldi, Y. Chauvin, T. Hunkapiller and M.A. McClure (1994): Hidden Markov Models of biological primary sequence information. *Proceedings of the National Academy of Science of USA*. 91, February, pp. 1059–1063.

P. Baldi and S. Brunak (1998): *Bioinformatics. The Machine Learning Approach.* A Bradford Book. The MIT Press. Cambridge, Massachusetts; London.

P. Baldi and Y. Chauvin (1994): Hidden Markov Models of the G-Protein-Coupled Receptor Family. *Journal of Computational Biology*, 1, pp. 311–336.

R. Durbin, S. Eddy, A. Krogh and G. Mitchison (1998): *Biological sequence analysis. Probabilistic models of proteins and nucleic acids.* Cambridge University Press. Cambridge.

S. Eddy (1996): Hidden Markov Models. *Current Opinion in Structural Biology*, 6, pp. 361–365.

S. Eddy (1998): HMMER User's Guide. Biological sequence analysis using profile hidden Markov models.
`http://hmmer.wustl.edu/`.

W.N. Grundy, T.L. Bailey, C.P. Elkan and M.E. Baker (1997): Meta–MEME: Motif-based hidden Markov models of protein families. *Computer Applications in Biological Sciences* (CABIOS), vol. 13, pp. 397–406.

W.N. Grundy (1998): *A Bayesian Approach to Motif-based Protein Modeling.* Ph.D. Thesis in Computer and Cognitive Science, University of California, San Diego,
`http://www.cs.columbia.edu/ bgrundy/`.

R. Hughey and A. Krogh (1996): Hidden Markov models for sequence analysis: extensions and analysis of the basic method. *Computer Applications in Biological Sciences* (CABIOS), 12, pp. 95–107.

A. Krogh, M. Brown, I. Saira Mian, K. Sjölander and D. Haussler (1994): Hidden Markov Models in Computational Biology: Applications to Protein Modelling. *Journal of Molecular Biology*, 235, pp. 1501–1531.

A. Krogh, I.S. Mian and D. Haussler (1994): A hidden Markov model that finds genes in *E. Coli* DNA. *Nucleic Acids Research*, 22, pp. 4768–4778.

A. Krogh (1999): An introduction to hidden Markov models for biological sequences. In S.L. Salzberg, D.B. Searls and S. Kasif: *Computational Methods in Molecular Biology*, Elsevier, Amsterdam etc.

J.V. White, C.M. Schultz and T.F. Smith (1994): Protein Classification by Stochastic Modelling and Optimal Filtering of Amino-Acid Sequences. *Mathematical Biosciences*, 119, pp. 35–75.

- Profiles, Motifs and Alignments

L. Allison, C.S. Wallace and C.N. Lee (1992): Finite-State Models in the Alignment of Macromolecules. *Journal of Molecular Evolution*, 35, pp. 77–89.

T. L. Bailey (1995): *Discovering motifs in DNA and protein sequences: The approximate common substring problem.* Ph.D. Thesis, Department of Computer Science and Engineering, University of University of California, San Diego.

T. L. Bailey and C. Elkan (1995): Unsupervised Learning of Multiple Motifs in Biopolymers using EM, *Machine Learning*, 21(1-2): pp. 51–80.

M. Gribskov, A.D. McLachlan and D. Eisenberg (1987): Profile Analysis, detection of distantly related proteins. *Proceedings of the National Academy of Sciences USA*, 84, pp. 4355–4358.

D. Gusfield (1997): *Algorithms on Strings, Trees and Sequences. Computer Science and Computational Biology.* Cambridge University Press. Cambridge.

Y. Tateno, K. Ikeo, T. Imanishi, H. Watanabel, T. Endo, Y. Yamaguchi, Y. Suzuki, K. Takahashi, K. Tsunoyama, M. Kawai, Y. Kawanishi, K. Naitou and T. Gojobori (1997): Evolutionary Motif and Its Biological and Structural Significance. *Journal of Molecular Evolution*, 44, pp. 38–43

M.S. Waterman (1995): *Introduction to Computational Biology. Maps, sequences and genomes.* Chapman and Hall, London, Glasgow, Weinheim, New York, Melbourne, Madras.

- Transient Markov chains

  J.G. Kemeny, J.L. Sell and A.W. Knapp (1976): *Denumerable Markov Chains.* Second Edition. Springer Verlag, New York et al. .

- Quasi stationarity

  B.R. Bhat (1962): Some Properties of Regular Markov Chains. *Annals of Mathematical Statistics*, 32, pp. 59–71.

  J.N. Darroch and E. Seneta (1965): On Quasi-stationary Distributions in Absorbing Discrete-Time Finite Markov Chains. *Journal of Applied Probability*, 2, pp. 88 - 100.

  H. Kesten (1995): A ratio limit theorem for (sub) Markov chains. *Advances in Applied Probability*, 27, pp. 652–691.

  J. Matthews (1970): A central limit theorem for absorbing Markov chains. *Biometrika*, 57, pp. 129–139.

- Scoring

  C. Barrett, R. Hughey and K. Karplus (1997): Scoring Hidden Markov Models. *Computer Applications in Biological Sciences* (CABIOS), 13, pp. 191 - 197.

  S. Eddy (1998): HMMER User's Guide. Biological sequence analysis using profile hidden Markov models.
  `http://hmmer.wustl.edu/`.

- Kimura's neutral theory of evolution

  W-H. Li (1997): *Molecular Evolution.* Sinauer Associates, Inc., Publishers, Sunderland Massachusetts.

Chapter 13

Derin's Algorithm

13.1 Introduction

Any mathematically precise treatment of the scoring, alignment, and training problems of HMM as stated in the preceding requires a few auxiliary factorizations and conditional independence statements inherent in a standard HMM to be made explicit. These results are consequences of the Markov property and the postulated conditional independence of emission variables given a sequence of states of the hidden Markov chain.

We exploit these properties to prove *Derin's backwards recursive formula for smoothing posterior probabilities* (Askar and Derin 1981), which is a first instance of an algorithmic probability connected with HMM to be presented in this text. Additional such schemes will be based on the forward and backward variables established in this chapter. These assertions about conditional distributions yield, also, sufficient conditions for a function of a Markov chain to be a Markov chain. The results will also be needed in the analysis of consistency of the maximum likelihood estimate for HMM in Chapter 17 below. The present chapter has a technical nature. The statements are in most cases intuitively obvious, and can be visualized by suitably separating the graph in Figure 13.1 but some of the proofs, if carried out in detail, are laborious and can be omitted at a first reading.

The main trick in the computations to follow is to calculate a probability by including suitable extra random variables in the argument of a probability so as to obtain a desired probability by marginalization. In this chapter we shall follow (Lucke 1996) and use the notation $\bigotimes$ in several of the marginal-

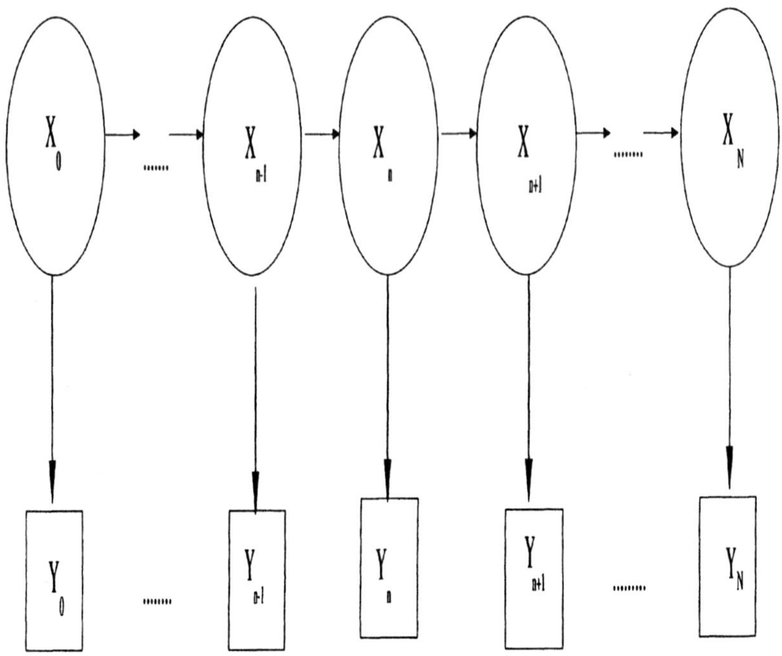

Figure 13.1: HMM: Influence diagram

izations when these require summations over a set of variables. In addition, the chain rule (2.16) in Chapter 1 is used in various extended forms, sometimes without explicit reference.

13.2 Derin's Formula

The goal is to find a backwards recursion with $n \leq m$ for the *smoothing posterior probability* defined as

$$\widehat{\pi}_j\left(n|m\right) = P\left(X_n = j | Y_0 = o_0, \ldots, Y_m = o_m\right) \tag{2.1}$$

for a standard HMM. The result below is found in (Askar and Derin 1981) and seems to be known as Derin's formula (see, e.g., Devijver 1985).

Proposition 13.2.1 *For* $n = 0, \ldots, N-1$

$$\widehat{\pi}_j(n|N) = \widehat{\pi}_j(n|n) \cdot \sum_{k=1}^{J} \frac{a_{j|k}}{\widehat{\pi}_k(n+1|n)} \widehat{\pi}_k(n+1|N). \tag{2.2}$$

■

This proposition will be proved in Section 13.7 using the results on conditional distributions established in the sections preceding Section 13.7. A forward recursion for the *filtering posterior probability* $\widehat{\pi}_j(n|n)$ is given in Exercise 3 in equation (8.1) and a forward recursion for the *prediction posterior probability* $\widehat{\pi}_k(n+1|n)$ in Exercise 3 in (8.2).

Thus there is a computational scheme in (2.2): $\widehat{\pi}_j(n|n)$ and $\widehat{\pi}_k(n+1|n)$ are computed recursively forwards and then stored to compute $\widehat{\pi}_j(n|N)$ recursively backwards. This has also bearing on a numerical implementation issue to be made explicit later. Derin's algorithm works with quantities normalized at each step that eliminate the underflow problem inherent in Baum's forward–backward procedure for computation of likelihoods.

In (Guédon and Cocozza-Thivent 1990) it is shown that Derin's formula can be extended to derive a learning algorithm for HMM with duration (hidden semi-Markov Model).

13.3 Various Sufficiency Properties

Here the typesetting is simplified, e.g., by writing a conditional probability

$$P(Y_m = o_m, \ldots, Y_N = o_N \mid X_n = j_n \ldots, X_N = j_N)$$

simply as

$$P(Y_m, \ldots, Y_N \mid X_n, \ldots, X_N)$$

without a large risk of confusion in the proofs to follow. Thus conditional independence becomes

$$P(Y_0, \ldots, Y_n \mid X_0, \ldots, X_n) = \prod_{i=0}^{n} P(Y_i \mid X_i). \tag{3.1}$$

The next two propositions in this section have the common feature of showing that the probability of a finite length emitted sequence conditioned on a state sequence of the hidden Markov chain depends only on a subsequence of the state sequence. The relevant subsequence can be seen as a sufficient reduction of the hidden Markov chain. The general idea of sufficiency is defined and described in exercises to Chapter 4 above. The properties are implicit in a statement of some five lines' length in the basic work (Baum et al. 1970) and are made explicitly available in (MacDonald and Zucchini 1997). Precautions with regard to some conditioning event having zero probability are not explicitly taken. In those cases both sides of the expressions will be taken as zero.

Proposition 13.3.1 *For all integers n and m such that $0 \leq n \leq m \leq N$*

$$P\left(Y_m, \ldots, Y_N \mid X_n \ldots, X_N\right) = P\left(Y_m, \ldots, Y_N \mid X_m, \ldots, X_N\right). \tag{3.2}$$

Proof: The left hand side of (3.2) can be expressed as

$$\frac{1}{P\left(X_n, \ldots, X_N\right)} \bigotimes P\left(Y_m, \ldots, Y_N \mid X_0 \ldots, X_N\right) \cdot P\left(X_0 \ldots, X_N\right),$$

where $\bigotimes$ designates the summation over $j_0, \ldots, j_{n-1}$ (which are the values of $X_0, \ldots, X_{n-1}$). If $n = 0$ there is no summation. By assumption (3.1) (and a marginalization argument)

$$P\left(Y_m, \ldots, Y_N \mid X_0 \ldots, X_N\right) = P\left(Y_m \mid X_m\right) \cdot \ldots \cdot P\left(Y_N \mid X_N\right).$$

This factor can be taken outside the summation sign $\bigotimes$, since $m \geq n$. This leaves us with

$$\prod_{l=m}^{N} P\left(Y_l \mid X_l\right) \frac{1}{P\left(X_n, \ldots, X_N\right)} \bigotimes P\left(X_0 \ldots, X_N\right),$$

where the sum is equal to $P\left(X_n, \ldots, X_N\right)$, since we are summing over $j_0, \ldots, j_{n-1}$. Thus the whole last expression is equal to

$$= \prod_{l=m}^{N} P\left(Y_l \mid X_l\right),$$

which is independent of n. Since the right hand side of (3.2) is a special case of the left hand side for $n = m$, this proves the assertion as claimed. ∎

We shall also in the rest of the proofs write for convenience of expression

$$X^{(t)} = \left(X_0, \ldots, X_t\right), Y^{(t)} = \left(Y_0, \ldots, Y_t\right),$$

where we use again the convention of dropping the $X_n = j_n$'s and $Y_n = o_n$'s.

Proposition 13.3.2 *For all integers $n = 0, \ldots, N-1$*

$$P(Y_{n+1}, \ldots, Y_N \mid X_0, \ldots, X_n) = P(Y_{n+1}, \ldots, Y_N \mid X_n). \tag{3.3}$$

Proof: The left hand side of (3.3) is

$$\frac{1}{P(X^{(n)})} \bigotimes P\left(X^{(N)}\right) \cdot P\left(Y_{n+1}, \ldots, Y_N \mid X^{(N)}\right), \tag{3.4}$$

where the summation $\bigotimes$ is over $j_{n+1}, \ldots, j_N$. By Proposition 13.3.1, equation (3.2) (with $m = n+1$, $n = 0$), we have

$$P\left(Y_{n+1}, \ldots, Y_N \mid X^{(N)}\right) = P(Y_{n+1}, \ldots, Y_N \mid X_{n+1}, \ldots, X_N),$$

and using the same proposition and equation once more (with $m = n+1$) we have

$$P(Y_{n+1}, \ldots, Y_N \mid X_{n+1}, \ldots, X_N) = P(Y_{n+1}, \ldots, Y_N \mid X_n, \ldots, X_N).$$

Thus we are dealing with

$$\begin{aligned} & \bigotimes P\left(X^{(N)}\right) \cdot P\left(Y_{n+1}, \ldots, Y_N \mid X^{(N)}\right) \\ = & \bigotimes P\left(X^{(N)}\right) P(Y_{n+1}, \ldots, Y_N \mid X_n, \ldots, X_N). \end{aligned}$$

By conditional probability $P\left(X^{(N)}\right) = P\left(X_{n+1}, \ldots, X_N | X^{(n)}\right) \cdot P\left(X^{(n)}\right)$ and by a consequence of *Markov property* (see Exercise 5.5.1 (c) in Chapter 7) we have

$$P\left(X_{n+1}, \ldots, X_N | X^{(n)}\right) = P(X_{n+1}, \ldots, X_N | X_n).$$

Thus the sum is, by an extension of the chain rule (2.16) in Chapter 1, equal to

$$\begin{aligned} & \bigotimes P\left(X^{(N)}\right) \cdot P(Y_{n+1}, \ldots, Y_N \mid X_n, \ldots, X_N) \\ = & \bigotimes \frac{P(Y_{n+1}, \ldots, Y_N, X_n, \ldots, X_N) \cdot P\left(X^{(n)}\right)}{P(X_n)}, \end{aligned} \tag{3.5}$$

and since we are in $\bigotimes$ summing over $j_{n+1}, \ldots, j_N$ we obtain this as

$$P\left(X^{(n)}\right) \bigotimes \frac{P(Y_{n+1}, \ldots, Y_N, X_n, \ldots, X_N)}{P(X_n)}$$

and

$$\begin{aligned}\bigotimes \frac{P(Y_{n+1},\dots,Y_N,X_n,\dots,X_N)}{P(X_n)} &= \frac{P(Y_{n+1},\dots,Y_N,X_n)}{P(X_n)} \\ &= P(Y_{n+1},\dots,Y_N|X_n).\end{aligned}$$

Thus in (3.4) we have found that

$$\begin{aligned}&\frac{1}{P(X^{(n)})}\bigotimes P\left(X^{(N)}\right)\cdot P\left(Y_{n+1},\dots,Y_N \mid X^{(N)}\right) \\ = &\frac{1}{P(X^{(n)})}P(Y_{n+1},\dots,Y_N|X_n)\cdot P\left(X^{(n)}\right),\end{aligned}$$

which proves the assertion. ∎

Proposition 13.3.3 *For all integers $n = 0,\dots,N$*

$$P(Y_0,\dots,Y_n \mid X_0,\dots,X_N) = P(Y_0,\dots,Y_n \mid X_0,\dots,X_n). \tag{3.6}$$

Proof: The left hand side of (3.6) is, with our notational conventions, written as

$$P\left(Y_0,\dots,Y_n \mid X^{(N)}\right) = \bigotimes P\left(Y_0,\dots,Y_N \mid X^{(N)}\right),$$

where the sum $\bigotimes$ is over $o_{n+1}, o_{n+2},\dots,o_N$. But then the conditional independence assumption (3.1) gives

$$\bigotimes P\left(Y_0,\dots,Y_N \mid X^{(N)}\right) = \prod_{l=0}^{n} P(Y_l|X_l) \bigotimes \prod_{l=n+1}^{N} P(Y_l|X_l)$$

where the summation gives $\bigotimes \prod_{l=n+1}^{N} P(Y_l|X_l) = 1$ or

$$P\left(Y_0,\dots,Y_n \mid X^{(N)}\right) = \prod_{l=0}^{n} P(Y_l|X_l),$$

as was to be proved. ∎

13.4 Lumping of Markov Chains

It was observed in an exercise of Chapter 10 that the emitted sequence of an HMM can be seen as a function of a Markov chain. Next, using a result obtained in the preceding section, we study the question of whether a function of a Markov chain is another Markov chain.

Now let g be a map from S onto the alphabet $\mathcal{O} = \{o_1, \ldots, o_K\}$. The process $\{Y_n\}_{n=0}^{\infty}$ is defined by

$$Y_n = g(X_n). \tag{4.7}$$

If $g(\cdot)$ is not one-to-one, i.e.,$K < J$, we are obviously lumping (or aggregating) the state space of the Markov chain. Lumping of the state space is often of interest in probabilistic modelling of DNA sequences, for example, $\{A, T, C, G\}$ can be lumped into the purine-pyriminide groups $\{A/G, C/T\}$ (Raftery and Tavaré 1994).

Simple examples show that $\{Y_n\}_{n=0}^{\infty}$ is not, in general, a Markov chain. There has been a considerable research in the characterization of those functions of Markov chains that are themselves Markov chains. The following proposition (Kelly 1982) gives a straightforward and natural set of sufficient conditions for functions of Markov chains to be Markovian functions. An extra process

$$Z_n = \phi(X_n) \tag{4.8}$$

will play the role of a kind of *sufficient statistic* in the sense that it is assumed to contain all the information in X_n relevant to the next state of the process Y_n.

Proposition 13.4.1 *Let* $\{X_n\}_{n=0}^{\infty} \in \text{Markov}(P, p_{X_0})$. *If*

$$P(Y_{n+1} = o_{n+1} \mid X_n = j) = P(Y_{n+1} = o_{n+1} \mid Z_n = \phi(j)) \tag{4.9}$$

and

$$P(Z_n = z_n \mid Y_n, \ldots, Y_0) = P(Z_n = z_n \mid Y_n), \tag{4.10}$$

then $\{Y_n = g(X_n)\}_{n=0}^{\infty}$ *is a Markov chain.*

Proof: The Markov property will be verified. By marginalization and the chain rule of conditional probability we get

$$P(Y_{n+1} \mid Y_n, \ldots, Y_0) = \sum_{z_n} P(Y_{n+1}, Z_n = z_n \mid Y_n, \ldots, Y_0)$$

$$= \sum_{z_n} P(Y_{n+1} \mid Z_n = z_n, Y_n, \ldots, Y_0) \cdot P(Z_n = z_n \mid Y_n, \ldots, Y_0).$$

Let us again use marginalization, to the effect that

$$P(Y_{n+1} \mid Z_n = z_n, Y_n, \ldots, Y_0) = \bigotimes P(Y_{n+1}, X_n, \ldots, X_0 \mid Z_n = z_n, Y_n, \ldots, Y_0) \qquad (4.11)$$

where the summation $\bigotimes$ is over $j_0, \ldots, j_n$ (the values of $X_0, \ldots, X_n$). Then the right hand side is equal to

$$\bigotimes P(Y_{n+1}|X_n, \ldots, X_0, Z_n = z_n, Y_n, \ldots, Y_0)\, P(X_n, \ldots, X_0 \mid Z_n = z_n, Y_n, \ldots, Y_0).$$

But since $Z_n = z_n, Y_n, \ldots, Y_0$ are, in view of (4.7) and (4.8), no longer random but exactly computable given $X_n, \ldots, X_0$, we have

$$P(Y_{n+1}|X_n, \ldots, X_0, Z_n = z_n, Y_n, \ldots, Y_0) = P(Y_{n+1}|X_n, \ldots, X_0).$$

Next we note that a function of a Markov chain satisfies the conditional independence property (3.1) in a trivial manner, as both sides of the equality are simultaneously either one or zero. Thus we are permitted to use (3.3) to obtain

$$P(Y_{n+1}|X_n, \ldots, X_0) = P(Y_{n+1}|X_n)$$

Then we use (4.9), so that

$$P(Y_{n+1}|X_n) = P(Y_{n+1} \mid Z_n = z_n).$$

But then

$$\bigotimes P(Y_{n+1}|X_n, \ldots, X_0, Z_n = z_n, Y_n, \ldots, Y_0)\, P(X_n, \ldots, X_0 \mid Z_n = z_n, Y_n, \ldots, Y_0)$$

$$\begin{aligned} &= P(Y_{n+1}|Z_n = z_n,) \cdot \bigotimes P(X_n, \ldots, X_0 \mid Z_n = z_n, Y_n, \ldots, Y_0) \\ &= P(Y_{n+1}|Z_n = z_n,). \end{aligned}$$

Thus we have obtained that

$$P(Y_{n+1} \mid Z_n = z_n, Y_n \ldots, Y_0) = P(Y_{n+1} \mid Z_n = z_n) = P(Y_{n+1} \mid Z_n = z_n, Y_n).$$

Using this and the condition (4.10) it follows that

$$\begin{aligned} & P\left(Y_{n+1} \mid Y_n, \ldots, Y_0\right) \\ = & \sum_{z_n} P\left(Y_{n+1} \mid Z_n = z_n, Y_n\right) \cdot P\left(Z_n = z_n \mid Y_n\right) \\ = & P\left(Y_{n+1} \mid Y_n\right), \end{aligned}$$

which was to be proved. ■

If Y_n and Z_n are functions of each other, $Y_n \equiv Z_n$, then (4.9) and (4.10) are reduced to one simple sufficient condition

$$P\left(Y_{n+1} = o_{n+1} \mid X_n = j\right) = P\left(Y_{n+1} = o_{n+1} \mid Y_n = g(j)\right). \tag{4.12}$$

13.5 Factorizations

The statements in this section are factorizations which demonstrate a kind of renewal structure for the HMM. For example, Proposition 13.5.1 below shows that if at some instants we actually were able to see the state of the hidden chain, then the past and future of the emitted process are independent, conditioned on this one value of the Markov chain. Most of the proofs are based on (MacDonald and Zucchini 1997).

Proposition 13.5.1 *For all integers* $n = 0, \ldots, N$

$$P\left(Y_0, Y_1, \ldots, Y_N \mid X_n\right) = P\left(Y_0, Y_1, \ldots, Y_n \mid X_n\right) \cdot P\left(Y_{n+1}, \ldots, Y_N \mid X_n\right). \tag{5.1}$$

Proof: The left hand side of (5.1) is, by definition of conditional expectation, equal to

$$P\left(Y^{(N)} \mid X_n\right) = \frac{P\left(Y^{(N)}, X_n\right)}{P\left(X_n\right)}.$$

Here we again obtain by marginalization that

$$P\left(Y^{(N)}, X_n\right) = \bigotimes_1 \bigotimes_2 P\left(Y_0, \ldots, Y_N, X_0, \ldots, X_{n-1}, X_n, X_{n+1}, \ldots, X_N\right), \tag{5.2}$$

where $\bigotimes_1$ is a summation over $j_0, \ldots, j_{n-1}$ and $\bigotimes_2$ is in its turn a summation over $j_{n+1}, \ldots, j_N$. The definition of conditional probability tells that

$$P\left(Y_0, \ldots, Y_N, X_0, \ldots, X_{n-1}, X_n, X_{n+1}, \ldots, X_N\right)$$

$$= P\left(Y_0, \ldots, Y_N | X^{(N)}\right) \cdot P\left(X^{(N)}\right)$$

and by the basic assumption (3.1) we have

$$P\left(Y_0, \ldots, Y_N | X^{(N)}\right) = P\left(Y_0, \ldots, Y_n | X^{(n)}\right) \cdot P\left(Y_{n+1}, \ldots, Y_N | X_{n+1}, \ldots, X_N\right).$$

Next we obtain from (3.2) in Proposition 13.3.1 (set $m = n + 1$) that

$$P\left(Y_{n+1}, \ldots, Y_N | X_{n+1}, \ldots, X_N\right) = P\left(Y_{n+1}, \ldots, Y_N | X_n, X_{n+1}, \ldots, X_N\right).$$

By conditional probability and the Markov property, see Exercise 5.5.1 (c) in Chapter 7, it follows that

$$\begin{aligned} P\left(X^{(N)}\right) &= P\left(X^{(n)}\right) P\left(X_{n+1}, \ldots, X_N | X^{(n)}\right) \\ &= P\left(X^{(n)}\right) P\left(X_{n+1}, \ldots, X_N | X_n\right). \end{aligned}$$

Inserting from the preceding we obtain

$$P\left(Y_0, Y_1, \ldots, Y_N | X^{(N)}\right) \cdot P\left(X^{(N)}\right) = T_1 \cdot T_2,$$

where

$$\begin{aligned} T_1 &= P\left(Y_0, \ldots, Y_n | X^{(n)}\right) P\left(X^{(n)}\right) \\ T_2 &= P\left(Y_{n+1}, \ldots, Y_N | X_n, X_{n+1}, \ldots, X_N\right) P\left(X_{n+1}, \ldots, X_N | X_n\right). \end{aligned}$$

Using an extension of the chain rule (2.16) in Chapter 1 we then rewrite T_2 to obtain

$$T_1 \cdot T_2 = P\left(Y_0, \ldots, Y_n | X^{(n)}\right) P\left(X^{(n)}\right) \cdot P\left(Y_{n+1}, \ldots, Y_N, X_{n+1}, \ldots, X_N | X_n\right)$$

Therefore in (5.2) we have, since $\bigotimes_2$ is a summation over $j_{n+1}, \ldots, j_N$,

$$\begin{aligned} &\bigotimes_1 \bigotimes_2 P\left(Y_0, \ldots, Y_N, X_0, \ldots, X_{n-1}, X_n, X_{n+1}, \ldots, X_N\right) \\ &= \bigotimes_1 P\left(Y_0, \ldots, Y_n | X^{(n)}\right) P\left(X^{(n)}\right) \\ &\quad \bigotimes_2 P\left(Y_{n+1}, \ldots, Y_N, X_{n+1}, \ldots, X_N | X_n\right) \\ &= \bigotimes_1 P\left(Y_0, \ldots, Y_n | X^{(n)}\right) \cdot P\left(X^{(n)}\right) \cdot P\left(Y_{n+1}, \ldots, Y_N | X_n\right) \end{aligned}$$

and therefore

$$P\left(Y^{(N)} \mid X_n\right) = \frac{\bigotimes_1 P(Y_0,\ldots,Y_n,X_0,\ldots,X_n)\cdot P(Y_{n+1},\ldots,Y_N|X_n)}{P(X_n)}. \quad (5.3)$$

By summing over $j_0,\ldots,j_{n-1}$, in (5.3) we obtain

$$\begin{aligned} &\bigotimes_1 P(Y_0,\ldots,Y_n,X_0,\ldots,X_n)\cdot P(Y_{n+1},\ldots,Y_N|X_n) \\ &= P(Y_0,\ldots,Y_n,X_n)\cdot P(Y_{n+1},\ldots,Y_N|X_n) \\ &= P(Y_0,\ldots,Y_n|X_n)\cdot P(X_n)\cdot P(Y_{n+1},\ldots,Y_N|X_n), \end{aligned} \quad (5.4)$$

and after division by $P(X_n)$ this gives $P\left(Y^{(N)},X_n\right)/P(X_n)$ as

$$P(Y_0,\ldots,Y_n|X_n)\cdot P(Y_{n+1},\ldots,Y_N|X_n),$$

which is equal to the right hand side of (5.1), as was claimed. ∎

The conditional probability $P(Y_{n+1},\ldots,Y_N \mid X_n)$ will be later called the **backward variable**. The next proposition is used to find a recursion for this backward variable.

Proposition 13.5.2 *For all integers* $n = 0,\ldots,N$

$$P(Y_n,Y_{n+1},\ldots,Y_N \mid X_n) = P(Y_n \mid X_n)\cdot P(Y_{n+1},\ldots,Y_N \mid X_n). \quad (5.5)$$

Proof: This is obtained from (5.1) by summing over $o_0,o_1,\ldots,o_{n-1}$. ∎

The following proposition is a variant of Proposition 13.5.1 and is used in the proof of Derin's backwards recursion below.

Proposition 13.5.3 *For all integers* $n = 0,\ldots,N-1$

$$P(Y_0,\ldots,Y_N \mid X_{n+1}) = P(Y_0,\ldots,Y_n \mid X_{n+1})\,P(Y_{n+1},\ldots,Y_N \mid X_{n+1}). \quad (5.6)$$

Proof: The proof is a simple modification of the proof of Proposition 13.5.1 above. We start with

$$P\left(Y^{(N)} \mid X_{n+1}\right) = \frac{P\left(Y^{(N)},X_{n+1}\right)}{P(X_{n+1})}$$

to obtain by marginalization

$$P\left(Y^{(N)},X_n\right) = \bigotimes_1\bigotimes_2 P(Y_0,\ldots,Y_N,X_0,\ldots,X_n,X_{n+1},X_{n+2},\ldots,X_N),$$

where $\bigotimes_1$ is now a summation over $j_0, \ldots, j_n$ and $\bigotimes_2$ is a summation over $j_{n+2}, \ldots, j_N$. As in the proof above we obtain

$$P\left(Y_0, Y_1, \ldots, Y_N | X^{(N)}\right) \cdot P\left(X^{(N)}\right) = T_1 \cdot T_2$$

with

$$\begin{aligned} T_1 &= P\left(Y_0, \ldots, Y_n | X^{(n)}\right) P\left(X^{(n+1)}\right) \\ T_2 &= P\left(Y_{n+1}, \ldots, Y_N | X_{n+1}, \ldots, X_N\right) \cdot P\left(X_{n+2}, \ldots, X_N | X_{n+1}\right). \end{aligned}$$

Then

$$\begin{aligned} T_1 \cdot T_2 &= P\left(Y_0, \ldots, Y_n | X^{(n)}\right) P\left(X^{(n+1)}\right) \\ &\cdot P\left(Y_{n+1}, \ldots, Y_N, X_{n+2}, \ldots, X_N | X_{n+1}\right). \end{aligned}$$

Therefore, since $\bigotimes_2$ is a summation over $j_{n+2}, \ldots, j_N$, we have

$$\begin{aligned} &\bigotimes_1 \bigotimes_2 P\left(Y_0, \ldots, Y_N, X_0, \ldots, X_{n-1}, X_n, X_{n+1}, \ldots, X_N\right) \\ &= \bigotimes_1 P\left(Y_0, \ldots, Y_n | X^{(n)}\right) P\left(X^{(n+1)}\right) \cdot \\ &\bigotimes_2 P\left(Y_{n+1}, \ldots, Y_N, X_{n+2}, \ldots, X_N | X_{n+1}\right) \\ &= \bigotimes_1 P\left(Y_0, \ldots, Y_n | X^{(n)}\right) \cdot P\left(X^{(n+1)}\right) \cdot P\left(Y_{n+1}, \ldots, Y_N | X_{n+1}\right). \end{aligned}$$

Owing to (3.6) Proposition 13.3.3, we obtain

$$P\left(Y_0, \ldots, Y_n | X^{(n)}\right) = P\left(Y_0, \ldots, Y_n | X^{(n+1)}\right).$$

Therefore

$$P\left(Y^{(N)} \mid X_{n+1}\right) = \frac{\bigotimes_1 P\left(Y_0, .., Y_n, X_0, .., X_{n+1}\right) P\left(Y_{n+1}, .., Y_N | X_{n+1}\right)}{P\left(X_{n+1}\right)}.$$

The desired result follows by summing over $j_0, \ldots, j_n$ in the expression above as in the proof of Proposition 13.5.1. ■

The next proposition turns out to be directly instrumental in finding the properties of what will be called **forward variables**.

Proposition 13.5.4 *For all integers $n = 0, \dots, N-1$*

$$P(Y_0, \dots, Y_N \mid X_n, X_{n+1}) = P(Y_0, \dots Y_n \mid X_n) \cdot P(Y_{n+1}, \dots, Y_N \mid X_{n+1}). \quad (5.7)$$

Proof: We start as before by writing the left hand side of (5.7) as

$$\frac{1}{P(X_n, X_{n+1})} \bigotimes_1 \bigotimes_2 P(Y_0, Y_1, \dots, Y_N, X_0, X_1, \dots, X_n, X_{n+1}, X_{n+2} \dots, X_N),$$

where $\bigotimes_1$ is a summation over $j_0, \dots, j_n$ and $\bigotimes_2$ is a summation over $j_{n+2}, \dots, j_N$. Then, as in the proof of Proposition 13.5.1, we have

$$\begin{aligned} & P(Y_0, Y_1, \dots, Y_N, X_0, X_1, \dots, X_{n-1}, X_n, X_{n+1}, \dots, X_N) \\ & = P\left(X^{(N)}\right) \cdot P\left(Y_0, \dots, Y_n | X^{(n)}\right) \cdot P(Y_{n+1}, \dots, Y_N | X_{n+1}, \dots, X_N), \end{aligned}$$

where we have used assumption (3.1). By definition of conditional probability we get

$$\begin{aligned} & P\left(X^{(N)}\right) \cdot P\left(Y_0, \dots, Y_n | X^{(n)}\right) \\ & = P\left(X^{(n)}\right) \cdot P\left(Y_0, \dots, Y_n | X^{(n)}\right) \cdot P\left(X_{n+1}, \dots, X_N | X^{(n)}\right). \end{aligned}$$

Inserting these in the double sum we have

$$\bigotimes_1 \bigotimes_2 P(Y_0, \dots, Y_N, X_0, \dots, X_{n-1}, X_n, X_{n+1}, \dots, X_N) = S_1 \cdot S_2,$$

where

$$\begin{aligned} S_1 &= \bigotimes_1 P\left(X^{(n)}\right) \cdot P\left(Y_0, .., Y_n | X^{(n)}\right) \\ S_2 &= \bigotimes_2 P(Y_{n+1}, .., Y_N | X_{n+1}, .., X_N) P\left(X_{n+1}, .., X_N | X^{(n)}\right) \end{aligned} \quad (5.8)$$

recalling the nature of the summation $\bigotimes_2$. We evaluate next the sum S_2. We have for the generic term in $\bigotimes_2$

$$\begin{aligned} & P(Y_{n+1}, \dots, Y_N | X_{n+1}, \dots, X_N) \cdot P\left(X_{n+1}, \dots, X_N | X^{(n)}\right) \\ = & \frac{P(Y_{n+1}, \dots, Y_N, X_{n+1}, \dots, X_N) \cdot P(X_{n+1}, \dots, X_N | X_n)}{P(X_{n+1}, \dots, X_N)}, \end{aligned}$$

where we have also used the Markov property. Now

$$\frac{P(X_{n+1},\ldots,X_N|X_n)}{P(X_{n+1},\ldots,X_N)}=\frac{P(X_n,X_{n+1},\ldots,X_N)}{P(X_n)\cdot P(X_{n+1},\ldots,X_N)}$$

and the Markov property gives this as

$$\frac{P(X_n)\prod_{l=n}^{N-1}P(X_{l+1}|X_l)}{P(X_n)P(X_{n+1})\prod_{l=n+1}^{N-1}P(X_{l+1}|(X_l)}$$

which is clearly equal to

$$\frac{P(X_{n+1}|X_n)}{P(X_{n+1})}.$$

This last ratio is independent of the summation variables in $\bigotimes_2$. Hence

$$\begin{aligned}&\bigotimes_2 P(Y_{n+1},\ldots,Y_N|X_{n+1},\ldots,X_N)\cdot P\left(X_{n+1},\ldots,X_N|X^{(n)}\right)\\&=\frac{P(X_{n+1}|X_n)}{P(X_{n+1})}\bigotimes_2 P(Y_{n+1},\ldots,Y_N,X_{n+1},\ldots,X_N)\\&=\frac{P(X_{n+1}|X_n)}{P(X_{n+1})}\cdot P(Y_{n+1},\ldots,Y_N,X_{n+1}).\end{aligned}$$

This expression is also independent of the summation variable in $\bigotimes_1$. Thus

$$S_1\cdot S_2=\frac{P(X_{n+1}|X_n)P\left(Y_{n+1}^N,X_{n+1}\right)}{P(X_{n+1})}\bigotimes_1 P\left(X^{(n)}\right)P\left(Y_0,\ldots,Y_n|X^{(n)}\right).$$

Obviously, by the definition of $\bigotimes_1$,

$$\begin{aligned}\bigotimes_1 P\left(X^{(n)}\right)\cdot P\left(Y_0,\ldots,Y_n|X^{(n)}\right)&=\bigotimes_1 P(Y_0,\ldots,Y_n,X_0,X_1,\ldots,X_n)\\&=P(Y_0,\ldots,Y_n,X_n).\end{aligned}$$

In summary we have obtained the left hand side of (5.7) as

$$\frac{1}{P(X_n,X_{n+1})}\cdot P(Y_0,\ldots,Y_n,X_n)\cdot\frac{P(X_{n+1}|X_n)}{P(X_{n+1})}\cdot P(Y_{n+1},\ldots,Y_N,X_{n+1}).$$

Since

$$\frac{1}{P(X_n,X_{n+1})}\cdot\frac{P(X_{n+1}|X_n)}{P(X_{n+1})}=\frac{1}{P(X_n)\cdot P(X_{n+1})}$$

we obtain the left hand side of (5.7) as

$$\frac{P(Y_0,\ldots,Y_n,X_n)}{P(X_n)}\cdot\frac{P(Y_{n+1},\ldots,Y_N,X_{n+1})}{P(X_{n+1})},$$

which is equal to the right hand side of (5.7). ∎

13.6 Final Time Sufficiency

The next proposition, also found in (MacDonald and Zucchini 1997), will be invoked later in finding the recursion for the backward variables in both the Baum–Petrie formulation and in Devijver's rescaling found in the next chapter.

Proposition 13.6.1 *For all integers n and m such that $0 \le n \le m \le N$*

$$P(Y_m,\ldots,Y_N \mid X_n,\ldots,X_m) = P(Y_m,\ldots,Y_N \mid X_m). \tag{6.1}$$

Proof: We start again by writing the left hand side of (5.7) as

$$\frac{1}{P(X_n,\ldots,X_m)}\bigotimes_1\bigotimes_2 P\left(Y_m,\ldots,Y_N|X^{(N)}\right)\cdot P\left(X^{(N)}\right)$$

where $\bigotimes_1$ designates summation over $j_0,\ldots,j_{n-1}$, and $\bigotimes_2$ designates summation over $j_{m+1},\ldots,j_N$. Proposition 13.3.1 gives by (3.2) $(n=0)$

$$P\left(Y_m,\ldots,Y_N|X^{(N)}\right) = P(Y_m,\ldots,Y_N|X_m,\ldots,X_N).$$

This is independent of the summation variable in $\bigotimes_1$. We evaluate first the summation $\bigotimes_1$, obtaining

$$\frac{1}{P(X_n,\ldots,X_m)}\cdot\bigotimes_1 P\left(X^{(N)}\right) = \frac{P(X_n,\ldots,X_N)}{P(X_n,\ldots,X_m)}$$
$$= P(X_{m+1},\ldots,X_N|X_n,\ldots,X_m).$$

Therefore the overall sum to be computed is

$$\bigotimes_2 P(Y_m,\ldots,Y_N|X_m,\ldots,X_N)\cdot P(X_{m+1},\ldots,X_N|X_n,\ldots,X_m).$$

We use the Markov property once more to obtain

$$P\left(X_{m+1},\ldots,X_N|X_n,\ldots,X_m\right)=P\left(X_{m+1},\ldots,X_N|X_m\right),$$

so that

$$\begin{aligned}
&\bigotimes_2 P\left(Y_m,\ldots,Y_N|X_m,\ldots,X_N\right)\cdot P\left(X_{m+1},\ldots,X_N|X_n,\ldots,X_m\right)\\
&=\bigotimes_2 \frac{P\left(Y_m,\ldots,Y_N,X_m,\ldots,X_N\right)}{P\left(X_m,\ldots,X_N\right)}\frac{P\left(X_m,\ldots,X_N\right)}{P\left(X_m\right)}\\
&=\frac{1}{P\left(X_m\right)}\bigotimes_2 P\left(Y_m,\ldots,Y_N,X_m,\ldots,X_N\right).
\end{aligned}$$

Since $\bigotimes_2$ is a summation over the values of $X_{m+1},\ldots,X_N$, we obtain

$$\frac{1}{P\left(X_m\right)}\bigotimes_2 P\left(Y_m,\ldots,Y_N,X_m,\ldots,X_N\right)=\frac{1}{P\left(X_m\right)}P\left(Y_m,\ldots,Y_N,X_m\right),$$

which is the assertion as claimed. ∎

13.7 Proof of Derin's Backwards Recursion

We finally prove Proposition 13.2.1. In order to simplify notation we write the conditional probability in (2.1) as

$$\widehat{\pi}_j\left(n|N\right)=P\left(X_n=j|Y^{(N)}\right).$$

By another marginalization and an extended version of the chain rule (2.16) in Chapter 1 we obtain

$$\begin{aligned}
\widehat{\pi}_j\left(n|N\right) &= \sum_{k=1}^{J} P\left(X_n=j,X_{n+1}=k|Y^{(N)}\right)\\
&= \sum_{k=1}^{J} P\left(X_n=j|X_{n+1}=k,Y^{(N)}\right)P\left(X_{n+1}=k|Y^{(N)}\right)\\
&= \sum_{k=1}^{J} P\left(X_n=j|X_{n+1}=k,Y^{(N)}\right)\widehat{\pi}_k\left(n+1|N\right)\\
&= \sum_{k=1}^{J} \frac{P\left(X_n=j,X_{n+1}=k,Y^{(N)}\right)}{P\left(X_{n+1}=k,Y^{(N)}\right)}\widehat{\pi}_k\left(n+1|N\right). \qquad (7.2)
\end{aligned}$$

In the numerator inside the summation in (7.2) we have

$$\begin{aligned}
&P\left(X_n = j, X_{n+1} = k, Y^{(N)}\right)\\
&= P\left(Y^{(N)} | X_n = j, X_{n+1} = k\right) P\left(X_n = j, X_{n+1} = k\right)\\
&= P\left(Y^{(n)} | X_n = j\right) P\left(Y_{n+1}, \ldots, Y_N | X_{n+1} = k\right) \cdot P\left(X_n = j\right) \cdot a_{j|k}
\end{aligned}$$

using the factorization (5.1) in Proposition 13.5.1 and the definition of $a_{j|k}$. Then, since $P\left(Y^{(n)}|X_n = j\right) P\left(X_n = j\right) = P\left(X_n = j|Y^{(n)}\right) P\left(Y^{(n)}\right)$, we have obtained

$$\begin{aligned}
&P\left(X_n = j, X_{n+1} = k, Y^{(N)}\right)\\
&= P\left(X_n = j | Y^{(n)}\right) P\left(Y^{(n)}\right) P\left(Y_{n+1}, \ldots, Y_N | X_{n+1} = k\right) \cdot a_{j|k}. \quad (7.3)
\end{aligned}$$

For the denominator inside the summation in (7.2) it holds that

$$\begin{aligned}
&P\left(X_{n+1} = k, Y^{(N)}\right) = P\left(Y^{(N)} | X_{n+1} = k\right) P\left(X_{n+1} = k\right)\\
= \;&P\left(Y^{(n)} | X_{n+1} = k\right) P\left(Y_{n+1}, \ldots, Y_N | X_{n+1} = k\right) P\left(X_{n+1} = k\right), (7.4)
\end{aligned}$$

where we have invoked (5.6) in Proposition 13.5.3. Next, from the identity

$$P\left(Y^{(n)} | X_{n+1} = k\right) P\left(X_{n+1} = k\right) = P\left(X_{n+1} = k | Y^{(n)}\right) P\left(Y^{(n)}\right)$$

we obtain

$$\begin{aligned}
&P\left(X_{n+1} = k, Y^{(N)}\right)\\
&= P\left(X_{n+1} = k | Y^{(n)}\right) P\left(Y^{(n)}\right) P\left(Y_{n+1}, \ldots, Y_N | X_{n+1} = k\right). \quad (7.5)
\end{aligned}$$

Therefore we obtain in from (7.2)–(7.5) and (2.1) that

$$\begin{aligned}
\widehat{\pi}_j (n|N) &= \sum_{k=1}^{J} \frac{P\left(X_n = j | Y^{(n)}\right) \cdot a_{j|k}}{P\left(X_{n+1} = k | Y^{(n)}\right)} \widehat{\pi}_k (n+1|N)\\
&= \widehat{\pi}_j (n|n) \cdot \sum_{k=1}^{J} \frac{a_{j|k}}{\widehat{\pi}_k (n+1|n)} \widehat{\pi}_k (n+1|N),
\end{aligned}$$

which is the result claimed in (2.2). ∎

13.8 Exercises:

In the exercises below a fixed HMM model λ is assumed but this is omitted from the notation.

1. Consider the burst noise in Gilbert's channel, Example 3.2 in Chapter 10.

   (a) Show that for any $o_{j_l} \in \{0,1\}$, $l = 0, 1, \ldots, n-1, n+1, \ldots, N$

   $$\begin{aligned} P(Y_{n+1} &= o_{j_{n+1}}, \ldots, Y_N = o_{j_N} | Y_0 = o_{j_0}, \ldots, Y_n = 1) \\ &= P(Y_{n+1} = o_{j_0}, \ldots, Y_N = o_{j_N} | Y_n = 1). \end{aligned}$$

   (b) Show that for any $o_{j_l} \in \{0,1\}$, $l = 0, 1, .., n-1, n+1, .., N$

   $$\begin{aligned} P(Y_{N+1} = & \; o_{j_{N+1}} | Y_0 = o_{j_0}, .., Y_n = 1, Y_{n+1} = o_{j_{n+1}}, .., Y_N = o_{j_N}) \\ = & \; P\left(Y_{N+1} = o_{j_{N+1}} | Y_n = 1, Y_{n+1} = o_{j_{n+1}}, .., Y_N = o_{j_N}\right). \end{aligned}$$

2. Compute the probability of error-free run in Gilbert's channel or the probability

   $$P(Y_1 = 0, \ldots, Y_{n-1} = 0, Y_n = 1 \mid Y_0 = 1)$$

   for the burst noise.

3. We present here forward recursions for filtering and prediction posterior probabilities. The notation is given in (2.1).

   (a) **Filtering posterior probability for an HMM**
   Show that

   $$\widehat{\pi}_j(n|n) = \frac{\left[\sum_{i=1}^{J} \widehat{\pi}_i(n-1|n-1) \cdot a_{i|j}\right] \cdot b_j(o_n)}{\sum_{j=1}^{J} \sum_{i=1}^{J} \widehat{\pi}_i(n-1|n-1) \cdot a_{i|j} \cdot b_j(o_n)}, \qquad j = 1, \ldots, J. \tag{8.1}$$

   This recursive filtering formula is perhaps first given in (Åström 1965), see also (Abend 1968, Raviv 1967). The limiting properties of $\widehat{\pi}_j(n|n)$ as $n \to \infty$ and $\{X_n\}_{n=0}^{\infty}$ is ergodic are established in (Kaijser 1975).

   (b) Apply the formula (8.1) to burst noise in Gilbert's channel.

(c) **Prediction posterior probability for an HMM**
Derive the following straightforward recursive formula for the one-step prediction probability by aid of the filtering probability as

$$\widehat{\pi}_j\left(n|n-1\right) = \sum_{i=1}^{J} \widehat{\pi}_i\left(n-1|n-1\right) \cdot a_{i|j}, \qquad j = 1, \ldots, J, \quad (8.2)$$

(Rudemo 1973). Thus (8.1) can be written as

$$\widehat{\pi}_j\left(n|n\right) = \frac{\widehat{\pi}_j\left(n|n-1\right) \cdot b_j\left(o_n\right)}{\sum_{j=1}^{J} \widehat{\pi}_j\left(n|n-1\right) \cdot b_j\left(o_n\right)}, \qquad j = 1, \ldots, J. \quad (8.3)$$

4. Show formally using a suitable property from the above the following (intuitively obvious) equality

$$P\left(X_n = j \mid Y_0 = o_0, \ldots, Y_{n-1} = o_{n-1}, X_0 = j_0, \ldots, X_{n-1} = i\right) = a_{i|j}.$$

5. Show that the log likelihood function for the sequence $\mathbf{o} = o_{j_0} \ldots o_{j_n}$ with respect to the HMM model family is

$$\log P\left(Y_0 = o_{j_0}, \ldots, Y_n = o_{j_n}\right) = \sum_{i=0}^{n} \log f\left(Y_i = o_{j_i} \mid o_{j_0}, \ldots, o_{j_{i-1}}\right), \quad (8.4)$$

where

$$f\left(Y_i = o_{j_i} \mid o_{j_0}, \ldots, o_{j_{i-1}}\right) = \sum_{l=1}^{J} \widehat{\pi}_l\left(i|i-1\right) b_l\left(o_{j_i}\right). \quad (8.5)$$

(Lindgren 1978).

6. Consider a Markov chain $(X_k)_{k=0}^{\infty}$ with the state space $\{0, 1\}$ and with the transition probability matrix

$$A = \begin{pmatrix} 1-p & p \\ q & 1-q \end{pmatrix}$$

and with the initial distribution

$$\pi(0) = (1-a, a).$$

The emitted sequence $(Y_k)_{k=0}^{\infty}$ is given by

$$Y_k = \begin{cases} 1 & \text{if } X_k + V_k \geq 1, \\ \\ 0 & \text{if } X_k + V_k \leq 0, \end{cases}$$

where $(V_k)_{k=0}^{\infty}$ is a sequence of independent, identically distributed discrete random variables, which are independent of of $(X_k)_{k=0}^{\infty}$, too. The variables V_k assume values in the alphabet $\{-1, 0, 1\}$ with the probabilities

$$1 - p_0 - p_1, \quad p_0, \quad p_1$$

respectively.

(a) Show that this is a hidden Markov model in the sense of our definition. Give the emission probability matrix B.

(b) For $j = 0, 1$ let

$$\widehat{\pi}_j\,(n|m) = P\,(X_n = j|Y_0, \ldots, Y_m) \tag{8.6}$$

be the prediction ($n > m$) or filtering ($n = m$) probability. Show that

$$\widehat{\pi}_1\,(n+1|n) = p - (p - (1-q)) \cdot \widehat{\pi}_1\,(n|n)\,.$$

We take $\widehat{\pi}_j\,(0|-1) = a \cdot (1-j) + (1-a) \cdot j$. Give a recursion for $\widehat{\pi}_1\,(n|n)$, too. (Askar and Derin 1981)

7. Let $Y_n = g\,(X_n)$ be a function of a Markov chain as in Section 13.4 above. Let us first introduce for any $o \in \mathcal{O}$, the level set

$$S_o = g^{-1}(o) = \{j \in S | g\,(j) = o\}. \tag{8.7}$$

This gives us a partition of S as

$$S = \cup_{i=1}^{K} S_{o_i}. \tag{8.8}$$

Prove the following sufficient condition for a Markovian function of a Markov chain with a finite state space found in (Burke and Rosenblatt 1958). This gives an intuitive criterion for the lumpability of a Markov chain.

Proposition 13.8.1 *Let $\{X_n\}_{n=0}^{\infty}$ be a Markov chain, with the stationary transition probability matrix $A = \left(a_{i|j}\right)_{i=1,j=1}^{J,J}$. If for any fixed b in $\mathcal{O}$*

$$\sum_{j \in S_b} a_{i|j} = P\left(X_{n+1} \in S_b | X_n = i\right) \tag{8.9}$$

has the same value, say $C(o_k, b)$, for all states i in any given level set S_{o_k}, $k = 1, \ldots, K$, then $\{Y_n = g\left(X_n\right)\}_{n=0}^{\infty}$ is a Markov chain. ∎

8. Show formally for $n = 0, 1, \ldots, N-1$, using a suitable property from the above the following (intuitively obvious) equality

$$P\left(Y_{n+1}, \ldots, Y_N \mid X_{n+1}, Y_0 \ldots, Y_k\right) = P\left(Y_{n+1}, \ldots, Y_N \mid X_{n+1}\right).$$

13.9 References and Further Reading:

K. Abend (1968): Compound decision procedures for unknown distributions and for dependent states of nature. *Pattern Recognition.* L.E. Kanal (ed.), Thompson, Washington, DC. pp. 207–249.

M. Askar and H. Derin (1981): A Recursive Algorithm for the Bayes Solution of the Smoothing Problem. *IEEE Transactions on Automatic Control*, 26, pp. 558–561.

L.E. Baum, T. Petrie, G. Soules and N. Weiss (1970): A Maximization Technique Occurring in the Statistical Analysis of Probabilistic Functions of Markov Chains. *The Annals of Mathematical Statistics*, 41, pp. 164–171.

C.J. Burke and M. Rosenblatt (1958): A Markovian Function of a Markov Chain. *The Annals of Mathematical Statistics*, 29, pp. 1112–1122.

P.A. Devijver (1985): Baum's forward–backward algorithm revisited. *Pattern Recognition Letters*, 3, pp. 369–373.

Y. Guédon and C. Cocozza-Thivent (1990): Explicit state occupancy modelling by hidden semi-Markov models: application of Derin's scheme. *Computer Speech and Language*, 4, pp. 167–192.

T. Kaijser (1975): A Limit Theorem for Partially Observed Markov Chains. *The Annals of Probability*, 3, pp. 677–696.

F.P. Kelly (1982): Markovian Functions of a Markov Chain. *Sankhya*, 44, Series A, pp. 372–379.

G. Lindgren (1978): Markov Regime Models for Mixed Distributions and Switching Regressions. *Scandinavian Journal of Statistics*, 5, pp. 81–91.

H. Lucke (1996): Which Stochastic Models Allow Baum–Welch Training? *IEEE Transactions on Signal Processing*, 44, pp. 2746–2756.

I.L. MacDonald and W. Zucchini (1997): *Hidden Markov and Other Models for Discrete-valued Time Series.* Chapman & Hall. Monographs on Statistics and Applied Probability 70, London, Weinheim, New York, Tokyo, Melbourne, Madras.

A.E. Raftery and S. Tavaré (1994): Estimation and Modelling Repeated Patterns in High Order Markov Chains with the Mixture Distribution Model. *Applied Statistics*, 43, pp. 179–199.

J. Raviv (1967): Decision making in Markov chains applied to the problem of pattern recognition. *IEEE Transactions on Information Theory*, 3, pp. 536–551.

M. Rudemo (1973): State Estimation for Partially Observed Markov Chains. *Journal of Mathematical Analysis and Applications*, 44, pp. 581–611.

K.J. Åström (1965): Optimal Control of Markov Processes with Incomplete State Information. *Journal of Mathematical Analysis and Applications*, 10, pp. 174–205.

Chapter 14

Forward–Backward Algorithm

14.1 Introduction

A procedure that facilitates all computational thinking with a hidden Markov model known as the forward-backward algorithm will now be derived. This algorithm provides a computationally efficient solution to the scoring (evaluation) problem for a standard HMM. In addition the algorithm is the means for deriving other algorithms needed in learning of HMM, e.g., in the computation of variance of the parameter estimates (Hughes 1997). The procedure is formally derived by use of the propositions established in Chapter 13.

The original forward–backward procedure is owed to (Baum and Petrie 1970). The learning/training problem in HMM can and will be formulated in terms of the quantities introduced in the forward–backward procedure, although there are some computational and numerical weaknesses. *Here and in the next two chapters we assume, unless otherwise stated, that there are no silent states, in the sense of the profile HMM in Chapter 12.*

Following (Devijver 1985) we will also establish the forward–backward procedure for the smoothing posterior probabilities. This resolves an implementation issue troubling in the forward-backward procedure.

14.2 Forward–Backward Algorithm and Evaluation

14.2.1 An Outline

The scoring (evaluation) problem is to compute the simultaneous probability for a sequence of emitted symbols, $\mathbf{o} = o_0 \ldots o_N$, conditioned on some model $\lambda = (A, B, \pi(0))$,

$$L_N = P(Y_0, \ldots, Y_n; \lambda) = \sum_{j_o=1}^{J} \cdots \sum_{j_N=1}^{J} \pi_{j_0}(0)\, b_{j_0}(0) \prod_{l=1}^{N} a_{j_{l-1}|j_l} b_{j_l}(l), \quad (2.1)$$

so that the exponential growth of operations in N involved in the summation is avoided. In order to simplify typesetting we omit the notational references to the model λ in the sequel.

The **forward–backward algorithm**, to be presented immediately, allows the evaluation of the probability (2.1) so that the computational requirement is linear in the sequence length $N+1$. The computation is based on the factorizations given in the preceding chapter and has the following expression:

$$L_N = P(Y_0 = o_0, \ldots, Y_N = o_N) = \sum_{j=1}^{J} \alpha_n(j) \cdot \beta_n(j), \quad (2.2)$$

where we define the *forward variable* $\alpha_n(j)$ as the simultaneous probability of the emitted sequence up to time $n \leq N$ and of the hidden Markov chain being in the state j at time n by

$$\alpha_n(j) = P(Y_0 = o_0, \ldots, Y_n = o_n, X_n = j) \quad (2.3)$$

and the *backward variable* $\beta_n(j)$ is defined as the probability of the emitted subsequence from time $n+1$ till the end N conditioned on the hidden Markov chain being in the state j at time n or

$$\beta_n(j) = P(Y_{n+1} = o_{n+1}, \ldots, Y_N = o_N | X_n = j). \quad (2.4)$$

We take $\beta_N(j) = 1$ for every j arbitrarily. To verify (2.2) let us consider

$$P(Y_0, \ldots, Y_N, X_n = j) = P(X_n = j)\, P(Y_0, \ldots, Y_N | X_n = j).$$

In view of Proposition 13.5.1 in Chapter 13 the right hand side is factorized as

$$P(X_n = j)\, P(Y_0 = o_0, .., Y_n = o_n | X_n = j)\, P(Y_{n+1} = o_{n+1}, .., Y_N = o_N | X_n = j)$$
$$= P(Y_0 = o_0, .., Y_n = o_n, X_n = j)\, P(Y_{n+1} = o_{n+1}, .., Y_N = o_N | X_n = j)\,.$$

Since

$$P(Y_0 = o_0, \ldots, Y_N = o_N) = \sum_{j=1}^{J} P(Y_0 = o_0, \ldots, Y_N = o_N, X_n = j)$$

we obtain (2.2). ■

The preceding would be merely a formally valid manipulation, if it did not turn out that there are recursive algorithms for $\alpha_n(j)$ and $\beta_n(j)$.

14.2.2 Forward Recursion

The idea is to express $\alpha_{n+1}(j)$ as a function of $\alpha_n(i)$s:

$$\begin{aligned}
\alpha_{n+1}(j) &= P(Y_0 = o_0, \ldots, Y_{n+1} = o_{n+1}, X_{n+1} = j) \\
&= \sum_{i=1}^{J} P(Y_0 = o_0, \ldots, Y_{n+1} = o_{n+1}, X_n = i, X_{n+1} = j) \\
&= \sum_{i=1}^{J} P(X_n = i, X_{n+1} = j) \\
&\quad \cdot P(Y_0 = o_0, \ldots, Y_{n+1} = o_{n+1} | X_n = i, X_{n+1} = j)\,.
\end{aligned}$$

Here we invoke Proposition 13.5.4 of Chapter 13 and obtain (with $N = n+1$) this as

$$\sum_{i=1}^{J} P(X_n = i, X_{n+1} = j)\, P(Y_0, \ldots, Y_n | X_n = i)\, P(Y_{n+1} = o_{n+1} | X_{n+1} = j)\,.$$

With the notations introduced

$$P(X_n = i, X_{n+1} = j) \cdot P(Y_{n+1} = o_{n+1} | X_{n+1} = j) = a_{i|j} \cdot b_j(o_{n+1}) \cdot P(X_n = i)\,.$$

Hence we have

$$\sum_{i=1}^{J} P(X_n = i, X_{n+1} = j) \cdot P(Y_0, \ldots, Y_n | X_n = i) \cdot P(Y_{n+1} = o_{n+1} | X_{n+1} = j$$

$$= \sum_{i=1}^{J} P(Y_0 = o_0, \ldots, Y_n = o_n, X_n = i) \cdot a_{i|j} \cdot b_j(o_{n+1}).$$

This last expression is, as the result of the stated definition of the forward variable, equal to

$$\sum_{i=1}^{J} \alpha_n(i) \cdot a_{i|j} \cdot b_j(o_{n+1}) = \left[\sum_{i=1}^{J} \alpha_n(i) \cdot a_{i|j}\right] \cdot b_j(o_{n+1}).$$

This completes the derivation of the forward algorithm. We summarize the result in a formal manner.

The Forward Algorithm

Consider the forward variable $\alpha_n(j)$ defined as

$$\alpha_n(j) = P(Y_0 = o_0, \ldots, Y_n = o_n, X_n = j), \tag{2.5}$$

additionally conditioned by the model λ.

Start:

$$\alpha_0(j) = b_j(o_0)\, \pi_j(0), \qquad j = 1, \ldots, J.$$

Recursion:

$$\alpha_{n+1}(j) = \left[\sum_{i=1}^{J} \alpha_n(i) \cdot a_{i|j}\right] \cdot b_j(o_{n+1}), \quad j = 1, \ldots, J, \quad 1 \le n \le N-1. \blacksquare$$

A trellis for the forward procedure is given in Figure 14.1, an adaptation of (Juang & Rabiner 1991 p. 256).

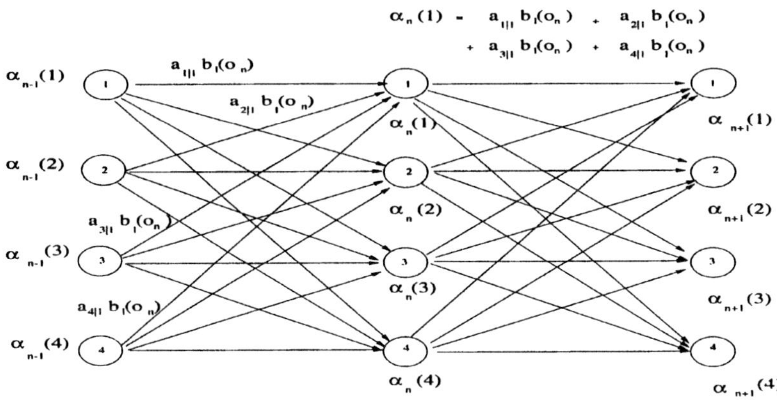

Figure 14.1: A trellis calculation of the forward probabilities

14.2.3 Backward Recursion

Next we treat the backward variable in a similar manner. By definition we get

$$\begin{aligned}\beta_n(j) &= \sum_{i=1}^{J} \frac{P\left(Y_{n+1}=o_{n+1},..,Y_N=o_N,X_n=j,X_{n+1}=i\right)}{P\left(X_n=j\right)}\\ &= \sum_{i=1}^{J} \frac{P\left(Y_{n+1}=o_{n+1},..,Y_N=o_N|X_n=j,X_{n+1}=i\right)P\left(X_n=j,X_{n+1}=i\right)}{P\left(X_n=j\right)}.\end{aligned}$$

Here $\dfrac{P\left(X_n=j,X_{n+1}=i\right)}{P\left(X_n=j\right)}=a_{j|i}$ and Proposition 13.6.1 (with $m=n+1$) from Chapter 13 entails

$$P\left(Y_{n+1},\ldots,Y_N|X_n=j,X_{n+1}=i\right)=P\left(Y_{n+1},\ldots,Y_N|X_{n+1}=i\right).$$

Applying furthermore the factorization in Proposition 13.5.2 from Chapter 13 we obtain

$$\begin{aligned}P\left(Y_{n+1}\right. &= \left.o_{n+1},\ldots,Y_N=o_N|X_{n+1}=i\right)\\ &= P\left(Y_{n+1}=o_{n+1}|X_{n+1}=i\right)\cdot P\left(Y_{n+2}=o_{n+2},\ldots,Y_N=o_N|X_{n+1}=i\right).\end{aligned}$$

Hence it follows that

$$\beta_n(j) = \sum_{i=1}^{J} P\left(Y_{n+1} = o_{n+1} | X_{n+1} = i\right) P\left(Y_{n+2} = o_{n+2}, \ldots, Y_N = o_N | X_{n+1} = i\right) \cdot a_{j|i}$$

Recalling the definition of the backward variable and the emission probability $b_i\left(o_{n+1}\right) = P\left(Y_{n+1} = o_{n+1} | X_{n+1} = i\right)$ we have thus obtained

$$\beta_n(j) = \sum_{i=1}^{J} b_i\left(o_{n+1}\right) \cdot \beta_{n+1}(i) \cdot a_{j|i}.$$

This completes the derivation of the backward algorithm.

The Backward Algorithm

Consider the backward variable $\beta_n(j)$ defined as

$$\beta_n(j) = P\left(Y_{n+1} = o_{n+1}, \ldots, Y_N = o_N | X_n = j\right), \tag{2.6}$$

also conditioned by the model λ.

Start:

$$\beta_N(j) = 1 \quad j = 1, \ldots, J.$$

Recursion :

$$\beta_n(j) = \sum_{i=1}^{J} b_i\left(o_{n+1}\right) \cdot \beta_{n+1}(i) \cdot a_{j|i}, \quad j = 1, \ldots, J \quad n = N-1, N-2, \ldots, 0. \blacksquare$$

The Forward and Backward procedures can be stated more compactly in a matrix notation that will be treated in a set of exercises below.

14.2.4 The Scoring (Evaluation) Problem

In view of (2.1)

$$L_N = P\left(Y_0 = o_0, \ldots, Y_N = o_N\right) = \sum_{j=1}^{J} \alpha_n(j) \cdot \beta_n(j). \tag{2.7}$$

we have for any $n = 0, \ldots, N$ a respective way of computing L_N. For example, with $n = N$ we have

$$L_N = \sum_{j=1}^{J} \alpha_N(j) \tag{2.8}$$

from the convention $\beta_N(j) = 1$. This is the solution to the evaluation problem used in speech processing (Juang and Rabiner 1991, p. 255). The scoring formulas can be cast in a very useful matrix notation that will be given in the exercises below.

14.3 Recursions for Posterior Probabilities

14.3.1 Filtering, Smoothing, Prediction

All probabilistic information about X_n given a sequence of observations $o_0 \ldots o_N$ is contained in the posterior probabilities

$$\widehat{\pi}_j\left(n|N\right) = P\left(X_n = j|Y_0 = o_0, \ldots, Y_N = o_N\right), j = 1, \ldots, J. \tag{3.1}$$

This probability is additionally conditioned by the model λ not made explicit so as not to burden the notation.

For $n > N$ the probability $\widehat{\pi}_j\left(n|N\right)$ deals with **prediction** posterior probability, $n = N$ the probability $\widehat{\pi}_j\left(n|n\right)$ is a **filtering** posterior probability. For $n < N$ we talk about a **smoothing** posterior probability. The terminology about prediction, filtering, and smoothing coincides with the one used in the completely analogous situations in signal processing (Elliott et al. 1995). The Figures 14.2, 14.3 and 14.4 attempt to illustrate these notions.

In (Devijver 1985) a forward–backward type of recursion is derived for the smoothing posterior probabilities. We start from

$$\widehat{\pi}_j\left(n|N\right) = P\left(X_n = j|Y_0 = o_0, \ldots, Y_n = o_N\right)$$

and simplify the notation

$$\begin{aligned}\frac{P\left(Y_0, \ldots, Y_N, X_n\right)}{P\left(Y_0, \ldots, Y_N\right)} &= \frac{P\left(Y_0, \ldots, Y_n, Y_{n+1}, \ldots, Y_N, X_n\right)}{P\left(Y_0, \ldots, Y_n\right) P\left(Y_{n+1}, \ldots, Y_N \mid Y_0, \ldots, Y_n\right)} \\ &= \frac{P\left(Y_0, Y_1, \ldots, Y_n, X_n\right) \cdot P\left(Y_{n+1}, \ldots, Y_N \mid X_n\right)}{P\left(Y_0, \ldots, Y_n\right) P\left(Y_{n+1}, \ldots, Y_N \mid Y_0, \ldots, Y_n\right)},\end{aligned}$$

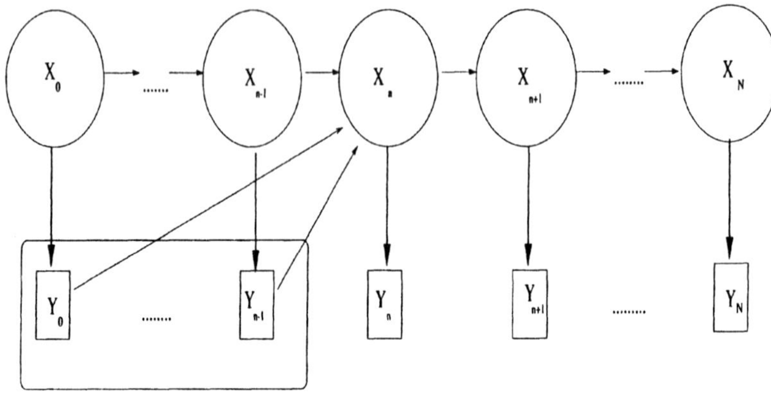

Figure 14.2: Prediction

where we have used the factorization (5.1) in Proposition 13.5.1 and the definition of conditional probability. Here Devijver (1985) introduces

$$\widehat{\pi}_j\,(n|N) = \tilde{\alpha}_n(j) \cdot \tilde{\beta}_n(j) \tag{3.2}$$

with

$$\tilde{\alpha}_n(j) = \frac{P\,(Y_0, \ldots, Y_n, X_n = j)}{P\,(Y_0, \ldots, Y_n)} \tag{3.3}$$

and

$$\tilde{\beta}_n(j) = \frac{P\,(Y_{n+1}, \ldots, Y_N \mid X_n = j)}{P\,(Y_{n+1}, \ldots, Y_N \mid Y_0, \ldots, Y_n)}. \tag{3.4}$$

These are evidently scalings of the forward and backward variables (2.3) and (2.4) of the Baum–Petrie recursions, respectively. The important property of these particular scalings is that there turn out to exist recursions for the scaled forward and backward variables $\tilde{\alpha}_n(j)$ and $\tilde{\beta}_n(j)$. These will be established next.

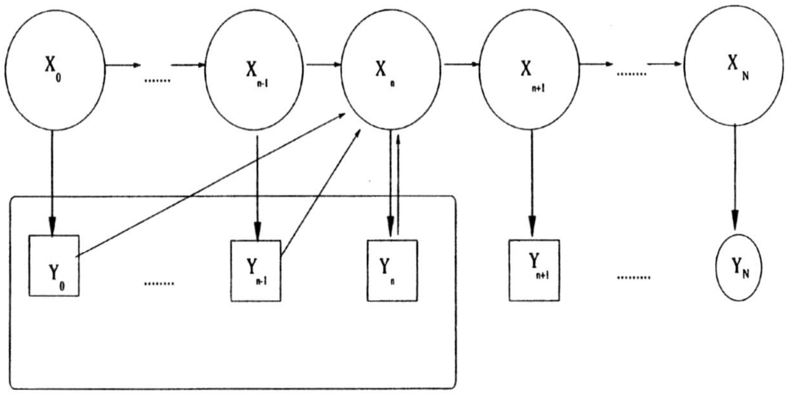

Figure 14.3: Filtering

14.3.2 Scaled Recursions

First, here $\tilde{\alpha}_n(j)$ is equal to the filtering posterior probability $\tilde{\alpha}_n(j) = \widehat{\pi}_j\,(n|n)$. By (8.1) in Exercise 3 in Chapter 13 we have

$$\tilde{\alpha}_n(j) = \frac{\left[\sum_{i=1}^{J} \tilde{\alpha}_{n-1}(i) \cdot a_{i|j}\right] \cdot b_j\,(o_n)}{\sum_{j=1}^{J} \sum_{i=1}^{J} \tilde{\alpha}_{n-1}(i) \cdot a_{i|j} \cdot b_j\,(o_n)}, \tag{3.5}$$

which is the desired forward recursion. Next we obtain

$$\tilde{\beta}_n(j) = \sum_{k=1}^{J} \frac{P\,(Y_{n+1}, Y_{n+2} \ldots, Y_N, X_{n+1} = k \mid X_n = j)}{P\,(Y_{n+1}, Y_{n+2} \ldots, Y_N \mid Y_0, \ldots, Y_n)}. \tag{3.6}$$

After some rearranging we have

$$\tilde{\beta}_n(j) = \sum_{k=1}^{J} \frac{P\left(Y_{(n+1)}^{(N)} \mid X_{n+1} = k, X_n = j\right) P\,(X_{n+1} = k|X_n = j)}{P\,(Y_{n+1}, Y_{n+2} \ldots, Y_N \mid Y_0, \ldots, Y_n)}. \tag{3.7}$$

By (6.1) $(m = n + 1)$ in Proposition 13.6.1 of the preceding Chapter 13 we obtain

$$P\,(Y_{n+1}, Y_{n+2}, \ldots, Y_N \mid X_{n+1} = k, X_n = j) = P\,(Y_{n+1}, Y_{n+2} \ldots, Y_N \mid X_{n+1} = k)$$

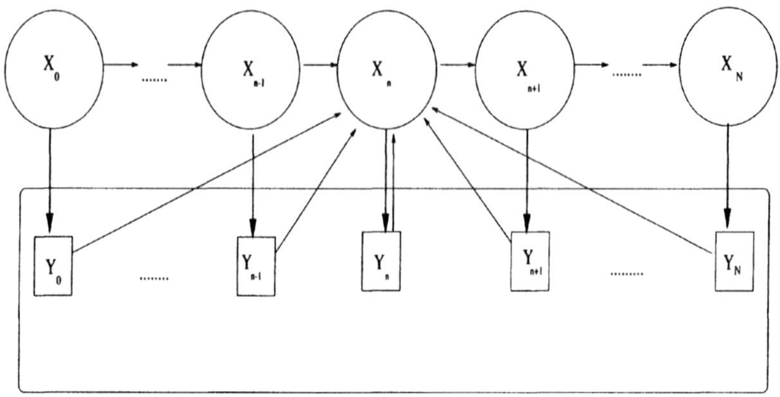

Figure 14.4: Smoothing

and by (5.5) in Proposition 13.5.2 in Chapter 13

$$\begin{aligned} & P\left(Y_{n+1}, Y_{n+2}, \ldots, Y_N \mid X_{n+1}=k\right) \\ & =P\left(Y_{n+1} \mid X_{n+1}=k\right) \cdot P\left(Y_{n+2}, \ldots, Y_N \mid X_{n+1}\right). \end{aligned} \tag{3.8}$$

Since $P\left(Y_{n+1} \mid X_{n+1}=k\right) = b_k\left(o_{j_k}\right)$ and $P\left(X_{n+1}=k | X_n=j\right) = a_{j|k}$ we have obtained in (3.7)

$$\begin{aligned} & P\left(Y_{n+1}, Y_{n+2}, \ldots, Y_N \mid X_{n+1}=k, X_n=j\right) P\left(X_{n+1}=k | X_n=j\right) \\ & =b_k\left(o_{j_k}\right) P\left(Y_{n+2}, \ldots, Y_N \mid X_{n+1}\right) a_{j|k}. \end{aligned}$$

In the denominators of (3.7) we have

$$\begin{aligned} & P\left(Y_{n+1}, \ldots, Y_N \mid Y_0, \ldots, Y_n\right) \\ & =P\left(Y_{n+1} \mid Y_0, \ldots, Y_n\right) \cdot P\left(Y_{n+2} \ldots, Y_N \mid Y_0, \ldots, Y_{n+1}\right). \end{aligned}$$

Thus we have

$$\tilde{\beta}_n(j)=\sum_{k=1}^J \frac{b_k\left(o_{j_k}\right) P\left(Y_{n+2}, \ldots, Y_N \mid X_{n+1}\right) a_{j|k}}{P\left(Y_{n+1} \mid Y_0, \ldots, Y_n\right) \cdot P\left(Y_{n+2}, \ldots, Y_N \mid Y_0, \ldots, Y_{n+1}\right)},$$

which by the definition of $\tilde{\beta}_n(j)$ in (3.4) gives

$$\tilde{\beta}_n(j)=\sum_{k=1}^J \frac{b_k\left(o_{j_k}\right) a_{j|k}}{P\left(Y_{n+1} \mid Y_0, \ldots, Y_n\right)} \cdot \frac{P\left(Y_{n+2}, \ldots, Y_N \mid X_{n+1}=k\right)}{P\left(Y_{n+2}, \ldots, Y_N \mid Y_0, \ldots, Y_{n+1}\right)},$$

which is equal to

$$\tilde{\beta}_n(j) = \frac{1}{P(Y_{n+1} \mid Y_0, \ldots, Y_n)} \sum_{k=1}^{J} b_k(o_{j_k})\, a_{j|k} \cdot \tilde{\beta}_{n+1}(k).$$

Finally,

$$\begin{aligned} P(Y_{n+1} \mid Y_0, \ldots, Y_n) &= \sum_{j=1}^{J}\sum_{k=1}^{J} P(Y_{n+1}, X_n = j, X_{n+1} = k \mid Y_0, \ldots, Y_n) \\ &= \sum_{j=1}^{J}\sum_{k=1}^{J} a_{j|k} b_j(o_{l_j})\, \tilde{\alpha}_{n-1}(j) \end{aligned}$$

using the results in the preceding Chapter in an obvious manner. We set

$$N_n = \frac{1}{\sum_{j=1}^{J}\sum_{k=1}^{J} a_{j|k} b_j(o_{l_j})\, \tilde{\alpha}_{n-1}(j)}$$

and obtain

$$\tilde{\beta}_n(j) = N_n \sum_{k=1}^{J} b_k(o_{j_k})\, a_{j|k} \cdot \tilde{\beta}_{n+1}(k).$$

14.3.3 Implementation

First, there is the summary:.

A Forward–Backward Algorithm for Posterior Smoothing

The forward variable $\tilde{\alpha}_n(j)$ and the backward variable $\tilde{\beta}_n(j)$ are defined as

$$\tilde{\alpha}_n(j) = \frac{P(Y_0, Y_1, \ldots, Y_n, X_n = j)}{P(Y_0, \ldots, Y_n)} \tag{3.9}$$

and

$$\tilde{\beta}_n(j) = \frac{P(Y_{n+1}, \ldots, Y_N \mid X_n = j)}{P(Y_{n+1}, \ldots, Y_N \mid Y_0, \ldots, Y_n)}. \tag{3.10}$$

Start:

$$\tilde{\alpha}_0(j) = N_0 b_j(o_0)\, \pi_j(0), \quad j = 1, \ldots, J.$$

$$\tilde{\beta}_N(j) = 1.$$

Recursions:

$$\tilde{\alpha}_{n+1}(j) = N_n \left[\sum_{i=1}^{J} \tilde{\alpha}_n(i) \cdot a_{i|j} \right] \cdot b_j\left(o_{n+1}\right), \quad j = 1, \ldots, J, \quad 1 \leq n \leq N-1,$$

and

$$\tilde{\beta}_n(j) = N_n \sum_{k=1}^{J} b_k\left(o_{j_k}\right) a_{j|k} \cdot \tilde{\beta}_{n+1}(k).$$

where in both cases

$$N_n = \frac{1}{\sum_{j=1}^{J} \sum_{k=1}^{J} a_{j|k} b_j\left(o_{l_j}\right) \tilde{\alpha}_{n-1}(j)}.$$

■

The forward recursions do not lead to an efficient implementation, but can be written in an equivalent form that can be effectively implemented. The scaled recursions above are immune to the *underflow problem*, namely that the forward and backward variables in the Baum–Petrie recursions are small, if $N - n$ is large.

Derin's backward recursion (2.2) in Proposition 13.2.1 in the preceding chapter is another solution to the underflow problem, but with a different scaling.

14.4 The Alignment Problem and the Viterbi Algorithm

14.4.1 Formulation

The alignment or decoding problem could be solved by finding for each $n = 0, 1, 2, \ldots, N$ the value $j_*(n)$ such that

$$j_*(n) = \arg\max_{1 \leq j \leq J} \widehat{\pi}_j\left(n|N\right) \tag{4.1}$$

The rule in (4.1) is optimal, in the sense of minimizing the expected number of false estimates of states at time n, assuming that λ is the true model, but it might provide a sequence $\mathbf{x}^* = j_*(0) \ldots j_*(N)$ with a low overall probability. In order to solve the alignment problem we maximize instead the probability

$$P\left(Y_0 = o_0, \ldots, Y_N = o_N, X_0 = j_0, \ldots, X_N = j_N\right)$$

by selection of $j_0 \dots j_N$, when the sequence $o_0 \dots o_N$ is fixed and the model λ is known and subsequently omitted in the notation.

Let us set

$$\delta_n(j) = \max_{j_0 \cdots j_{n-1}} P\left(Y_0 = o_0, \dots, Y_n = o_n, X_0 = j_0, \dots, X_n = j\right), \tag{4.2}$$

which is the highest probability (*score*) along a single subsequence of states which at time n is in state j and accounts for the first $n + 1 \leq N$ emitted symbols. But then conditional independence properties valid for an HMM yield a multiplicative structure for the score, which yields an algorithm based on the *Bellman optimality principle.*

14.4.2 The Bellman Optimality Principle

Proposition 14.4.1

$$\delta_n(j) = \left[\max_{i=1,\dots,J} \delta_{n-1}(i) \cdot a_{i|j}\right] \cdot b_j\left(o_n\right). \tag{4.3}$$

Proof: Using the notational conventions in Chapter 13 we set

$$P\left(Y_0 = o_0, \dots, Y_n = o_n, X_0 = j_0, \dots, X_n = j\right) = P\left(Y^{(n)} \mid X^{(n)}\right) \cdot P\left(X^{(n)}\right).$$

By the conditional independence of the emitted symbols and the definition of conditional probability this can be decomposed as

$$\prod_{j=1}^{n} P\left(Y_j \mid X_j\right) \cdot P\left(X_n \mid X^{(n-1)}\right) P\left(X^{(n-1)}\right).$$

The Markov property of the hidden chain and some reorganization entail

$$\begin{aligned} &\prod_{j=1}^{n} P\left(Y_j \mid X_j\right) \cdot P\left(X_n \mid X^{(n-1)}\right) P\left(X^{(n-1)}\right) \\ &= b_j\left(o_n\right) \prod_{j=1}^{n-1} P\left(Y_j \mid X_j\right) \cdot P\left(X_n \mid X_{n-1}\right) P\left(X^{(n-1)}\right) \\ &= a_{j_{n-1}|j} \cdot b_j\left(o_n\right) \cdot \prod_{j=1}^{n-1} P\left(Y_j \mid X_j\right) \cdot P\left(X^{(n-1)}\right). \end{aligned} \tag{4.4}$$

Reverting back to the fullyfledged notation this is equal to

$$b_j\left(o_n\right) a_{i|j} P\left(Y_0 = o_0, \ldots, Y_{n-1} = o_{n-1}, X_0 = j_0, \ldots, X_{n-1} = i\right).$$

If we now pause to think of what we are requested to do when finding $\delta_n(j)$ in (4.2), we should consider the trellis of the hidden Markov chain as given in Figure 14.5. For each $j \in S$ at time n (say $n = 4$ in Figure 14.5) we have

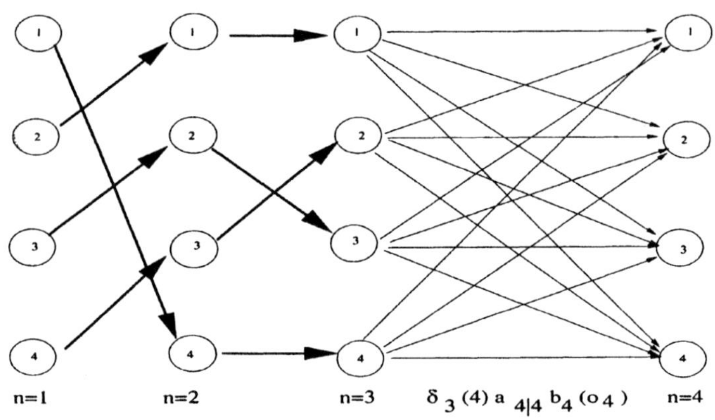

Figure 14.5: A trellis with the best paths

to find the transition to that state from a state $i \in S$ at time $n - 1(= 3)$ giving the best score. There are many paths leading to i at time $n - 1$. But we see that the score depending on the current transition is factorized as the product

$$a_{i|j} P\left(Y_0 = o_0, \ldots, Y_{n-1} = o_{n-1}, X_0 = j_0, \ldots, X_{n-1} = i\right).$$

This shows clearly that if we do not choose at time $n-1$ for every $i \in S$ that special path in the trellis leading to i with maximal probability, we cannot obtain

$$\max_{j_0 \ldots j_{n-1}} a_{i|j} P\left(Y_0 = o_0, \ldots, Y_{n-1} = o_{n-1}, X_0 = j_0, \ldots, X_{n-1} = i\right).$$

Hence

$$\max_{j_0 \ldots j_{n-1}} a_{j_{n-1}|j} P\left(Y_0 = o_0, \ldots, Y_{n-1} = o_{n-1}, X_0 = j_0, \ldots, X_{n-1} = j_{n-1}\right)$$

$$= \left[\max_{i=1,\ldots,J} \delta_{n-1}(i) \cdot a_{i|j} \right]$$

as was to be proved. ■

The subsequence yielding $\delta_n(j)$ is called a *survivor* (illustrated by a thick arrow in Figure 14.5) and denoted by

$$\psi_n(j) = \arg\max_{i=1,\ldots,J} \delta_{n-1}(i) \cdot a_{i|j}$$

and consists of the prefix yielding $\delta_{n-1}(i)$ concatenated by the best scoring transition between times $n-1$ and n. Hence we need at any j and any n only remember the survivor and no other path leading to this state through the trellis.

The complete procedure yielding the best decoded state sequence (path) is now summarized in the following manner (Forney 1973).

14.4.3 Dynamic Programming for the Alignment (Decoding) Problem

Viterbi Algorithm (VA)

Storage: n time index, for each $j \in S$ the survivor $\psi_n(j)$ and the corresponding scores $\delta_n(j)$, $j \in S$.

Start: $n = 0$. Compute for each $j \in S$

$$\delta_0(j) = \pi_j(0) \cdot b_j(o_0),$$

$$\psi_0(j) = \emptyset.$$

Recursion: Compute

$$\delta_{n+1}(j) = \left[\max_{i=1,\ldots,J} \delta_n(i) \cdot a_{i|j} \right] \cdot b_j(o_{n+1}).$$

Store the survivors

$$\psi_n(j) = \arg\max_{i=1,\ldots,J} \delta_{n-1}(i) \cdot a_{i|j}.$$

for $j = 1, \ldots, N$. Set $n+1$ to n unless $n = N$ and repeat.

Termination:

$$P^* = \max_{i=1,\ldots,J} \delta_N(i)$$

$$j_*(N) = \arg\max_{i=1,\ldots,J} \delta_N(i).$$

Traceback: The best path is found by

$$j_*(n) = \psi_{n+1}\left(j_*\left(n+1\right)\right), \qquad n = N-1, N-2, \ldots, 0. \qquad \blacksquare$$

14.4.4 Viterbi Learning

The Viterbi algorithm can, in addition, be used to solve the training/learning problem by an algorithm from (Waterman 1995, p. 248). One can thus learn from a set of training sequences both the model and a multiple alignment in the sense of alignment to the model in Chapter 12.

1) **Start**: Choose an initial HMM model of the given architecture. If no prior information is available make all transitions equally likely.

2) **Alignment**: Use the Viterbi algorithm to align $\mathbf{o}^{(1)}, \ldots, \mathbf{o}^{(r)}$ to the current HMM.

3) **Statistics**: Collect the count statistics

$$\begin{aligned} n_i(k) &= \text{the number of times state } i \text{ occurs in the alignment of } \mathbf{o}^{(k)}, \\ n_{i|j}(k) &= \text{the number of transitions (excluding the final transition)} \\ &\quad\ \text{from } i \text{ to } j \text{ in the alignment of } \mathbf{o}^{(k)}, \\ m_{j|l}(k) &= \text{the number of times the symbol } o_l \text{ is emitted in state } j \text{ by} \\ &\quad\ \text{the sequence } \mathbf{o}^{(k)}, \end{aligned}$$

compute

$$e_{i|j} = \sum_{k=1}^{r} n_{i|j}(k), \quad e_i = \sum_{k=1}^{r} n_i(k)$$

$$d_{j|l} = \sum_{k=1}^{r} m_{j|l}(k),$$

and standardize these to relative frequencies to re-estimate the model.

4) **Repeat**: Repeat steps 2) and 3) until the parameter estimates converge. The properties of this algorithm will be treated in a set of exercises. ■

14.5 Exercises

1. In the sequel we will need

$$\widehat{\pi}_{j|k}\left(n|N\right) = P\left(X_n = j, X_{n+1} = k|Y_0 = o_0, \ldots, Y_N = o_N\right), \qquad (5.1)$$

which is the conditional posterior probability that a transition has taken place between any two states j and k at time $n+1$. Show that

$$\widehat{\pi}_{j|k}\left(n|N\right) = \frac{\widehat{\pi}_j\left(n|N\right) a_{j|k} b_k\left(o_{j_{n+1}}\right) \beta_{n+1}(k)}{\beta_n(j)}.$$

2. Show that

$$\widehat{\pi}_j\left(n|N\right) = \frac{\alpha_n(j) \cdot \beta_n(j)}{\sum_{j=1}^{J} \alpha_n(j) \cdot \beta_n(j)}.$$

3. Let us introduce

$$v_l(o) = \begin{cases} 1 & \text{if } o = o_l \in \mathcal{O}, \\ 0 & \text{otherwise.} \end{cases}$$

Then for any $o \in \mathcal{O}$ we have the identity

$$P\left(Y_n = o|X_n = j\right) = \prod_{l=1}^{K} b_j\left(o_l\right)^{v_l(o)}.$$

We introduce also the $J \times J$ matrix $\mathcal{B}_n$

$$= \begin{pmatrix} \prod_{l=1}^{K} b_1\left(o_l\right)^{v_l\left(o_{j_n}\right)} & 0 & 0 & \cdots & 0 \\ 0 & \prod_{l=1}^{K} b_2\left(o_l\right)^{v_l\left(o_{j_n}\right)} & 0 & \cdots & 0 \\ \vdots & \cdots & \vdots & \cdots & \vdots \\ 0 & 0 & 0 & \cdots & \prod_{l=1}^{K} b_J\left(o_l\right)^{v_l\left(o_{j_n}\right)} \end{pmatrix}.$$

In other words, the elements on the diagonal are the emission probabilities $P\left(Y_n = o_{j_n}|X_n = j\right)$ with $j = 1, \ldots, j$.

(a) Let $\underline{\alpha}_{n+1}$ be a column vector of forward variables, i.e.,

$$\underline{\alpha}_{n+1} = \begin{pmatrix} \alpha_n(1) \\ \alpha_n(2) \\ \vdots \\ \alpha_n(J) \end{pmatrix}.$$

Show that

$$\underline{\alpha}_{n+1} = \mathcal{B}_{n+1} \cdot A^T \underline{\alpha}_n,$$

where A^T is the transposed matrix.

(b) Let $\underline{\beta}_{n+1}$ be a column vector of backward variables, i.e.,

$$\underline{\beta}_{n+1} = \begin{pmatrix} \beta_n(1) \\ \beta_n(2) \\ \vdots \\ \beta_n(J) \end{pmatrix}.$$

Show that

$$\underline{\beta}_n = A\mathcal{B}_{n+1} \cdot \underline{\beta}_{n+1}.$$

4. Show that if

$$\underline{\mathbf{1}} = \begin{pmatrix} 1 \\ 1 \\ \vdots \\ 1 \end{pmatrix}$$

is a $J \times 1$ (column) vector of ones and

$$\pi(0) = \begin{pmatrix} \pi_1(0) \\ \vdots \\ \pi_J(0) \end{pmatrix}$$

then

$$P(Y_0 = o_0, \ldots, Y_N = o_N; A, B, \pi(0)) = \underline{\mathbf{1}}^T \left[\prod_{l=1}^{N} \mathcal{B}_l \cdot A^T \right] \cdot \mathcal{B}_0 \pi(0),$$

where we have the matrix product

$$\prod_{l=1}^{N} \mathcal{B}_l \cdot A^T = \left(\mathcal{B}_N \cdot A^T\right) \cdots \left(\mathcal{B}_1 \cdot A^T\right).$$

The next two exercises are meant as demonstrations of the advantages of this matrix formula, as well as of computing and modelling by HMM.

5. **Missing an emitted symbol**
Suppose that we have an emitted sequence with a missing jth symbol or $\mathbf{o}' = o_0 o_1 \ldots o_{j-1} o_{j+1} \ldots o_n$. This leads to the *pattern completion problem* (Kung 1993). Assume that the missing of the symbol is due to a random erasure. Show that

$$\begin{aligned} & P\left(Y_0 = o_0, \ldots, Y_{j-1} = o_{j-1}, Y_{j+1} = o_{j+1}, \ldots, Y_n = o_n; A, B, \pi(0)\right) \\ = \; & P_2 \cdot P_1, \end{aligned} \tag{5.2}$$

where

$$P_2 = \underline{1}^T \left[\prod_{l=j+1}^{n} \mathcal{B}_l \cdot A^T \right]$$

$$P_1 = A^T \left[\prod_{l=1}^{j-1} \mathcal{B}_l \cdot A^T \right] \cdot \mathcal{B}_0 \pi(0).$$

6. **Linkage analysis**
Consider again Example 9.2.1 in chapter 9. The inheritance patterns at each locus are hidden. The observed data consist of the disease status or phenotype of each member in a pedigree. The probability distributions of these data conditioned on the inheritance vector can be computed using observed allele frequencies summed over all genotype configurations compatible with the phenotype and the inheritance vector. The details of these computations are not really necessary in this exercise.

Give a formula for the probability of the phenotypes determined by the M loci if the inheritance patterns and the phenotypes are treated as a hidden (non-homogeneous) Markov model.

The algorithm for effective computation of probability using the desired formula is known as the *Lander–Green* algorithm (Lander and Green 1987). Clearly the algorithmic developments are relying on the special structure of the transition matrix and of the inheritance probabilities. Faster versions of this algorithm have been derived in (Kruglyak et al. 1995,1996), implemented in the computer software package GENEHUNTER, as well as in (Idury and Elston 1997).

7. **VA for a second order Markov chain**
$\{X_n\}_{n=0}^{\infty}$ is a random sequence that assumes values in a finite state

space $S = \{1, 2, \ldots, J\}$ with J states and has a second order Markov property o

$$\begin{aligned} & P(X_n = j | X_{n-1} = i, X_{n-2} = k, \ldots, X_0 = i_0) \\ = \; & P(X_n = j | X_{n-1} = i, X_{n-2} = k), \quad n \geq 1, \quad i, j, k \in S. \end{aligned}$$

Define a new sequence Z_n (see Exercise 12 in Chapter 7) as

$$Z_0 = X_0, Z_i = [X_{i-1} X_i],$$

where $[X_{i-1} X_i]$ denotes a string of two elements from S. We have $P(Z_0 = j) = \pi_j(0)$ and

$$P(Z_n = [ij] \,| Z_{n-1} = [kj]) = P(X_n = j | X_{n-1} = i, X_{n-2} = k).$$

$\{Z_n\}_{n \geq 1}$ is a first order Markov chain with S adjoined to the set of all possible two-strings as state space.

An observed random process $\{Y_n\}_{n=0}^{\infty}$ has the finite state space $\mathcal{O} = \{o_l\}_{l=1}^{K}$ and for any sequence of states $X_0 = j_0, \ldots, X_n = j_n$ the probability of the sequence $o_0 \ldots o_n$ is

$$P(Y_0 = o_0 \ldots, Y_n = o_n \mid X_0 = j_0, \ldots, X_n = j_n) = \prod_{l=0}^{n} b_{j_l}(l). \quad (5.3)$$

(a) Then we verify that

$$b_j(o_k) = P(Y_n = o_k | Z_n = jl).$$

and

$$P(Y_0 = o_0 \ldots, Y_n = o_n \mid Z_0 = j_0, \ldots, Z_n = j_n j_{n-1}) = \prod_{l=0}^{n} b_{j_l}(l).$$

We observe only the process $\{Y_n\}_{n=0}^{\infty}$. The process $\{X_n\}_{n=0}^{\infty}$ is hidden.

(b) Set

$$\begin{aligned} a_{i|j}^{(1)} &= P(X_n = j, X_{n-1} = i | X_{n-1} = i), \\ a_{ik|j}^{(2)} &= P(X_n = j, X_{n-1} = i | X_{n-1} = i, X_{n-2} = k). \end{aligned}$$

For $1 \leq j \leq J$ with $Y_0 = o_0 \ldots, Y_n = o_n$

$$d_0(j) = \pi_j(0) \cdot b_j(o_0)$$

and for $1 \leq j \leq J$ and $1 \leq i \leq J$

$$d_1(i,j) = d_0(i)\, a_{i|j}^{(1)} \cdot b_j(o_1).$$

(c) Use the standard Viterbi algorithm for the HMM $\{Z_n, Y_n\}_{n=0}^{\infty}$ to argue that the recursion step in the decoding problem for $\{X_n, Y_n\}_{n=0}^{\infty}$ is for $2 \leq l \leq n$ and for $1 \leq j \leq J$ and $1 \leq i \leq J$ given by

$$d_l(i,j) = \max_{1 \leq k \leq J} \left[d_{l-1}(k,i)\, a_{ik|j}^{(2)} \cdot b_j(o_l) \right].$$

(d) Set

$$c_l(i,j) = \arg\max_{1 \leq k \leq J} \left[d_{l-1}(k,i)\, a_{ik|j}^{(2)} \right].$$

Show that the traceback of the Viterbi algorithm must be given by

$$X_l = c_l(X_{l+1}, X_{l+2}).$$

(Kundu et. al. 1989).

8. **Delete States and the Forward Recursion**
The profile HMM architecture outlined in Chapter 12 contains *delete* states. These states do not emit any symbols. This feature is not present in the standard framework in (Baum et al. 1970, Levinson et al. 1983, Rabiner 1989) and the forward and backward procedures need to be modified. Consider the following idea (owed to Timothy Leek) of a forward recursion taking into account the existence of delete states. For $0 \leq n \leq N-1$, $j = 1, \ldots, J$

$$\begin{aligned}
\alpha_{n+1}(j) &= \left[\sum_{i=1}^{J} I_i \alpha_n(i) \cdot a_{i|j} \right] \cdot b_j(o_{n+1}) \\
&+ \sum_{i=1}^{J} |1 - I_i| \left[\sum_{k=1}^{J} I_k \alpha_n(k) a_{k|i} \right] a_{i|j} b_j(o_{n+1}) \\
&+ \sum_{i=1}^{J} |1 - I_i| \sum_{k=1}^{J} |1 - I_k| \sum_{l=1}^{J} I_l \alpha_n(l) a_{l|k} a_{k|i} a_{i|j} b_j(o_{n+1}),
\end{aligned}$$

where

$$I_i = \begin{cases} 1 & \text{if } i \text{ is an emitting state,} \\ \\ 0 & \text{if } i \text{ is a delete state.} \end{cases}$$

Note that this reduces to the standard recursion if there are no delete states. Can this be derived in the same manner as the standard recursion? Explain by considering a trellis how this is supposed to work. What is the restriction on the transitions from and to delete states in a profile HMM for this to be appropriate?

9. **Forward variables for HMM with duration**
This exercise refers to the hidden semi-Markov model in Example 11.3.1 in Chapter 11 above. This differs from the standard model in that if the hidden MC is in state j, $X_n = j$, a discrete duration d_n is chosen from the length distribution ϕ_j

$$\underline{\phi} = (\phi_1, \phi_2, \ldots \phi_K),$$

where each ϕ_j is a probability distribution on the positive integers. A random process $\{Y_n\}_{n\geq 0}$ with the finite state space $\mathcal{O} = \{o_1, o_2, \ldots o_K\}$ is generated by as a segment $s_n = o_{k_1} \ldots o_{k_{d_n}}$ with d_n symbols of $\mathcal{O}$ with the conditional distribution

$$\prod_{r=1}^{d_n} b_j(k_r)$$

where the emission probability is

$$b_j(k_r) = P\left(Y_{d_0+\ldots d_{n-1}+r} = o_{k_r} | X_n = j\right).$$

We introduce a new forward variable

$$\alpha_n(j) = P(Y_0 = o_0, \ldots, Y_n = o_n, \text{state } j \text{ ends at } n) \qquad (5.4)$$

(a) Show that

$$\alpha_n(j) = \sum_{i=1}^{J} \sum_{d=1}^{T} \alpha_{n-d}(i) a_{i|j} \phi_j(d) \prod_{r=n-d+1}^{n} b_j(k_r),$$

where T is the maximum duration in any state. This is initiated by

$$\begin{aligned}\alpha_1(j) &= \pi_j(0)\phi_j(1)\, b_j(k_1) \\ \alpha_2(j) &= \pi_j(0)\phi_j(2)\prod_{r=1}^{2} b_j(k_r) + \sum_{i=1, i\neq j}^{J} \alpha_1(i) a_{i|j}\phi_j(1)\, b_j(k_2)\end{aligned}$$

etc., until $\alpha_D(j)$ has been computed.

(b) Show that these forward variables solve the evaluation problem for the duration model $\lambda = \left(A, B, \underline{\phi}, \pi(0)\right)$ in the sense that

$$P(Y_0 = o_0, \ldots, Y_N = o_N) = \sum_{j=1}^{J} \alpha_N(j).$$

14.6 References and Further Reading:

- Theory of HMM

L.E. Baum, T. Petrie, G. Soules and N. Weiss (1970): A Maximization Technique Occurring in the Statistical Analysis of Probabilistic Functions of Markov Chains. *The Annals of Mathematical Statistics*, 41, pp. 164–171.

P.A. Devijver (1985): Baum's forward–backward algorithm revisited. *Pattern Recognition Letters*, 3, pp. 369–373.

R.J. Elliott, L. Aggoun and J.B. Moore: *Hidden Markov Models. Estimation and Control.* Applications of Mathematics. Stochastic Modelling and Applied Probability. 29. Springer Verlag,1995, New York, Berlin.

D. Forney (1973): Viterbi Algorithm. *Proceedings of the IEEE*, March, 61, n:o 3, pp. 268–278.

J.P. Hughes (1997): Computing the observed information in the hidden Markov model using the EM algorithm. *Statistics & Probability Letters*, 12, pp. 107–114.

B.H. Juang and L.R. Rabiner (1991): Hidden Markov Models for Speech Recognition. *Technometrics*, 33, pp. 251–272.

A. Kundu, Y. He and P. Bahl (1989): Recognition of Handwritten Word: First and Second Order Written Markov Model Based Approach. *Pattern Recognition*, 22, pp. 283–297.

S.Y. Kung (1993): *Digital Neural Networks*, Chapter 7, PTR Prentice Hall, Englewood Cliffs, New Jersey, 07632.

S.E. Levinson, L.R. Rabiner and M.M. Sondhi (1983): An Introduction to the Applications of the Theory of Probabilistic Functions of a Markov Chain to Automatic Speech Recognition. *The Bell System Technical Journal*, 62, pp. 1053–1074.

I.L. MacDonald and W. Zucchini (1997): *Hidden Markov and Other Models for Discrete-valued Time Series*. Chapman & Hall. Monographs on Statistics and Applied Probability 70, London, Weinheim, New York, Tokyo, Melbourne, Madras.

L.R. Rabiner (1989): A Tutorial on Hidden Markov Models and Selected Applications in Speech Recognition. *Proceedings of the IEEE*, 77, pp. 257–286.

M.S. Waterman (1995): *Introduction to Computational Biology. Maps, sequences and genomes*. Chapman and Hall, London, Glasgow, Weinheim, New York, Melbourne, Madras.

- HMM and Linkage Analysis

R.M. Idury and R.C. Elston (1997): A Faster and More General Hidden Markov Model Algorithm for Multipoint Likelihood Calculations. *Human Heredity*, 47, pp. 197–202.

L. Kruglyak, M.J. Daly and E.S. Lander (1995): Rapid Multipoint Linkage Analysis of Recessive Traits in Nuclear Families, Including Homozygosity Mapping. *American Journal of Human Genetics*, 56, pp. 519–527.

L. Kruglyak, M.J. Daly, M.P. Reeve–Daly and E.S. Lander (1996): Parametric and Nonparametric Linkage Analysis: A Unified Multipoint Approach. *American Journal of Human Genetics*, 58, pp. 1347–1363.

E.S. Lander and P. Green (1987): Construction of multilocus genetic linkage maps in humans. *Proceedings of the National Academy of Sciences USA*, 84, pp. 2363–2367.

Chapter 15

Baum–Welch Learning Algorithm

15.1 The Learning Problem

15.1.1 Statement of the Learning Problem

Given an observed sequence $\mathbf{o} = o_0 \dots o_n$, the learning/training/estimation problem for HMM asks for the

$$\lambda = (A, B, \pi(0))$$

which specifies the most probable hidden Markov model for the given sequence $\mathbf{o}$, also called the *training sequence*. The estimation problem is first studied in terms of the method of maximum likelihood (ML). In other words, we want to find the model $\widehat{\lambda}_{\mathrm{ML}}$ defined by

$$\widehat{\lambda}_{\mathrm{ML}} = \arg\max_{\lambda} P\left(Y_0 = o_0, \dots, Y_n = o_n; \lambda\right).$$

The joint probability of any sequence of emitted (observed) symbols $\mathbf{o}$ with respect to λ is

$$\begin{aligned} & P\left(Y_0 = o_0 \dots, Y_n = o_n; \lambda\right) \\ & = \sum_{j_o=1}^{J} \cdots \sum_{j_n=1}^{J} \left[\pi_{j_0}(0)\, b_{j_0}(0) \cdot \prod_{l=1}^{n} a_{j_{l-1}|j_l} b_{j_l}(o_l) \right]. \end{aligned} \tag{1.1}$$

One might find it surprising, considering the complex nature of (1.1), that (local) maximization of $P(Y_0 = o_0 \ldots, Y_n = o_n; \lambda)$ can be done by a simple explicit procedure, as will be shown in this and the next chapter. The procedure is known as the *Baum–Welch re-estimation procedure.* We need a number of preparations to explain what it looks like and how it works as well as to elucidate its theoretical properties. The text (Kung 1993) derives and motivates the Baum–Welch algorithm using the notion of *backpropagation* as found in the theory of training of multilayer feedforward neural networks.

The learning problem under study is concerned with estimating the tables of probabilities $A, B, \pi(0)$ as such: the parametric problem, where some vector-valued θ indexing $A(\theta), B(\theta), \pi(0)(\theta)$ is estimated, is not treated here.

There are versions of the algorithm based on Devijver's recursion and Derin's recursion in Chapter 14. The paper (Roberts et al. 1993) gives a stochastic version, a Gibbs sampler for the Bayesian estimation procedure that eliminates the forward–backward recursions.

15.1.2 Plan for the Study of the Learning Problem

Since this is a problem with hidden variables, an idea like the EM algorithm, as treated in Chapter 4, seems natural. Indeed we are going to define a procedure based on an auxiliary function which somehow involves an estimate of the missing information about the state sequence. The pertinent auxiliary function is to be called the *quasi-log likelihood function for HMM.* This is derived in Subsections 15.2.1 to 15.2.3. The quasi-log likelihood depends on a previous estimate and is maximized to yield the explicit expressions for Baum–Welch re-estimates. These re-estimates will finally be written by means of forward and backward variables. This completes part 1 of the presentation. Part 2, i.e., the next chapter, will be concerned with the question of the relation of the re-estimates and the stationary (critical) points of the likelihood function.

15.2 The Quasi-log likelihood for HMM

15.2.1 A Lower Bound for the Log Likelihood Ratio

Let now t denote the number of state sequences $\mathbf{x} = j_0 j_1 \ldots j_n$ of length $n+1$ *which have positive probability w.r.t. the model* λ *for the given sequence of emitted symbols* $\mathbf{o}$. We enumerate the state sequences $(j_0, \ldots, j_n)$ by the index s, $s = 1, \ldots, t$. Then we set

$$u_s = P\left(Y_0 = o_0 \ldots, Y_n = o_n, X_0 = j_0, \ldots, X_n = j_n; \lambda\right) \text{ if } (j_0 \ldots j_n) \mapsto s.$$

For any other model λ^* we set

$$v_s = P\left(Y_0 = o_0 \ldots, Y_n = o_n, X_0 = j_0, \ldots, X_n = j_n; \lambda^*\right) \text{ if } (j_0 \ldots j_n) \mapsto s.$$

Note that some v_s may, in fact, be equal to zero, since we are checking state paths with positive probability with regard to λ. We have to exclude the converse situation and thus make the following assumption. The terminology about absolute continuity is taken from measure theory.

Assumption 15.2.1 (Absolute Continuity) We assume that the model λ^* does not assign a positive probability, conditioned on the given $\mathbf{o}$, to a state path in S^{n+1} that has probability zero with regard to the model λ or, if we have $\mathbf{x}^\dagger = j_0^\dagger \ldots j_n^\dagger$ such that

$$P\left(Y_0 = o_0 \ldots, Y_n = o_n, X_0 = j_0^\dagger, \ldots, X_n = j_n^\dagger; \lambda\right) = 0$$

then

$$P\left(Y_0 = o_0 \ldots, Y_n = o_n, X_0 = j_0^\dagger, \ldots, X_n = j_n^\dagger; \lambda^*\right) = 0. \qquad \blacksquare$$

We shall next give a lower bound for the *log likelihood ratio*

$$\ln \frac{P\left(Y_0 = o_0 \ldots, Y_n = o_n; \lambda^*\right)}{P\left(Y_0 = o_0 \ldots, Y_n = o_n; \lambda\right)}, \tag{2.1}$$

which by the ideas of Bayesian inference compares the plausibility of the two models for the fixed sequence of emitted symbols. Here ln designates the natural logarithm.

Proposition 15.2.1 *Under the assumption that for* $s = 1, \ldots, t$ *we have*

$$u_s > 0$$

and under the assumption 15.2.1 about absolute continuity we have

$$\ln \frac{P(Y_0 = o_0 \ldots, Y_n = o_n; \lambda^*)}{P(Y_0 = o_0 \ldots, Y_n = o_n; \lambda)} \geq \frac{Q(\lambda, \lambda^*) - Q(\lambda, \lambda)}{P(Y_0 = o_0 \ldots, Y_n = o_n; \lambda)}, \tag{2.2}$$

where we have put

$$Q(\lambda, \lambda^*) = Q(\lambda, \lambda^* \mid \mathbf{o}) = \sum_{s=1}^{t} u_s \ln v_s \tag{2.3}$$

and

$$Q(\lambda, \lambda) = Q(\lambda, \lambda \mid \mathbf{o}) = \sum_{s=1}^{t} u_s \ln u_s. \tag{2.4}$$

Proof: By our definition in (1.1) we have

$$\sum_{s=1}^{t} u_s = P(Y_0 = o_0 \ldots, Y_n = o_n; \lambda) \tag{2.5}$$

since summing over s exhausts all of the probability and by the assumption on absolute continuity

$$\sum_{s=1}^{t} v_s = P(Y_0 = o_0 \ldots, Y_n = o_n; \lambda^*). \tag{2.6}$$

Then we get by an appropriate manipulation of the log sum inequality in (2.5.4) in Chapter 2 that

$$\ln \frac{\sum_{s=1}^{t} v_s}{\sum_{s=1}^{t} u_s} \geq \frac{1}{\sum_{s=1}^{t} u_s} \left[\sum_{s=1}^{t} (u_s \ln v_s - u_s \ln u_s) \right], \tag{2.7}$$

as was claimed. ∎

Thus if we can find a model λ^* for which the right hand side of (2.2) is positive, then we have found a better model, in the sense of higher likelihood, for the sequence $\mathbf{o}$. The largest guaranteed improvement, cf. (2.2), results when we maximize $Q(\lambda, \lambda^*)$ in (2.3) as a function of λ. The function $Q(\lambda, \lambda^*)$ is sometimes called a *quasi-log likelihood function.*

If there is a path such that $v_s = 0$, then we take $Q(\lambda, \lambda^*) = -\infty$. Hence a model λ^* with this property with regard to λ will never be a candidate for improvement of likelihood in Baum–Welch learning.

From this point of advantage we see that the situation is like in the EM procedures discussed in Chapter 4. At a first stage a model λ has been handed down to us as a first approximation of the probability mechanism of the HMM supposedly generating the observed seqeunces. The next step is to try to improve this by maximization of $Q(\lambda, \lambda^*)$. We shall say that

'λ is the **current model**'

and that

'λ^* is the **re-estimated model**'.

15.2.2 Expressions for the Quasi-log likelihood Function

We next find suitable expressions for the quasi-log likelihood $Q(\lambda, \lambda^*)$ defined in (2.3). By the properties of standard HMM we have

$$\begin{aligned} v_s &= P\left(Y_0 = o_0 \ldots, Y_n = o_n, X_0 = j_0, \ldots, X_n = j_n; \lambda^*\right) \\ &= \pi^*_{j_0}(0) \cdot \prod_{l=0}^{n} b^*_{j_l}(o_l) \cdot \prod_{l=1}^{n} a^*_{j_{l-1}|j_l}, \end{aligned}$$

if $(j_0 \ldots j_n) \mapsto s$. Therefore

$$\ln v_s = \ln \pi^*_{j_0}(0) + \sum_{l=0}^{n} \ln b^*_{j_l}(o_l) + \sum_{l=1}^{n} \ln a^*_{j_{l-1}|j_l}, \tag{2.8}$$

where we have to keep in mind in the right hand side that we are dealing with the state sequence enumerated with s.

Let us now introduce, similarly as in the learning theory for Markov chains in Chapter 8, the frequency counts

$n_{i|j}(s) =$ the number of l such that $1 \leq l \leq n, \quad j_{l-1} = i, j_l = j, \quad 0 \leq l \leq n-1,$

i.e., the number of transitions of in the state path s from state i to state j, and

$$m_{j|k}(s) = \text{ the number of times the symbol } k \text{ is emitted in state } j \text{ in path } s$$

We set also

$$r_j(s) = \begin{cases} 1 & \text{if the initial state } j_0 = j \\ 0 & \text{otherwise.} \end{cases} \tag{2.9}$$

These pathwise frequency counts are clearly just agents of computation, not actually observable entities.

Using (2.8) and regrouping terms in the summations according to the state transitions and emitted symbols we can write (2.3) as

$$\begin{aligned} Q(\lambda, \lambda^*) &= \sum_{s=1}^{t} u_s \ln v_s \\ &= \sum_{s=1}^{t} u_s \left[\sum_{j=1}^{J} r_j(s) \ln \pi_j^*(0) + \sum_{j=1}^{J} \sum_{k=1}^{K} m_{j|k}(s) \ln b_j^*(o_k) \right. \\ &\quad \left. + \sum_{j=1}^{J} \sum_{i=1}^{J} n_{i|j}(s) \ln a_{i|j}^* \right], \end{aligned} \tag{2.10}$$

which is equal to

$$\begin{aligned} = \sum_{j=1}^{J} \left[\sum_{s=1}^{t} u_s r_j(s) \right] \ln \pi_j^*(0) &+ \sum_{j=1}^{J} \sum_{k=1}^{K} \left[\sum_{s=1}^{t} u_s m_{j|k}(s) \right] \ln b_j^*(o_k) \\ &+ \sum_{j=1}^{J} \sum_{i=1}^{J} \left[\sum_{s=1}^{t} u_s n_{i|j}(s) \right] \ln a_{i|j}^*. \end{aligned}$$

Here we set

$$e_j = \sum_{s=1}^{t} u_s r_j(s), \tag{2.11}$$

$$c_{i|j} = \sum_{s=1}^{t} u_s n_{i|j}(s) \tag{2.12}$$

and

$$d_{j|k} = \sum_{s=1}^{t} u_s m_{j|k}(s). \tag{2.13}$$

By this we obtain

$$Q(\lambda, \lambda^*) = \sum_{j=1}^{J} e_j \ln \pi_j^*(0) + \sum_{j=1}^{J} \sum_{k=1}^{K} d_{j|k} \ln b_j^*(o_k) + \sum_{i=1}^{J} \sum_{j=1}^{J} c_{i|j} \ln a_{i|j}^*. \tag{2.14}$$

Let us hasten to point out here that this function actually depends on the current model λ, since e_j, $c_{i|j}$ and $d_{j|k}$ are computed using

$$u_s = P(Y_0 = o_0 \ldots, Y_n = o_n, X_0 = j_0, \ldots, X_n = j_n; \lambda) \text{ if } (j_0, \ldots, j_n) \mapsto s.$$

Hence the quantities in (2.11), (2.12) and (2.13) are *expectations* in a sense, since they are sums of numbers weighted by a *non-normalized* sequence of positive numbers $(u_1, \ldots, u_t)$.

15.2.3 Maximization of the Quasi-log likelihood Function

It is not difficult to maximize $Q(\lambda, \lambda^*)$ in (2.14), since this is a sum of a finite number of expressions separated in the optimization variables. Hence we are confronted with several maximizations of a function of the form

$$\Phi(x) = \sum_{i=1}^{M} c_i \ln x_i,$$

where $c_i > 0$ and the optimization variables x_i are subject to the constraint $\sum_{i=1}^{M} x_i = 1$. Maximum likelihood estimates for multinomial processes in Chapter 3 and for Markov chains in Chapter 8 are already familiar instances. The unique global maximum of $\Phi(x)$ is attained when

$$x_i = \frac{c_i}{\sum_{i=1}^{M} c_i} \qquad \text{for } i = 1, \ldots, M. \tag{2.15}$$

Therefore $Q(\lambda, \lambda^*)$ is maximized if we choose λ^* using for all j and k and i

$$\pi_j^*(0) = \frac{e_j}{\sum_{j=1}^{J} e_j}, \tag{2.16}$$

$$b_j^*(o_k) = \frac{d_{j|k}}{\sum_{k=1}^{K} d_{j|k}}, \tag{2.17}$$

$$a_{i|j}^* = \frac{c_{i|j}}{\sum_{j=1}^{J} c_{i|j}}. \tag{2.18}$$

Because each path s starts somewhere, we obtain $\sum_{j=1}^{J} r_j(s) = 1$, and we note that

$$\sum_{j=1}^{J} e_j = \sum_{s=1}^{t} u_s \sum_{j=1}^{J} r_j(s) = \sum_{s=1}^{t} u_s = P\left(Y_0 = o_0 \ldots, Y_n = o_n; \lambda\right),$$

since the numbers u_s exhaust the positive probabilities (under λ) of the state sequences given the fixed emitted sequence $\mathbf{o}$. Also,

$$\sum_{k=1}^{K} d_{j|k} = \sum_{s=1}^{t} u_s \sum_{k=1}^{K} m_{j|k}(s) = \sum_{s=1}^{t} u_s n_j(s) = n_j, \tag{2.19}$$

since $\sum_{k=1}^{K} m_{j|k}(s)$ tells us just the number of times the path s visits j, so that

$$\sum_{k=1}^{K} m_{j|k}(s) = n_j(s) = \text{ the number of } l \text{ such that } j_l = j,\ 0 \leq l \leq n$$

and n_j is proportional to the expected number of visits to j conditioned on the emitted sequence and the model λ.

15.2.4 An Intermediate Summary of Re-estimation

We have by the reformulations above:

1. For $j = 1, \ldots, J$,

$$\pi_j^*(0) = \frac{e_j}{P\left(Y_0 = o_0 \ldots, Y_n = o_n; \lambda\right)}. \tag{2.20}$$

2. For $j = 1, \ldots, J$ and for $k = 1, \ldots, K$

$$b_j^*(o_k) = \frac{d_{j|k}}{n_j}. \tag{2.21}$$

3. For $j = 1, \ldots, J$ and for $i = 1, \ldots, J$

$$a^*_{i|j} = \frac{c_{i|j}}{\sum_{j=1}^{J} c_{i|j}}. \tag{2.22}$$

This means that $\pi^*(0)$ and $b^*_j(o_k)$ are probability distributions on their respective alphabets and $\left(a^*_{i|j}\right)_{i=1,j=1}^{J,J}$ is a transition matrix for a Markov chain. Hence (2.20)–(2.22) automatically give a well defined new model $\lambda^* = (A^*, B^*, \pi(0)^*)$. It is clear why (2.20)–(2.21) are called *re-estimation formulas.* We have performed the two steps:

'E' step: We form

$$Q(\lambda, \lambda^*) = \sum_{j=1}^{J} e_j \ln \pi^*_j(0) + \sum_{j=1}^{J}\sum_{k=1}^{K} d_{j|k} \ln b^*_j(o_k) + \sum_{i=1}^{J}\sum_{j=1}^{J} c_{i|j} \ln a^*_{i|j}. \tag{2.23}$$

Strictly speaking, expectation here means sums of logarithms with non-normalized weights.

M step: We maximize $Q(\lambda, \lambda^*)$ as a function of $\pi^*_j(0)$, $b^*_j(o_k)$ and $a^*_{i|j}$. ∎

In the form given above these natural re-estimates of the model are not computationally effective, since they involve summations over a possibly very large number of paths. Next we plan to recapitulate the re-estimates in (2.20)–(2.22) using the forward and backward probabilities introduced earlier.

In (Roberts et al. 1993) the **'E' step** is replaced by sampling (Markov Chain Monte Carlo) from a posterior distribution for λ thereby canceling the forward and backward variables.

Chapter 2 of (Elliott et al. 1995) introduces a technique of *reference probability* or an *absolutely continuous change of measure* for hidden Markov models. By this method it is also possible to compute the Baum–Welch re-estimates without the forward and backward variables.

15.3 Forward and Backward Probabilities

There are actually two rationales for using forward and backward variables: one is the algorithmic efficiency and the other is mathematical expediency in a proof of convergence to a stationary point of the log likelihood function to be established eventually. *In this section all probability computations are with respect to the current model λ, though this is not always reflected in the notation.*

15.3.1 Smoothing Probabilities

We consider the *smoothing probability* ($0 \leq l \leq n$) (here we take $N = n, \quad n = l$ with respect to the notations in Chapter 14)

$$\widehat{\pi}_j(l|n) = P(X_l = j | Y_0 = o_0, \ldots, Y_n = o_n). \tag{3.1}$$

We shall also need a *smoothing transition probability* (cf. Exercise 1 Chapter 14) to be defined as

$$\widehat{\pi}_{j|i}(l|n) = P(X_l = j, X_{l+1} = i | Y_0 = o_0, \ldots, Y_n = o_n). \tag{3.2}$$

By marginalization from (3.2) we have the following results.

Lemma 15.3.1

$$\widehat{\pi}_j(l|n) = \sum_{i=1}^{J} \widehat{\pi}_{j|i}(l|n). \tag{3.3}$$

■

Next we show how these smoothing probabilities can be invoked to reorganize the re-estimates. Let us consider the indicator function

$$I_{\{X_l=j, X_{l+1}=i\}} = \begin{cases} 1 & \text{if } X_l = i, X_{l+1} = j, \\ 0 & \text{otherwise.} \end{cases}$$

Then by the definition of expectation of a discrete random variable

$$E\left[I_{\{X_l=i, X_{l+1}=j\}} | Y_0, \ldots, Y_n\right] = P(X_l = i, X_{l+1} = j | Y_0, \ldots, Y_n).$$

By linearity of expectation we obtain in (3.3)

$$\widehat{\pi}_j(l|n) = \sum_{j=1}^{J} E\left[I_{\{X_l=j, X_{l+1}=i\}} | Y_0 = o_0, \ldots Y_n = o_n\right],$$

which is equal to

$$E\left[Z_l | Y_0 = o_0, \ldots, Y_n = o_n\right],$$

where

$$Z_l \equiv \sum_{i=1}^{J} I_{\{X_l=j, X_{l+1}=i\}}.$$

This shows explicitly that $\widehat{\pi}_j\left(l|n\right)$ is the expected number of times the state j is visited at time l by the paths of the Markov chain given $o_0 \ldots o_n$ and conditioned on the current model. Here Z_l is a random variable, whose generic outcome is the number of times any of the permitted state paths visits j at l, or equivalently, transits from j given $o_0 \ldots o_n$.

The random variable Z_l is equivalently described using the enumeration of the state sequences $(j_0 \ldots j_n)$ by the index s, $s = 1, \ldots, t$ used above. In fact Z_l has t outcomes of positive probability corresponding to

$$z_l(s, j) = \sum_{i=1}^{J} I_{\{j_l=j, j_{l+1}=i\}} \text{ if } (j_0 \ldots j_n) \mapsto s,$$

respectively. But then expectation is by definition $E\left[Z_l | Y_0 = o_0, \ldots Y_n = o_n\right] = C \cdot \sum_{s=1}^{t} z_l(s, j) \cdot u_s$, where C is the normalization constant for the numbers $u_1, \ldots, u_t$ or

$$C = \frac{1}{P\left(Y_0 = o_0, \ldots, Y_n = o_n\right)} = \frac{1}{\sum_{s=1}^{t} u_s}. \tag{3.4}$$

Since

$$u_s = P\left(Y_0 = o_0 \ldots, Y_n = o_n, X_0 = j_0, \ldots, X_n = j_n\right) \text{ if } (j_0 \ldots j_n) \mapsto s.$$

we have that

$$C \cdot u_s = \frac{u_s}{\sum_{s=1}^{t} u_s} = P\left(X_0 = j_0, \ldots, X_n = j_n \mid Y_0 = o_0 \ldots, Y_n = o_n,\right).$$

With $l = 0$ we obtain

$$\sum_{s=1}^{t} z_0(s, j) \cdot u_s = \sum_{s=1}^{t} r_j(s) \cdot u_s,$$

where $r_j(s)$ is defined in (2.9). Hence from (2.11)

$$\widehat{\pi}_j\left(0|n\right) = \frac{e_j}{P\left(Y_0 = o_0, \ldots, Y_n = o_n\right)}$$

and therefore (2.20) is given the following very reasonable interpretation of π_j^*.

Lemma 15.3.2 *Under the current model* λ

$$\pi_j^* = \widehat{\pi}_j\left(0|n\right). \tag{3.5}$$

∎

Next we prove:

Lemma 15.3.3 *Under the current model* λ

$$b_j^*(o_k) = \frac{\sum_{l=0}^{n} I_{\{Y_l=k\}}\widehat{\pi}_j\left(l|n\right)}{\sum_{l=0}^{n} \widehat{\pi}_j\left(l|n\right)}. \tag{3.6}$$

Proof: Our strategy is to prove that:

(a)

$$\sum_{l=0}^{n} P\left(X_l = j, Y_l = o_k|Y_0 = o_0, \ldots, Y_n = o_n\right) = \sum_{l=0}^{n} I_{\{Y_l=k\}}\widehat{\pi}_j\left(l|n\right);$$

(b)

$$C \cdot d_{j|k} = \sum_{l=0}^{n} P\left(X_l = j, Y_l = o_k|Y_0 = o_0, \ldots, Y_n = o_n\right).$$

We prove first part (**a**). Let us set

$$I_{\{X_l=j,Y_l=o_k\}} = \begin{cases} 1 & \text{if } X_l = j, Y_l = o_k \\ 0 & \text{otherwise.} \end{cases}$$

Then we have, as is readily seen,

$$I_{\{X_l=j,Y_l=o_k\}} = I_{\{X_l=j\}} \cdot I_{\{Y_l=o_k\}}.$$

Thus

$$\begin{aligned} &\sum_{l=0}^{n} P\left(X_l = j, Y_l = o_k|Y_0 = o_0, \ldots, Y_n = o_n\right) \\ &= \sum_{l=0}^{n} E\left(I_{\{X_l=j,Y_l=o_k\}}|Y_0 = o_0, \ldots Y_n = o_n\right) \\ &= \sum_{l=0}^{n} E\left(I_{\{X_l=j\}} \cdot I_{\{Y_l=o_k\}}|Y_0 = o_0, \ldots, Y_n = o_n\right). \end{aligned} \tag{3.7}$$

which is the expected number of occurrences of $X_l = j$ with $Y_l = o_k$, and is equal to

$$\sum_{l=0}^{n} I_{\{Y_l=o_k\}} \cdot E\left(I_{\{X_l=j\}}|Y_0 = o_0, \ldots, Y_n = o_n\right),$$

using a standard property of conditional expectation, since each Y_l is included in the conditioning $Y_0 = o_0, \ldots, Y_n = o_n$. This becomes

$$\begin{aligned}\sum_{l=0}^{n} I_{\{Y_l=o_k\}} \quad &\cdot \quad E\left(I_{\{X_l=j\}}|Y_0 = o_0, \ldots, Y_n = o_n\right) \\ &= \sum_{l=0}^{n} I_{\{Y_l=o_k\}} \cdot P\left(X_l = j|Y_0 = o_0, \ldots, Y_n = o_n\right),\end{aligned}$$

which by (3.1) is equal to

$$\sum_{l=0}^{n} I_{\{Y_l=o_k\}} \widehat{\pi}_j \left(l|n\right),$$

as was desired. To prove (**b**) we consider the random variable

$$\sum_{l=0}^{n} I_{\{X_l=j, Y_l=o_k\}},$$

the outcomes of positive probability of which are

$$z(s, j, o_k) = \sum_{l=0}^{n} I_{\{j_l=j, o_l=o_k\}} \text{ if } (j_0 \ldots j_n) \mapsto s.$$

Here o_l is the emitted symbol at time l. Thus

$$\sum_{l=0}^{n} P\left(X_l = j, Y_l = o_k|Y_0 = o_0, \ldots Y_n = o_n\right) = C \cdot \sum_{s=1}^{t} z(s, j, o_k) \cdot u_s,$$

but by (2.13) we see that the expected number of visits to state j with emission of o_k computed in terms of the non-normalized conditional path probabilities is

$$d_{j|k} = \sum_{s=1}^{t} u_s m_{j|k}(s).$$

Thus

$$C \cdot \sum_{s=1}^{t} z(s, j, o_k) \cdot u_s = C \cdot d_{j|k}.$$

But then we have obtained

$$\sum_{l=0}^{n} I_{\{Y_l = o_k\}} \widehat{\pi}_j (l|n) = C \cdot d_{j|k} \tag{3.8}$$

with C from (3.4).

Furthermore the sum of $\widehat{\pi}_j (l|n)$ over l gives the expected frequency of visits in the state j, given the observed symbols and conditioned on the current model, since by (3.3)

$$\sum_{l=0}^{n} \widehat{\pi}_j (l|n) = \sum_{i=1}^{J} \sum_{l=0}^{n} \widehat{\pi}_{j|i} (l|n) . \tag{3.9}$$

But by the same path enumeration argument that was used above and the interpretation of $\widehat{\pi}_j (l|n)$ and $\widehat{\pi}_{j|i} (l|n)$) discussed after Lemma 15.3.1, this is equal to the expected number of times the chain visits the state j, or by (2.19),

$$\sum_{l=0}^{n} \widehat{\pi}_j (l|n) = C \cdot n_j, \tag{3.10}$$

where C is the normalization constant from (3.4). In view of (2.13) and (2.21) we have proved the lemma. ■

It remains to find a formula for $a^*_{i|j}$ in (2.22) by means of the smoothing probabilities. By the same reasoning that was used above we see immediately from (2.12) and (2.22) that $a^*_{i|j}$ can be expressed by sums of smoothing probabilities as:

Lemma 15.3.4 *Under the current model* λ

$$a^*_{i|j} = \frac{\sum_{l=0}^{n-1} \widehat{\pi}_{i|j} (l|n)}{\sum_{l=0}^{n-1} \widehat{\pi}_i (l|n)}. \tag{3.11}$$

Proof: The proof is left as an exercise. ■

15.3.2 Re-estimates in Forward and Backward Variables

Our immediate goal is to write the smoothing probabilities in terms of the forward and backward variables. After having done that the lemmas in the preceding subsection give the Baum–Welch re-estimates in a computationally effective form, disregarding here questions of numerical stability. Let us recall the forward variable $\alpha_l(j)$ defined for every $j \in S$ as

$$\alpha_l(j) = P\left(Y_0 = o_0, \ldots, Y_l = o_l, X_l = j\right), \tag{3.12}$$

which is computed under the current model λ. For every $j \in S$ the backwards probability is

$$\beta_l(j) = P\left(Y_{l+1} = o_{l+1}, \ldots, Y_n = o_n | X_l = j\right) \tag{3.13}$$

(under the current model λ). Then we have

$$\alpha_l(j) \cdot \beta_l(j) = P\left(Y_0 = o_0, \ldots, Y_n = o_n, X_l = j\right),$$

as shown in Chapter 14. Thus we have by the definition of conditional probability and (3.1):

Lemma 15.3.5

$$\widehat{\pi}_j\left(l|n\right) = \frac{\alpha_l(j) \cdot \beta_l(j)}{P\left(Y_0 = o_0, \ldots, Y_n = o_n\right)}. \tag{3.14}$$

■

Next

$$\begin{aligned} \sum_{l=0}^{n-1} \widehat{\pi}_{i|j}\left(l|n\right) &= \sum_{l=0}^{n-1} P\left(X_l = i, X_{l+1} = j | Y_0 = o_0, \ldots Y_n = o_n\right) \\ &= C \cdot \sum_{l=0}^{n-1} \left[P\left(Y_0 = o_0, \ldots, Y_n = o_n | X_l = i, X_{l+1} = j\right)\right. \\ &\quad \left. P\left(X_l = i, X_{l+1} = j\right)\right], \end{aligned}$$

which obtained by Bayes' formula and with C in (3.4) and is equal to

$$C \cdot \sum_{l=0}^{n-1} P\left(Y_0 = o_0, \ldots, Y_n = o_n | X_l = i, X_{l+1} = j\right) \cdot P\left(X_l = i\right) \cdot a_{i|j},$$

which follows by $P(X_l = i, X_{l+1} = j) = P(X_l = i) \cdot a_{i|j}$. But Proposition 13.5.4 in Chapter 13 implies that

$$\begin{aligned} &P(Y_0 = o_0, \ldots, Y_n = o_n | X_l = i, X_{l+1} = j) \\ &= P(Y_0 = o_0, \ldots, Y_l = o_l | X_l = i) \cdot P(Y_{l+1} = o_{l+1}, \ldots, Y_n = o_n | X_{l+1} = j) \end{aligned}$$

and Proposition 13.5.2 in Chapter 13 yields this as

$$P(Y_0, \ldots, Y_l | X_l = i) \, P(Y_{l+1} | X_{l+1} = j) \cdot P(Y_{l+2}, \ldots, Y_n | X_{l+1} = j).$$

Then we have obtained, by the definition of the forward and backward variables in Chapter 14

$$P(X_l = i) \cdot P(Y_0 = o_0, \ldots, Y_n = o_n | X_l = i, X_{l+1} = j) = \alpha_l(i) \cdot b_j(o_{l+1}) \cdot \beta_{l+1}(j).$$

Thus we have

$$\sum_{l=0}^{n-1} \widehat{\pi}_{j|i}(l|n) = \frac{\sum_{l=0}^{n-1} \alpha_l(i) \cdot a_{i|j} \cdot b_{l+1}(j) \cdot \beta_{l+1}(j)}{P(Y_0 = o_0, \ldots, Y_n = o_n)} \tag{3.15}$$

Hence we have in lemma 15.3.4 in formula (3.11) from (3.14) and (3.15) that

$$\begin{aligned} a^*_{i|j} &= \frac{\sum_{l=0}^{n-1} \widehat{\pi}_{i|j}(l|n)}{\sum_{l=0}^{n-1} \widehat{\pi}_i(l|n)} \\ &= \frac{\dfrac{\sum_{l=0}^{n-1} \alpha_l(i) \cdot a_{i|j} \cdot b_j(o_{l+1}) \cdot \beta_{l+1}(j)}{P(Y_0 = o_0, \ldots, Y_n = o_n)}}{\dfrac{\sum_{l=0}^{n-1} \alpha_l(i) \cdot \beta_l(i)}{P(Y_0 = o_0, \ldots Y_n = o_n)}} \\ &= \frac{a_{i|j} \cdot \sum_{l=0}^{n-1} \alpha_l(i) \cdot b_i(o_{l+1}) \cdot \beta_{l+1}(i)}{\sum_{l=1}^{n-1} \alpha_l(i) \cdot \beta_l(i)}. \end{aligned} \tag{3.16}$$

Here we see that the new model λ^* satisfies the assumption about absolute continuity, since any state transition with $a_{i|j} = 0$ in λ will be carried over as $a^*_{i|j}$ to λ^*. Also, by lemma 15.3.3 and (3.6) and (3.14) and (3.15)

$$b^*_j(o_k) = \frac{\sum_{l=0}^{n} I_{\{Y_l = k\}} \widehat{\pi}_j(l|n)}{\sum_{l=0}^{n} \widehat{\pi}_j(l|n)}$$

$$= \frac{\frac{\sum_{l=0}^{n} I_{\{Y_l=o_k\}}\alpha_l(j)\cdot\beta_l(j)}{P(Y_0=o_0,\ldots Y_n=o_n)}}{\frac{\sum_{l=0}^{n}\alpha_l(j)\cdot\beta_l(j)}{P(Y_0=o_0,\ldots Y_n=o_n)}}$$

$$= \frac{\sum_{l=0}^{n} I_{\{Y_l=o_k\}}\alpha_l(j)\cdot\beta_l(j)}{\sum_{l=0}^{n}\alpha_l(j)\cdot\beta_l(j)}. \tag{3.17}$$

Finally, by Lemma 15.3.2 and (3.14)

$$\pi_j^* = \widehat{\pi}_j\,(0|n)$$

$$= \frac{\alpha_0(j)\cdot\beta_0(j)}{P(Y_0=o_0,\ldots,Y_n=o_n)}. \tag{3.18}$$

In the preceding we have seen that $P(Y_0=o_0,\ldots,Y_n=o_n)=\sum_{j=1}^{J}\alpha_l(j)\cdot\beta_l(j)$. The equations (3.16)–(3.17) constitute thus the desired version of the Baum–Welch re-estimation formulas.

15.4 Baum–Welch Learning for MAP

15.4.1 Prior Density

When estimating a model from data, there is always the possibility that the estimated model will be overfitted data, in the sense that it models the training sequences well but will not fit other sequences from the same biological family of sequences.

If a profile HMM, as discussed in Chapter 12, is taken to have different parameters for all modules, there is a danger of overfitting even if the number of training sequences is large. In a profile HMM with $F-1$ match states there are $9\cdot(F-1)$ probabilities accounting for the transitions and $2\cdot(K-1)\cdot F$ for the emission probabilities and a few more for the transitions in the START and in the END. For amino acids with $K=20$ and, e.g., $F=50$ parametric learning feels like a formidable task. The Baum–Welch estimates are basically expectations of frequency counts conditioned on the current model and therefore many estimated parameters could be (close to) zero.

The Baum–Welch algorithm can be modified to allow for regularizers by addition of pseudo-counts in all counted frequencies of symbols and expected counts of transitions, conditioned by the current model (Hughey and Krogh

1996), and then normalising. Regularization is, as pointed out Chapter 3, a method of estimating the probability distributions by interpolating smoothly between reliance on the prior information concerning likely distributions and confidence in the observed frequencies.

For any standard HMM a prior is a probability distribution over the rows $\left(a_{i|j}\right)_{j=1}^{J}$ of the transition probability matrix A and the rows $\left(b_j(l)\right)_{l=1}^{K}$ of the emission probability matrix. We can once more use a product of multivariate Dirichlet densities. This gives us a Baum-Welch algorithm for the MAP (Maximum A Posteriori) estimate.

15.4.2 MAP re-estimates

For the ratio of posterior densities ($\propto$ likelihood $\times$ prior density) for the re-estimated model λ^* and the current model λ we obtain, as in the the preceding,

$$\ln \frac{P\left(\lambda^* | Y_0 = o_0 \ldots, Y_n = o_n\right)}{P\left(\lambda \mid Y_0 = o_0 \ldots, Y_n = o_n\right)} \geq \frac{Q\left(\lambda, \lambda^*\right) - Q\left(\lambda, \lambda\right)}{P\left(Y_0 = o_0 \ldots, Y_n = o_n; \lambda\right)} + \ln \frac{\phi\left(\lambda^*\right)}{\phi\left(\lambda\right)}.$$

Let us now recall the simultaneous prior density as a product of the **multivariate Dirichlet densities** for $\pi(0)^*$, B^* and A^* as given in section 8.4 in Chapter 8

$$\phi\left(\lambda^*\right) = P_1 \cdot P_2 \cdot P_3 \tag{4.1}$$

with

$$\begin{aligned} P_1 &= \operatorname{Dir}\left(\pi(0)^*; \alpha \cdot q_1, \ldots, \alpha \cdot q_J\right), \\ P_2 &= \prod_{j=1}^{J} \operatorname{Dir}\left(\underline{b}_j^*; \rho_j \cdot r_{j|1}, \ldots, \rho_j \cdot r_{J|K}\right), \\ P_3 &= \prod_{i=1}^{J} \operatorname{Dir}\left(\underline{a}_i^*; \alpha_i \cdot q_{i|1}, \ldots, \alpha_i \cdot q_{i|J}\right), \end{aligned}$$

with the respective hyperparameters. A generic example is

$$\operatorname{Dir}\left(\underline{a}_i^*; \alpha_i \cdot q_{i|1}, \ldots, \alpha_i \cdot q_{i|J}\right) = \frac{\Gamma\left(\alpha_i\right)}{\prod_{j=1}^{J} \Gamma\left(\alpha_i q_{i|j}\right)} \cdot \prod_{j=1}^{J} \left(a_{i|j}^*\right)^{\alpha_i q_{i|j} - 1}, \tag{4.2}$$

where

$$\alpha_i > 0, \qquad q_{i|j} > 0, \qquad \sum_{j=1}^{J} q_{i|j} = 1.$$

Let us set, for ease of writing,

$$L(\lambda) := P(Y_0 = o_0 \ldots, Y_n = o_n; \lambda).$$

Then we see that the maximum a posteriori estimation using the Baum–Welch procedure amounts to maximizing

$$\frac{Q(\lambda, \lambda^*)}{L(\lambda)} + \ln \phi(\lambda^*)$$

as a function of λ^*. This can, as is clear, be maximized explicitly as a function of each of $\pi_j^*(0)^*$, $b_j^*(o_k)$ and $a_{i|j}^*$, since the expression is additively separated with regard to the optimization variables and with regard to the corresponding constraints. Inserting from (4.2) we need to maximize, e.g.,

$$\sum_{j=1}^{J} \left[\frac{c_{i|j}}{L(\lambda)} + \left(\alpha_i q_{i|j} - 1\right) \right] \ln a_{i|j}^*$$

under the constraint $\sum_{j=1}^{J} a_{i|j}^* = 1$. But under the assumption

$$\frac{c_{i|j}}{L(\lambda)} + \left(\alpha_i q_{i|j} - 1\right) > 0$$

for all j this is formally no different from the situation dealt with above and thus we obtain

$$a_{i|j}^* = \frac{\dfrac{c_{i|j}}{L(\lambda)} + \alpha_i q_{i|j} - 1}{\displaystyle\sum_{j=1}^{J} \frac{c_{i|j}}{L(\lambda)} + \sum_{j=1}^{J} \alpha_i q_{i|j} - J} \tag{4.3}$$

for $j = 1, \ldots, J$ and for $i = 1, \ldots, J$. In the software suite SAM [1] these estimates seem to be implemented as

$$a_{i|j}^* = \frac{\dfrac{c_{i|j}}{L(\lambda)} + \alpha_i q_{i|j}}{\dfrac{\sum_{j=1}^{J} c_{i|j}}{L(\lambda)} + \displaystyle\sum_{j=1}^{J} \alpha_i q_{i|j}}. \tag{4.4}$$

[1] cf. `http://www.cse.ucsc.edu/research/compbio/sam.html`

From the preceding we recall that

$$\begin{aligned}\frac{c_{i|j}}{L(\lambda)} &= \frac{a_{i|j}\cdot\sum_{l=0}^{n-1}\alpha_l(i)\cdot b_j(o_{l+1})\cdot\beta_{l+1}(j)}{L(\lambda)},\\ \sum_{j=1}^{J}\frac{c_{i|j}}{L(\lambda)} &= \frac{\sum_{l=0}^{n-1}\alpha_l(i)\cdot\beta_l(i)}{L(\lambda)}.\end{aligned}$$

This involves an unfortunate source of notational confusion, but the distinction between the hyperparameters α_i and the forward variables $\alpha_l(i)$ can perhaps be claimed to be clear. The same argument gives for $j = 1,\ldots,J$ and for $k = 1,\ldots,K$ a recursion for $b_j^*(o_k)$, too.
Thus the **Baum-Welch iteration for MAP is**

$$a_{i|j}^* = \frac{\dfrac{a_{i|j}\cdot\sum_{l=0}^{n-1}\alpha_l(i)\cdot b_j(o_{l+1})\cdot\beta_{l+1}(j)}{L(\lambda)} + \alpha_i q_{i|j} - 1}{\dfrac{\sum_{l=0}^{n-1}\alpha_l(i)\cdot\beta_l(i)}{L(\lambda)} + \sum_{j=1}^{J}\alpha_i q_{i|j} - J} \tag{4.5}$$

and

$$b_j^*(o_k) = \frac{\dfrac{\sum_{l=0}^{n} I_{\{Y_l=o_k\}}\alpha_l(j)\cdot\beta_l(j)}{L(\lambda)} + \rho_i r_{i|j} - 1}{\dfrac{\sum_{l=0}^{n}\alpha_l(j)\cdot\beta_l(j)}{L(\lambda)} + \sum_{j=1}^{J}\rho_i r_{i|j} - J}. \tag{4.6}$$

When dealing with the profile HMM there may, e.g., be the prior belief that 'match to delete' transitions are more probable than to 'delete to delete' transitions which can be built into an HMM by setting

$$\alpha_{\text{delete}}q_{\text{delete|delete}} < \alpha_{\text{match}}q_{\text{match|delete}}$$

Note that we have not modified Baum–Welch to take into account deletion states, cf. (Hughey and Krogh 1996).

15.5 Exercises

1. **Parameter matching for Gilbert's burst noise**
   We have the binary state space $S = \{b, g\}$ and the transition matrix

$$A = \begin{pmatrix} 1-p & p \\ q & 1-q \end{pmatrix}.$$

The Markov chain is hidden in another process $\{Y_n\}_{n=0}^{\infty}$ with the state space $\mathcal{O} = \{0, 1\}$. The process $\{Y_n\}_{n=0}^{\infty}$ has the emission probability matrix

$$B = \begin{pmatrix} \rho & 1-\rho \\ 1 & 0 \end{pmatrix},$$

where $0 \leq \rho \leq 1$. The parameters to be estimated are ρ, p and q. The following probabilities a,b and c can be directly counted in terms of frequencies from observed data (a sequence of observed bits):

$$a = P(Y_n = 1), \qquad b = P(Y_n = 1 \mid Y_{n-1} = 1),$$

$$c = \frac{P(Y_n = 1, Y_{n-1} = 1, Y_{n-2} = 1)}{P(Y_n = 1, Y_{n-1} = 0, Y_{n-2} = 1) + P(Y_n = 1, Y_{n-1} = 1, Y_{n-2} = 1)}$$

(a) What is the meaning of c ?

(b) Show that

$$c = \frac{(1-\rho)(1-q)^2}{(1-q)^2 + qp}.$$

(c) Solve for ρ, p and q in terms of a,b and c.

This is a method of estimation called *parameter matching.*

2. Prove Lemma 15.3.4 above.

3. If the quasi-log likelihood function for the re-estimate row $\left(a^*_{i|j}\right)_{j=1}^{J}$ given a current model λ is defined by

$$Q\left(\lambda, \left(a^*_{i|j}\right)_{j=1}^{J}\right) = \sum_{j=1}^{J} \sum_{l=1}^{n} P(Y_0, \ldots, Y_n, X_{l-1} = i, X_l = j; \lambda) \log a^*_{i|j},$$

find the formula for the re-estimates $a^*_{i|j}$ obtained by maximization of this function.

4. **n independent emitted sequences**
Let us suppose that we have r training sequences $\mathbf{o}^{(l)}$ of letters o_{l_i}

$$\mathbf{o}^{(l)} = \left(o^{(l)}_{i_1} o^{(l)}_{i_2} \ldots o^{(l)}_{i_j} \ldots o^{(l)}_{i_{m_l}}\right), \qquad l = 1, \ldots, r,$$

where $m_l = |\mathbf{o}^{(l)}|$ and $o_{i_j}^{(l)}$ are symbols in the alphabet $\mathcal{O} = \{o_1, \ldots, o_K\}$. The question is to estimate (A, B) ignoring $\pi(0)$.

Let each training sequence be treated as an independent outcome. Thus the likelihood function is

$$\prod_{k=1}^{r} P_k = \prod_{k=1}^{r} P\left(Y_1^{(k)} = o_{i_1}^{(k)}, Y_2^{(k)} = o_{i_2}^{(k)} \ldots, Y_{m_l}^{(k)} = o_{i_{m_k}}^{(k)}; A, B\right),$$

if $Y_n^{(k)}$ is seen as the nth random variable emitted by the common HMM in the kth sequence. Each factor in this product has a familiar expression in terms of a sum over all permitted paths. Verify in sufficient detail that the Baum–Welch iterations are to be modified as follows

$$a_{i|j}^{*} = \frac{\sum_{k=1}^{r} \frac{1}{P_k} a_{i|j}^{(k)} \cdot \sum_{l=0}^{m_k-1} \alpha_l^{(k)}(i) \cdot b_j^{(k)}(o_{l+1}) \cdot \beta_{l+1}^{(k)}(j)}{\sum_{k=1}^{r} \frac{1}{P_k} \sum_{l=0}^{m_k-1} \alpha_l^{(k)}(i) \cdot \beta_l^{(k)}(i)}, \tag{5.1}$$

and

$$b_j^{*}(o_s) = \frac{\sum_{k=1}^{r} \frac{1}{P_k} \sum_{l=0}^{m_k} I_{\{Y_l^{(k)}=o_s\}} \alpha_l^{(k)}(j) \cdot \beta_l^{(k)}(j)}{\sum_{k=1}^{r} \frac{1}{P_k} \sum_{l=0}^{n} \alpha_l^{(k)}(j) \cdot \beta_l^{(k)}(j)}, \tag{5.2}$$

where $a_{i|j}^{(k)}$, etc., are the Baum–Welch estimates for the kth training sequence, respectively.

5. **Viterbi Learning. Part 1: Reestimation**
Let

$$\begin{aligned} &P(\mathbf{x}^{*}(\lambda'), \mathbf{o}; \lambda') \\ &= \max_{j_0 \cdots j_n} P(X_0 = j_0, \ldots, X_n = j_n, Y_0 = o_0 \ldots, Y_n = o_n; \lambda') \end{aligned}$$

be the joint probability of the best decoded path $\mathbf{x}^{*}(\lambda') = j_0^{*} \ldots j_n^{*}$ and the observed emitted sequence $\mathbf{o}$ conditioned by the current model $\lambda' = (A', B')$. Note that the Viterbi algorithm presupposes a given current model. Next we determine λ^{*} so that

$$\lambda^{*} = \arg\max_{\lambda} P(\mathbf{x}^{*}(\lambda'), \mathbf{o}; \lambda). \tag{5.3}$$

Derive the re-estimates of A^* and B^*. You may ignore the initial distribution.
Hint: Note that

$$P(\mathbf{x}^*(\lambda'), \mathbf{o}; \lambda) = \prod_{l=0}^{n} b_{j_l^*}(o_l) \cdot \prod_{l=0}^{n} a_{j_{l-1}^*|j_l^*},$$

and that the transition counts and emission counts can be retrieved from $\mathbf{x}^*(\lambda')$. (Juang and Rabiner 1990).

6. Let the emission alphabet $\mathcal{O}$ consist of K symbols which are real p - dimensional vectors. Set $N = p \cdot (n+1)$. Show that with

$$\log L(\lambda) = \log P(Y_0 = o_0 \ldots, Y_n = o_n; \lambda),$$

that

$$0 \leq \frac{1}{N} \arg\max_\lambda \log L(\lambda) - \frac{1}{N} \log P(\mathbf{x}^*, \mathbf{o}; \lambda^*) \leq \frac{1}{p} \log J, \qquad (5.4)$$

where λ^* is defined in (5.3). *Hint*: Show first that

$$0 \leq \frac{1}{N} \log L(\lambda) - \frac{1}{N} \log P(\mathbf{x}^*, \mathbf{o}; \lambda^*) \leq \frac{1}{p} \log J.$$

What is the interpretation of this result? What happens with the result when tested on a left to right HMM? (Merhav and Ephraim 1991).

7. **Mixtures of emission distributions**
We have the emission alphabet $\mathcal{O} = \{o_1, o_2, \ldots o_K\}$ and the processes $\{Y_n\}_{n=0}^{\infty}$ and $\{X_n\}_{n=0}^{\infty}$ are for any fixed n related by the *emission probability mixtures*

$$b_j(o_k) = \sum_{r=1}^{M} \mu_{jr} b_{jr}(o_k),$$

where $\mu_{jr} \geq 0$ and $\sum_{r=1}^{M} \mu_{jr} = 1$ for any j and $b_{jr}(o_k) \geq 0$, $\sum_{k=1}^{K} b_{jr}(o_k) = 1$. The generative model is a three stage procedure. First, the Markov chain assumes the state j, then $r \in \{1, \ldots, M\}$ is picked using the distribution $\{\mu_{jr}\}_{r=1}^{M}$ and then the emitted symbol is drawn from the distribution $b_{jr}(o_k)$. Let B_n denote the variable ('the switch') that has $\{1, \ldots, M\}$ as alphabet. Then we have

$$\mu_{jr} = P(B_n = r \mid X_n = j)$$

and

$$b_{jr}(o_k) = P(Y_n = o_k \mid B_n = r, X_n = j).$$

The conditional independence assumption is made for the pairs (Y_n, B_n) given the state sequence. Thus a model is

$$\lambda = \left(A, (\mu_{jr})_{j=1,r=1}^{J,M}, (b_{jr}(o_k))_{j=1,r=1,k=1}^{J,M,K}, \pi(0)\right).$$

(a) Show that for $l \leq n$

$$P(Y_0 = o_0, \ldots, Y_n = o_n, X_l = i, B_l = r; \lambda)$$
$$= \sum_{j=1}^{J} \alpha_{n-1}(j) \cdot a_{j|i}\mu_{jr}b_{jr}(o_l)\beta_n(i).$$

(b) If the quasi-log likelihood function for $(\mu_{jr}^*)_{r=1}^{M}$ given a current model λ is defined by

$$Q\left(\lambda, (\mu_{jr}^*)_{r=1}^{M}\right)$$
$$= \sum_{r=1}^{M}\sum_{l=0}^{n} P(Y_0 = o_0, \ldots, Y_n = o_n, X_l = i, B_l = r; \lambda) \log \mu_{jr}^*,$$

find the formula for the re-estimate μ_{jr}^* obtained by maximization of this function and interpret the result in words. (Juang 1985).

8. **Delete states and Re-estimation**
Reconsider the situation in Exercise 4 above. Suppose that i is a delete state. Explain the rationale for the following modification

$$a_{i|j}^* = \frac{\sum_{k=1}^{r} \frac{1}{P_k} a_{i|j}^{(k)} \cdot \sum_{l=0}^{m_k-1} \alpha_l^{(k)}(i) \cdot b_j^{(k)}(o_l) \cdot \beta_l^{(k)}(j)}{\sum_{k=1}^{r} \frac{1}{P_k} \sum_{l=0}^{m_k-1} \alpha_l^{(k)}(i) \cdot \beta_l^{(k)}(i)} \qquad (5.5)$$

of the re-estimate of the transition probability.

9. **A Bound for Kullback distance for HMM**
By definition the Kullback distance between two HMM's is

$$D(\lambda|\lambda^*) = \bigotimes_{\mathbf{o}} P(Y_0, \ldots, Y_n; \lambda) \log \frac{P(Y_0, \ldots, Y_n; \lambda)}{P(Y_0, \ldots, Y_n; \lambda^*)},$$

where $\bigotimes_{\mathbf{o}}$ denotes summation over all $n+1$ - strings of symbols in an emission alphabet $\mathcal{O}$. There is, in general, no explicit expression for $D(\lambda|\lambda^*)$. We study some bounds for this. Here the initial distribution is omitted so that

$$P(Y_0 = o_0, \ldots, Y_n = o_n; \lambda) = \sum_{s=1}^{t} p(\mathbf{o}, s; \lambda),$$

where

$$p(\mathbf{o}, s; \lambda) := \prod_{i=1}^{J} \prod_{j=1}^{J} a_{i|j}^{n_{i|j}(s)} \cdot \prod_{j=1}^{J} \prod_{k=1}^{K} b_j(k)^{m_{j|k}(s)}$$

using the frequency counts as in the preceding. Similar expressions hold for $P(Y_0 = o_0, \ldots, Y_n = o_n; \lambda^*)$. Show (use the log sum inequality) that

$$D(\lambda|\lambda^*) \leq D_1 + D_2,$$

where

$$D_1 = \sum_{i=1}^{J} \sum_{j=1}^{J} \log \frac{a_{i|j}}{a_{i|j}^*} \sum_{\mathbf{o}} \sum_{s=1}^{t} n_{i|j}(s) p(\mathbf{o}, s; \lambda)$$

and

$$D_2 = \sum_{j=1}^{J} \sum_{k=1}^{K} \log \frac{b_j(k)}{b_j^*(k)} \sum_{\mathbf{o}} \sum_{s=1}^{t} m_{j|k}(s) p(\mathbf{o}, s; \lambda).$$

10. Show that if $\lambda = (A, B)$ is an HMM with a left to right architecture, then

$$\frac{\sum_{\mathbf{o}} \sum_{s=1}^{t} n_{i|j}(s) p(\mathbf{o}, s; \lambda)}{\sum_{\mathbf{o}} \sum_{s=1}^{t} \sum_{j=1}^{J} n_{i|j}(s) p(\mathbf{o}, s; \lambda)} = a_{i|j}$$

and

$$\frac{\sum_{\mathbf{o}} \sum_{s=1}^{t} m_{j|k}(s) p(\mathbf{o}, s; \lambda)}{\sum_{\mathbf{o}} \sum_{s=1}^{t} \sum_{k=1}^{K} m_{j|k}(s) p(\mathbf{o}, s; \lambda)} = b_j(k)$$

where the notations are those used in the Exercise 9 above.

11. **A Bound for Kullback Distance for HMM (contnd)**
Use the result in Exercise 10 to rewrite the bound $D_1 + D_2$ for the Kullback distance in Exercise 9.

12. **Re-estimation formulas for HMM with duration (I)**
This exercise is a continuation of Exercise 9 in Chapter 14 above. In loc.cit. we introduced

$$\alpha_n(j) = P\left(Y_0 = o_0, \ldots, Y_n = o_n, \text{state } j \text{ ends at } n\right). \tag{5.6}$$

In addition to this we need three more forward and backward variables:

$$\begin{aligned}
\alpha_n^\star(j) &= P\left(Y_0 = o_0, \ldots, Y_n = o_n, \text{state } j \text{ begins at } n+1\right), \\
\beta_n(j) &= P\left(Y_{n+1} = o_0, \ldots, Y_N = o_N \mid \text{state } j \text{ ends at } n\right), \\
\beta_n^\star(j) &= P\left(Y_{n+1} = o_0, \ldots, Y_N = o_N \mid \text{state } j \text{ begins at } n+1\right).
\end{aligned}$$

Show that the relationships between these are

$$\begin{aligned}
\alpha_n^\star(j) &= \sum_{i=1}^{J} \alpha_n(i) a_{i|j} \\
\alpha_n(j) &= \sum_{d=1}^{T} \alpha_{n-d}^\star(j)\phi_j(d) \prod_{r=n-d+1}^{n} b_j(k_r), \\
\beta_n(i) &= \sum_{j=1}^{J} \beta_n^\star(i) a_{i|j}, \\
\beta_n^\star(i) &= \sum_{d=1}^{T} \beta_{n+d}(i)\phi_i(d) \prod_{r=n+1}^{n+d} b_i(k_r).
\end{aligned}$$

(Rabiner 1989).

13. **Re-estimation formulas for HMM with duration (II)**
Please check carefully and explain the meaning of the following formulas for Baum-Welch algorithms for HMM with duration, where the superscript † designates the re-estimates:

(a)

$$b_j^\dagger(o_k) = \frac{\sum_{l=0}^{n} I_{\{Y_l = o_k\}} \left[\sum_{\tau<l} \alpha_\tau^\star(j)\beta_\tau^\star(j) - \sum_{\tau<l} \alpha_\tau(j)\beta_\tau(j)\right]}{\sum_{k=1}^{K}\sum_{l=0}^{n} I_{\{Y=o_k\}} \left[\sum_{\tau<l} \alpha_\tau^\star(j)\beta_\tau^\star(j) - \sum_{\tau<l} \alpha_\tau(j)\beta_\tau(j)\right]}.$$

(b)

$$a^{\dagger}_{i|j} = \frac{a_{i|j} \cdot \sum_{l=0}^{n-1} \alpha_l(i) \cdot \beta_l^{\star}(j)}{\sum_{j=1}^{J} a_{i|j} \cdot \sum_{l=0}^{n-1} \alpha_l(i) \cdot \beta_l^{\star}(j)}.$$

(c)

$$\phi_j^{\dagger}(d) = \frac{\sum_{l=0}^{n} \alpha_l^{\star}(j) \phi_j(d) \beta_{n+d} \prod_{r=l+1}^{n+d} b_j(k_r)}{\sum_{d=1}^{T} \sum_{l=0}^{n} \alpha_l^{\star}(j) \phi_j(d) \beta_{n+d} \prod_{r=l+1}^{n+d} b_j(k_r)}.$$

(Rabiner 1989).

15.6 References and Further Reading:

L.E. Baum, T. Petrie, G. Soules and N. Weiss (1970): A Maximization Technique Occurring in the Statistical Analysis of Probabilistic Functions of Markov Chains. *The Annals of Mathematical Statistics*, 41, pp. 164–171.

R.J. Elliott, L. Aggoun and J.B. Moore: *Hidden Markov Models. Estimation and Control.* Applications of Mathematics. Stochastic Modelling and Applied Probability. 29. Springer Verlag,1995, New York, Berlin.

R. Hughey and A. Krogh (1996): Hidden Markov models for sequence analysis: extensions and analysis of the basic method. *Computer Applications in Biological Sciences* (CABIOS), 12, pp. 95–107.

B.H. Juang (1985): Maximum-likelihood Estimation for Mixture Multivariate Stochastic Observations of Markov Chains. *AT & T Technical Journal*, 64, pp. 1235–1249.

B.H. Juang and L.R. Rabiner (1990): The segmental K-means Algorithm for for estimating Parameters of Hidden Markov Models. *IEEE Transactions on Acoustics, Speech and Signal Processing*, 38, pp. 1639–1641.

S.Y. Kung (1993): *Digital Neural Networks*, Chapter 7, PTR Prentice Hall, Englewood Cliffs, New Jersey, 07632.

S.E. Levinson, L.R. Rabiner and M.M. Sondhi (1983): An Introduction to the Applications of the Theory of Probabilistic Functions of a Markov Chain to Automatic Speech Recognition. *The Bell System Technical Journal*, 62, pp. 1053–1074.

N. Merhav and Y. Ephraim (1991): Maximum Likelihood Hidden Markov Modelling Using a Dominant Sequence of States. *IEEE Transactions on Signal Processing*, 39, pp. 2111–2115.

L.R. Rabiner (1989): A Tutorial on Hidden Markov Models and Selected Applications in Speech Recognition. *Proceedings of the IEEE*, 77, pp. 257 - 286.

C.P. Roberts, G. Celeux and J. Diebolt (1993): Bayesian estimation for hidden Markov chains: a stochastic implementation. *Statistics and Probability Letters*, 16, pp. 77–83.

Chapter 16

Limit Points of Baum–Welch

16.1 Introduction

16.1.1 A Summary of Re-estimates

1. For $j = 1, \ldots, J$,

$$\pi_j^*(0) = \frac{\alpha_0(j) \cdot \beta_0(j)}{P\left(Y_0 = o_0, \ldots, Y_n = o_n\right)}. \tag{1.1}$$

π_j^* is the expected frequency of j at starting time given $o_0 \ldots o_n$ and conditioned on the current model λ.

2. For $j = 1, \ldots, J$ and for $k = 1, \ldots, K$

$$b_j^*(o_k) = \frac{\sum_{l=0}^{n} I_{\{Y_l = o_k\}} \alpha_l(j) \cdot \beta_l(j)}{\sum_{l=0}^{n} \alpha_l(j) \cdot \beta_l(j)}. \tag{1.2}$$

$b_j^*(o_k)$ is the expected number of visits in state j and emitting the symbol o_k divided by the expected number of transitions from state j, given $o_0 \ldots o_n$ and conditioned on the current model λ.

3. For $j = 1, \ldots, J$ and for $i = 1, \ldots, J$

$$a_{i|j}^* = \frac{a_{i|j} \cdot \sum_{l=0}^{n-1} \alpha_l(i) \cdot b_j\left(o_{l+1}\right) \cdot \beta_{l+1}(j)}{\sum_{l=0}^{n-1} \alpha_l(i) \cdot \beta_l(i)}. \tag{1.3}$$

$a^*_{i|j}$ is the ratio of the expected number of transitions from state i to state j divided by the expected number of transitions from state i given $o_0 \dots o_n$ and conditioned on the current model λ. ∎

Note that the expressions depend on the emitted symbols $\mathbf{o} = o_0 \dots o_n$ in a non-transparent manner, since the current model and the emitted sequence $\mathbf{o}$ influence the re-estimates also through the forward and backward probabilities.

16.1.2 Plan for the Further Study

The Baum–Welch algorithm deals with iterations of a map. Namely, if the current model λ is set to $\lambda^{(t)}$ and the re-estimate λ^* is set to $\lambda^{(t+1)}$, then we write (1.1)–(1.3) in one big comprehensive formula as

$$\lambda^{(t+1)} = \tau\left(\lambda^{(t)}\right), t = 0, 1, 2, \dots. \tag{1.4}$$

We say that the Baum–Welch algorithm has *converged*, if

$$\lambda^* = \tau\left(\lambda^*\right). \tag{1.5}$$

Then λ^* is a *fixed point of* τ.

Our purpose is now to analyze the properties of the iteration and to find out how the fixed points of the algorithm are related to finding $\widehat{\lambda}_{\text{ML}}$, the maximum likelihood hidden Markov model in

$$\widehat{\lambda}_{\text{ML}} = \arg\max_{\lambda} P\left(Y_0 = o_0 \dots, Y_n = o_n; \lambda\right).$$

The calculations in section 16.2 turn out to suggest that the right hand sides of the re-estimates (1.3)–(1.2) can be associated with the stationary points of the Lagrangians corresponding to $\arg\max_{\lambda} P\left(Y_0 = o_0 \dots, Y_n = o_n; \lambda\right)$. Then we can define a general transformation of the parameter space to itself and show that the likelihood increases strictly for iterations of this transformation unless the algorithm has converged. Then it can be proved that the *limit points of the iteration* (1.4) in the interior of the parameter space *are stationary points* (*local maxima* or *saddle points*) *of the likelihood.* For definition of stationary (or critical) points we refer to any text on multivariable calculus or to (Khuri 1993). In fact the analysis of algorithmic properties of Baum–Welch or EM is not a branch of probability theory but a case for the study of non-linear maximization of a differentiable function under constraints (Kuhn–Tucker theory, see Luenberger 1984) and can be cast in terms of the theory of algorithms owed to Zangwill (Luenberger 1984, Wu 1983). In the final section we outline the global convergence analysis based on Zangwill's work.

16.1.3 The Theorem on Limit Points of Re-estimates

The fundamental result to be treated in this chapter is as follows. We introduce the manifold Λ, defined as

$$\Lambda = \{\lambda = (A, B, \pi(0)) \mid \text{satisfying (I)–(III) in section 10.2 } \}. \qquad (1.6)$$

Theorem 16.1.1 (Baum–Petrie 1970) *Assume that all the limit points of a sequence of models $\tau^t\left(\lambda^{(0)}\right)$, chosen by Baum–Welch learning, for an arbitrary $\lambda^{(0)}$, stay in the interior of the manifold Λ in* (1.6). *Then the limit points are stationary points of the likelihood function*

$$L(\lambda) = P(Y_0 = o_0 \ldots, Y_n = o_n; \lambda)$$

for an HMM. ∎

A proof of this will be worked out in the sections below. First we find a formal expression for the stationary points of $L(\lambda)$. These findings lead to the definition of the main tool of the proof, which is the so called *growth transformation* (Baum and Sell 1968).

It should be recalled that Baum–Welch learning works under the absolute continuity assumption stated in the preceding chapter. Hence an initially chosen structure of zeroes in A will be unmoved during the iterations. Also, the general architecture of the HMM is fixed but not explicitly specified.

The results and techniques in the sequel are found in (Baum et al. 1970, Levinson et al. 1983, and Rabiner 1989).

16.2 On the Stationary Points of the Likelihood

16.2.1 A Formal Optimization Calculus

Let now $L(\lambda)$ designate

$$L(\lambda) = P(Y_0 = o_0 \ldots, Y_n = o_n; \lambda).$$

Let us consider first the straightforward maximization of this with respect to A. Since the constraints are for each row separately, we introduce for Q, the Lagrange function for $L(\lambda)$ with respect to the transition matrix, by

$$Q = L(\lambda) + \sum_{i=1}^{J} \rho_i \left(\sum_{j=1}^{J} a_{i|j} - 1 \right),$$

where ρ_i are as yet undetermined Lagrange multipliers. In a **stationary point** in the interior of the manifold defined by the constraints $a_{i|j} \geq 0$, $\sum_{j=1}^{J} a_{i|j} = 1$ it will hold that

$$\frac{\partial}{\partial a_{r|s}} Q = \frac{\partial}{\partial a_{r|s}} L(\lambda) + \rho_r = 0$$

Multiplying this equation by $a_{r|s}$ and summing over s gives

$$\sum_{s=1}^{J} a_{r|s} \frac{\partial}{\partial a_{r|s}} L(\lambda) = - \left[\sum_{s=1}^{J} a_{r|s} \right] \cdot \rho_r = -\rho_r.$$

Since $\frac{\partial}{\partial a_{r|s}} L(\lambda) = -\rho_r$ at a stationary point, we have thus that $L(\lambda)$ has a stationary point when

$$a_{r|s} = \frac{a_{r|s} \frac{\partial}{\partial a_{r|s}} L(\lambda)}{\sum_{s=1}^{J} a_{r|s} \frac{\partial}{\partial a_{r|s}} L(\lambda)}. \tag{2.1}$$

This is a nonlinear and coupled system of equations for $a_{r|s}$, $r, s \in S$ which we cannot solve analytically. It turns out, however, that this way of looking at finding the maximum likelihood model $\widehat{\lambda}_{\text{ML}}$ will give some useful insights about the Baum–Welch re-estimation formulas. This is seen after some further computational manipulations.

16.2.2 Some Computational Details

We now wish to demonstrate a computational formula given for clarity of presentation as a lemma. For convenience of reference we recapitulate here the recursions for forward and backward probabilities with

$$\alpha_0(j) = b_j(o_0)\,\pi_j(0), \quad j = 1, \ldots, J, \tag{2.2}$$

$$\alpha_{l+1}(j) = \left[\sum_{i=1}^{J} \alpha_l(i) \cdot a_{i|j}\right] \cdot b_j(o_{l+1}), \quad j = 1, .., J, 1 \leq l \leq n-1, \tag{2.3}$$

$$\beta_n(j) = 1 \quad j = 1, \ldots, J, \tag{2.4}$$

$$\beta_l(j) = \sum_{i=1}^{J} b_i(o_{n+1}) \cdot \beta_{l+1}(i) \cdot a_{j|i}, \quad j = 1, .., J, l = n-1, .., 0. \tag{2.5}$$

(With respect to the notations used in Chapter 14 we have $N = n$, $n = l$.) The steps verifying the next lemma are similar to the operations in backpropagation for training multi layer perceptrons.

Lemma 16.2.1

$$\frac{\partial}{\partial a_{r|s}} L(\lambda) = \sum_{l=0}^{n-1} \alpha_l(r) \cdot b_s(o_{l+1}) \cdot \beta_{l+1}(s). \tag{2.6}$$

Proof: By our analysis of the evaluation problem using the forward and backward variables we have $n+1$ rules for computing $L(\lambda)$ and we use the evaluation formula in the form

$$L(\lambda) = P(Y_0 = o_0, \ldots, Y_n = o_n; \lambda) = \sum_{j=1}^{J} \alpha_n(j) \cdot \beta_n(j). \tag{2.7}$$

By our choice $\beta_n(j) = 1$, but we let it remain in the expression. We use next the forward recursion (2.3) for $\alpha_n(j)$ to get

$$L(\lambda) = \sum_{j=1}^{J} \sum_{i=1}^{J} \alpha_{n-1}(i) \cdot a_{i|j} \cdot b_j(o_n) \cdot \beta_n(j). \tag{2.8}$$

Hence if we differentiate partially the right hand side of (2.8) with respect to $a_{r|s}$ we obtain by the rule for derivative of product that

$$\begin{aligned} \frac{\partial}{\partial a_{r|s}} L(\lambda) &= \alpha_{n-1}(r) \cdot b_s(o_n) \cdot \beta_n(s) \\ &\quad + \sum_{j=1}^{J} \sum_{i=1}^{J} \left[\frac{\partial}{\partial a_{r|s}} \alpha_{n-1}(i)\right] \cdot a_{i|j} \cdot b_j(o_n) \cdot \beta_n(j) \end{aligned}$$

$$+\sum_{j=1}^{J}\sum_{i=1}^{J}\alpha_{n-1}(i)\cdot a_{i|j}\cdot b_j\left(o_n\right)\cdot\frac{\partial}{\partial a_{r|s}}\beta_n(j). \tag{2.9}$$

Here $\frac{\partial}{\partial a_{r|s}}\beta_n(j) = 0$ for all r, s, j. Hence we need only to calculate the value of the middle term in the right hand side. The task is accomplished for the benefit of our purposes by using the forward recursion (2.3) once more:

$$\frac{\partial}{\partial a_{r|s}}\alpha_{n-1}(i) = \frac{\partial}{\partial a_{r|s}}\left[\sum_{l=1}^{J}\alpha_{n-2}(l)\cdot a_{l|i}\right]\cdot b_i\left(o_{n-1}\right). \tag{2.10}$$

If the differentiation in the right hand side is performed we obtain

$$\frac{\partial}{\partial a_{r|s}}\alpha_{n-1}(i) = \begin{cases} \alpha_{n-2}(r)b_s\left(o_{n-1}\right) + \sum_{l=1}^{J}\frac{\partial}{\partial a_{r|s}}\alpha_{n-2}(l)\cdot a_{l|s}\cdot b_s\left(o_{n-1}\right) & i = s, \\ \\ \sum_{l=1}^{J}\frac{\partial}{\partial a_{r|s}}\alpha_{n-2}(l)\cdot a_{l|i}\cdot b_i\left(o_{n-1}\right) & i \neq s. \end{cases}$$

Insertion of this partial derivative in the middle term in the right hand side of (2.9) gives

$$\begin{aligned} &\sum_{j=1}^{J}\sum_{i=1}^{J}\left[\frac{\partial}{\partial a_{r|s}}\alpha_{n-1}(i)\right]\cdot a_{i|j}\cdot b_j\left(o_n\right)\cdot\beta_n(j) \\ = &\sum_{j=1}^{J}\alpha_{n-2}(r)b_s\left(o_{n-1}\right)a_{s|j}\cdot b_j\left(o_n\right)\cdot\beta_n(j) \\ &+\sum_{j=1}^{J}\left[\sum_{l=1}^{J}\frac{\partial}{\partial a_{r|s}}\alpha_{n-2}(l)\cdot a_{l|s}\cdot b_s\left(o_{n-1}\right)\right]a_{s|j}b_j\left(o_n\right)\beta_n(j) \\ &+\sum_{j=1}^{J}\sum_{i=1,i\neq s}^{J}\sum_{l=1}^{J}\frac{\partial}{\partial a_{r|s}}\alpha_{n-2}(l)a_{l|i}b_i\left(o_{n-1}\right)a_{i|j}b_j\left(o_n\right)\beta_n(j). \end{aligned} \tag{2.11}$$

Now we compute the value of each of the three terms in the right hand side of (2.11). First,

$$\begin{aligned} &\sum_{j=1}^{J}\alpha_{n-2}(r)b_s\left(o_{n-1}\right)a_{s|j}\cdot b_j\left(o_n\right)\cdot\beta_n(j) \\ = &\ \alpha_{n-2}(r)b_s\left(o_{n-1}\right)\left[\sum_{j=1}^{J}\left(a_{s|j}\cdot b_j\left(o_n\right)\cdot\beta_n(j)\right)\right], \end{aligned}$$

where we now use the backward recursion (2.5) to obtain

$$\alpha_{n-2}(r) b_s(o_{n-1}) \beta_{n-1}(s). \tag{2.12}$$

The two last terms in the right hand side of (2.11) are put together in the triple sum

$$\Phi = \sum_{j=1}^{J} \sum_{i=1}^{J} \sum_{l=1}^{J} \frac{\partial}{\partial a_{r|s}} \alpha_{n-2}(l) \cdot a_{l|i} \cdot b_i(o_{n-1}) a_{i|j} \cdot b_j(o_n) \cdot \beta_n(j).$$

By interchanging the order of the summations we obtain

$$\Phi = \sum_{i=1}^{J} \sum_{l=1}^{J} \left[\frac{\partial}{\partial a_{r|s}} \alpha_{n-2}(l) \right] a_{l|i} \cdot b_i(o_{n-1}) \left[\sum_{j=1}^{J} a_{i|j} b_j(o_n) \beta_n(j) \right].$$

But the backward recursion (2.5) now yields

$$\left[\sum_{j=1}^{J} a_{i|j} b_j(o_n) \beta_n(j) \right] = \beta_{n-1}(i).$$

Hence the triple sum becomes nothing else but

$$\Phi = \sum_{i=1}^{J} \sum_{l=1}^{J} \left[\frac{\partial}{\partial a_{r|s}} \alpha_{n-2}(l) \right] a_{l|i} \cdot b_i(o_{n-1}) \beta_{n-1}(i).$$

By this we have rewritten the right hand side of (2.9) as

$$\begin{aligned} \frac{\partial}{\partial a_{r|s}} L(\lambda) &= \alpha_{n-1}(r) \cdot b_s(o_n) \cdot \beta_n(s) \\ &\quad + \alpha_{n-2}(r) \cdot b_s(o_{n-1}) \cdot \beta_{n-1}(s) \\ &\quad + \sum_{i=1}^{J} \sum_{l=1}^{J} \left[\frac{\partial}{\partial a_{r|s}} \alpha_{n-2}(l) \right] a_{l|i} \cdot b_i(o_{n-1}) \beta_{n-1}(i). \end{aligned} \tag{2.13}$$

But the double sum is seen to be exactly of the same form that we have computed in (2.9). Hence we may continue by differentiation of $\frac{\partial}{\partial a_{r|s}} \alpha_{n-2}(l)$ and reorganizing the resulting sums exactly in the same manner as above. Since $\alpha_j(0)$ in (2.2) does not involve any $a_{i|j}$, we see that when going downwards the differentiation stops at $l = 0$, as it should. Thus we see that (2.6) is valid, i.e.,

$$\frac{\partial}{\partial a_{r|s}} L(\lambda) = \sum_{l=0}^{n-1} \alpha_l(r) \cdot b_s(o_{l+1}) \cdot \beta_{l+1}(s).$$

as was to be proved. ∎

16.2.3 Re-estimation Formulas Rediscovered

Let us now reconsider (2.1), i.e.,

$$a_{r|s} = \frac{a_{r|s} \frac{\partial}{\partial a_{r|s}} L(\lambda)}{\sum_{s=1}^{J} a_{r|s} \frac{\partial}{\partial a_{r|s}} L(\lambda)}.$$

If we insert (2.6) in the denominator we have

$$\sum_{s=1}^{J} a_{r|s} \frac{\partial}{\partial a_{r|s}} L(\lambda) = \sum_{s=1}^{J} \sum_{l=0}^{n-1} a_{r|s} \alpha_l(r) \cdot b_s(o_{l+1}) \cdot \beta_{l+1}(s).$$

We interchange the order of summation and get

$$\sum_{s=1}^{J} \sum_{l=0}^{n-1} a_{r|s} \alpha_l(r) \cdot b_s(o_{l+1}) \cdot \beta_{l+1}(s) = \sum_{l=0}^{n-1} \alpha_l(r) \cdot \left[\sum_{s=1}^{J} a_{r|s} b_s(o_{l+1}) \cdot \beta_{l+1}(s) \right],$$

where the backward recursion (2.5) gives

$$\sum_{s=1}^{J} a_{r|s} b_s(o_{l+1}) \cdot \beta_{l+1}(s) = \beta_l(r).$$

Hence we have the equality

$$\sum_{s=1}^{J} a_{r|s} \frac{\partial}{\partial a_{r|s}} L(\lambda) = \sum_{l=0}^{n-1} \alpha_l(r) \cdot \beta_l(r).$$

Thus

$$a_{r|s} = \frac{a_{r|s} \sum_{l=0}^{n-1} \alpha_l(r) \cdot b_s(o_{l+1}) \cdot \beta_{l+1}(s)}{\sum_{l=0}^{n-1} \alpha_l(r) \cdot \beta_l(r)}. \tag{2.14}$$

This is a system of coupled non-linear equations in $a_{r|s}$. The right hand side of (2.14) is identical to re-estimation formula (1.3).

In other words, a fixed point of the Baum-Welch re-estimation iteration is a stationary point of $P(\lambda)$ in the interior of the domain of λ, since the fixed point solves the equations (2.1). With a similar piece of mathematics we can analyze $\frac{\partial}{\partial \pi_j(0)} L(\lambda)$ and $\frac{\partial}{\partial b_j(o_k)} L(\lambda)$ to obtain the analogous conclusion about the re-estimation formulas (1.2) and (1.1).

16.3 The Growth Transformation

16.3.1 Introduction

The preceding discussion of (2.1) or

$$a_{r|s} = \frac{a_{r|s} \frac{\partial}{\partial a_{r|s}} L(\lambda)}{\sum_{s=1}^{J} a_{r|s} \frac{\partial}{\partial a_{r|s}} L(\lambda)}$$

suggests that it should be useful to consider the following general type of transformations

$$\mathbf{T}_{ij}(\mathbf{x}) = \frac{x_{ij} \frac{\partial}{\partial x_{ij}} \phi(\mathbf{x})}{\sum_{j=1}^{q_i} x_{ij} \frac{\partial}{\partial x_{ij}} \phi(\mathbf{x})} \tag{3.1}$$

of a manifold to itself. The detailed definition of the pertinent manifold is given below. By the argument in the preceding subsection the right hand sides of the re-estimation formulas (1.3)–(1.1) are special cases of this transformation (in their respective parameter spaces). We shall now study this transformation and show why it can be called a *growth transformation* and why the Baum–Welch re-estimation procedure will (locally) maximize the likelihood. We need one more observation.

16.3.2 Polynomials Homogeneous in Degree n

Let us recall the probability formula

$$\begin{aligned} & P\left(Y_0 = o_0 \ldots, Y_n = o_n; \lambda\right) \\ = & \sum_{j_o=1}^{J} \cdots \sum_{j_n=1}^{J} [\pi_{j_0}(0)\, b_{j_0}(0) \cdot \prod_{l=1}^{n} a_{j_{l-1}|j_l} b_{j_l}(l)]. \end{aligned} \tag{3.2}$$

In (3.2) we effectively sum over the state sequences of length $n+1$ that have positive probability conditioned on the model λ and given the sequence of emission symbols. We enumerate, as done before, the state sequences $(j_0 \ldots j_n)$ by the index s, $s = 1, \ldots, t$ if the number of such sequences is t. Then there is the correspondence

$$(j_0 \ldots j_n) \mapsto s, s = 1, \ldots t.$$

Also, we recall

$$n_{i|j}(s) = \text{ the number of } l \text{ such that } 1 \leq l \leq n, \qquad j_{l-1} = i, j_l = j, \tag{3.3}$$

the number of transitions of the Markov chain from state i to state j,

$$m_{j|k}(s) = \text{ the number of times the symbol } k \text{ is emitted in state } j \tag{3.4}$$

and

$$r_j(s) = \begin{cases} 1 & \text{if the initial state } j_0 = j, \\ 0 & \text{otherwise.} \end{cases} \tag{3.5}$$

Then we can reorganize (3.2) to obtain

$$\sum_{s=1}^{t} \prod_{j=1}^{J} \pi_j(0)^{r_j(s)} \cdot \prod_{i=1}^{J} \prod_{j=1}^{J} a_{i|j}^{n_{i|j}(s)} \cdot \prod_{j=1}^{J} \prod_{k=1}^{K} b_j(k)^{m_{j|k}(s)}. \tag{3.6}$$

Let us regard this as a function of $a_{i|j}$, $i = 1, \ldots, J$, $j = 1, \ldots, J$. We substitute $\rho \cdot a_{i|j}$ for $a_{i|j}$, respectively, in this expression. Here ρ is a real number. Then

$$\prod_{i=1}^{J} \prod_{j=1}^{J} \left(\rho \cdot a_{i|j}\right)^{n_{i|j}(s)} = \rho^{\sum_i \sum_j n_{i|j}(s)} \cdot \prod_{i=1}^{J} \prod_{j=1}^{J} a_{i|j}^{n_{i|j}(s)}.$$

For any path s,

$$\sum_i \sum_j n_{i|j}(s) = \sum_i n_i(s) = n$$

since $\sum_j n_{i|j}(s) = n_i(s)$ is the number of times the path s visits the state i prior to time n and the path visits one and only one of the states at each instant, and there are n instants counted. Thus $\sum_i \sum_j n_{i|j}(s)$ has the same value for each path s and we have

$$\begin{aligned} &\sum_{s=1}^{t} \prod_{j=1}^{J} \pi_j(0)^{r_j(s)} \cdot \prod_{i=1}^{J} \prod_{j=1}^{J} \left(\rho a_{i|j}\right)^{n_{i|j}(s)} \cdot \prod_{j=1}^{J} \prod_{k=1}^{K} b_j(k)^{m_{j|k}(s)} \\ &= \rho^{n+1} \sum_{s=1}^{t} \prod_{j=1}^{J} \pi_j(0)^{r_j(s)} \cdot \prod_{j=1}^{J} \prod_{k=1}^{K} b_j(k)^{m_{j|k}(s)} \cdot \prod_{i=1}^{J} \prod_{j=1}^{J} a_{i|j}^{n_{i|j}(s)}. \end{aligned} \tag{3.7}$$

This shows that $P(Y_0 = o_0 \ldots, Y_n = o_n; \lambda)$ is, as a function of $a_{i|j}$, $i = 1, \ldots, J$, $j = 1, \ldots, J$, a homogeneous polynomial of degree $n+1$. The coefficients of this polynomial, composed of $\prod_{j=1}^{J} \pi_j(0)^{r_j(s)}$, $\prod_{i=1}^{J} \prod_{j=1}^{J} a_{i|j}^{n_{i|j}(s)}$, and $\prod_{k=1}^{K} b_j(k)^{m_{j|k}(s)}$ are positive.

A similar argument (note, e.g., $\sum_j \sum_k m_{j|k}(s) = n+1$) shows that $L(\lambda)$ taken in turn as a function of $b_j(k)$, $j = 1, \ldots, J$, $k = 1, \ldots, K$ and of $\pi(0)$ is a homogeneous polynomial of a certain degree with regard to each of the groups of variables in the expression. These observations give us the crucial property of the growth transformations to be found in the next section.

16.3.3 Baum–Eagon Inequality

The theorem in this section has been proven in (Baum and Eagon 1968). In (Baum and Sell 1968) the restriction in (Baum and Eagon 1968) to polynomials homogeneneous in degree d is relaxed to include all polynomials with positive coefficients. The proof in (Baum and Sell 1968) requires more advanced mathematics than what is reasonable to discuss here. We need some notations. $\mathcal{S}$ designates the manifold

$$\mathcal{S} = \left\{ \mathbf{x} = (x_{ij})_{i=1,j=1}^{p,q_i} \,|x_{ij} \geq 0, \sum_{j=1}^{q_i} x_{ij} = 1, i = 1, \ldots, p, j = 1, \ldots, q_i \right\}. \tag{3.8}$$

We also let

$$\mu = (\mu_{ij})_{i=1,j=1}^{p,q_i}$$

denote an array of non-negative integers. In addition we set

$$\mathbf{x}^{\mu} = \prod_{i=1}^{p} \prod_{j=1}^{q_i} x_{ij}^{\mu_{ij}}.$$

Similarly c_μ is an abbreviation for $c_{\mu_{ij}}$. A *polynomial* ϕ is written in these notations as

$$\phi(\mathbf{x}) = \sum_{\mu} c_\mu \mathbf{x}^{\mu}.$$

The notation $\frac{\partial}{\partial x_{ij}} \phi(\mathbf{x})$ means the partial derivative of ϕ evaluated at point $\mathbf{x} \in \mathcal{S}$.

Theorem 16.3.1 *Let $\phi(\mathbf{x})$ be a polynomial with nonnegative coefficients, which is homogeneous of degree d in its variables x_{ij}. Let $\mathbf{T}(\mathbf{x})$ denote the point of $\mathcal{S}$, whose i, j coordinate is*

$$\mathbf{T}_{ij}(\mathbf{x}) = \frac{x_{ij} \frac{\partial}{\partial x_{ij}} \phi(\mathbf{x})}{\sum_{j=1}^{q_i} x_{ij} \frac{\partial}{\partial x_{ij}} \phi(\mathbf{x})}. \tag{3.9}$$

Then

$$\phi(\mathbf{T}(\mathbf{x})) > \phi(\mathbf{x}) \tag{3.10}$$

unless $\mathbf{T}(\mathbf{x}) = \mathbf{x}$.

Proof: Using the notations introduced above we have that

$$\mathbf{T}_{ij}(\mathbf{x}) = \frac{\sum_{\mu} c_\mu \mu_{ij} \mathbf{x}^{\mu}}{\sum_{j=1}^{q_i} \sum_{\mu} c_\mu \mu_{ij} \mathbf{x}^{\mu}}.$$

In other words we wish to prove

$$\phi \equiv \phi(\mathbf{x}) = \sum_{\mu} c_\mu \mathbf{x}^{\mu} \le \sum_{\mu} c_\mu \prod_{i=1}^{p} \prod_{j=1}^{q_i} \mathbf{T}_{ij}(\mathbf{x})^{\mu_{ij}}. \tag{3.11}$$

To do this we note the identity

$$\phi = \sum_\mu \left[c_\mu \prod_{i=1}^{p} \prod_{j=1}^{q_i} \mathbf{T}_{ij}(\mathbf{x})^{\mu_{ij}} \right]^{1/(d+1)} \cdot \left[c_\mu^{d/(d+1)} \mathbf{x}^\mu \prod_{i=1}^{p} \prod_{j=1}^{q_i} \left(\frac{1}{\mathbf{T}_{ij}(\mathbf{x})} \right)^{\mu_{ij}/(d+1)} \right].$$

Now we apply Hölder's inequality, see (7.3) in Appendix 16.7, which gives

$$\phi \le \left[\sum_\mu c_\mu \prod_{i=1}^{p} \prod_{j=1}^{q_i} \mathbf{T}_{ij}(\mathbf{x})^{\mu_{ij}} \right]^{1/(d+1)} \left[\sum_\mu c_\mu \mathbf{x}^\mu \prod_{i=1}^{p} \prod_{j=1}^{q_i} \left(\frac{x_{ij}}{\mathbf{T}_{ij}(\mathbf{x})} \right)^{\mu_{ij}/d} \right]^{d/d+1} \tag{3.12}$$

In this last step we have used the identity

$$(\mathbf{x}^\mu)^{(d+1)/d} = \mathbf{x}^\mu \cdot \prod_{i=1}^{p} \prod_{j=1}^{q_i} x_{ij}^{\mu_{ij}/d}.$$

Since

$$\sum_{i=1}^{p} \sum_{j=1}^{q_i} \mu_{ij}/d \equiv 1$$

by homogeneity of ϕ, we can apply the Geometric Mean–Arithmetic Mean Inequality (7.1) proved in Appendix 16.7, to the double products within the second bracket in (3.12) and conclude that

$$\sum_\mu c_\mu \mathbf{x}^\mu \prod_{i=1}^{p} \prod_{j=1}^{q_i} \left(\frac{x_{ij}}{\mathbf{T}_{ij}(\mathbf{x})} \right)^{\mu_{ij}/d} \le \sum_\mu c_\mu \mathbf{x}^\mu \sum_{i=1}^{p} \sum_{j=1}^{q_i} \frac{\mu_{ij}}{d} \left(\frac{x_{ij}}{\mathbf{T}_{ij}(\mathbf{x})} \right). \tag{3.13}$$

Here we substitute the definition of $\mathbf{T}_{ij}(\mathbf{x})$ as given in (3.9) in the right hand side of (3.13) and obtain

$$\sum_\mu c_\mu \mathbf{x}^\mu \sum_{i=1}^{p} \sum_{j=1}^{q_i} \frac{\mu_{ij}}{d} \left(\frac{x_{ij}}{\mathbf{T}_{ij}(\mathbf{x})} \right)$$

$$= \frac{1}{d} \sum_\mu c_\mu \mathbf{x}^\mu \sum_{i=1}^{p} \sum_{j=1}^{q_i} \mu_{ij} x_{ij} \cdot \left(\frac{\sum_{j_0=1}^{q_i} \sum_{\mu'} c_{\mu'} \mu'_{ij_0} \mathbf{x}^{\mu'}}{\sum_{\mu'} c_{\mu'} \mu'_{ij} \mathbf{x}^{\mu'}} \right). \tag{3.14}$$

Then we interchange the order of the finite summations to obtain

$$\frac{1}{d} \sum_{i=1}^{p} \sum_{j=1}^{q_i} x_{ij} \left[\frac{\sum_\mu c_\mu \mu_{ij} \mathbf{x}^\mu}{\sum_{\mu'} c_{\mu'} \mu'_{ij} \mathbf{x}^{\mu'}} \right] \cdot \sum_{j_0=1}^{q_i} \sum_{\mu'} c_{\mu'} \mu'_{ij_0} \mathbf{x}^{\mu'},$$

for each i, j the expression within the brackets $[\cdot]$ is $= 1$ and by hypothesis $\sum_{j=1}^{q_i} x_{ij} = 1$ for each i. Hence the whole of the latter expression reduces to nothing more than

$$\frac{1}{d}\sum_{i=1}^{p}\sum_{j_0=1}^{q_i}\sum_{\mu'} c_{\mu'}\mu'_{ij_0}\mathbf{x}^{\mu'}.$$

This is, however, exactly equal to

$$\frac{1}{d}\sum_{i=1}^{p}\sum_{j_0=1}^{q_i} x_{ij}\frac{\partial\phi}{\partial x_{ij_0}},$$

which by Euler's theorem on homogeneous functions (7.5), see Appendix 16.7, is equal to

$$\sum_{\mu} c_{\mu}\mathbf{x}^{\mu}.$$

Thus if we now apply this upper bound $\sum_{\mu} c_{\mu}\mathbf{x}^{\mu}$ to the expression within the second brackets in (3.12), raise both sides of (3.12) to the $(d+1)$st power, and divide both sides of the resulting inequality by $\left(\sum_{\mu} c_{\mu}\mathbf{x}^{\mu}\right)^{d}$, we have the desired inequality (3.11). That

$$\phi\left(\mathbf{T}\left(\mathbf{x}\right)\right) > \phi\left(\mathbf{x}\right)$$

if $\mathbf{T}(\mathbf{x}) \neq \mathbf{x}$ follows by (3.13) and the case of strict Geometric Mean–Arithmetic Mean Inequality, if the terms in the sum are not all equal. ∎

16.4 Baum–Welch Learning and Maximum Likelihood

We have in the above introduced the comprehensive notation for Baum–Welch re-estimation as

$$\lambda^{(t+1)} = \tau\left(\lambda^{(t)}\right), \qquad t = 0, 1, 2, \ldots, \tag{4.1}$$

and said that the Baum-Welch-algorithm has *converged to* λ, if

$$\lambda^{*} = \tau\left(\lambda^{*}\right). \tag{4.2}$$

It is clear from the explicit expressions above that τ is a continuous map in the appropriate manifold. The likelihood function is

$$L(\lambda) = \sum_{s=1}^{t} \prod_{j=1}^{J} \pi_j(0)^{r_j(s)} \cdot \prod_{i=1}^{J} \prod_{j=1}^{J} a_{i|j}^{n_{i|j}(s)} \cdot \prod_{j=1}^{J} \prod_{k=1}^{K} b_j(k)^{m_{j|k}(s)}. \tag{4.3}$$

Application of theorem 16.3.1 to $L(\lambda)$ as ϕ (three times with A, B and $\pi(0)$ as respective variables and constants), as is possible, yields that

$$L\left(\tau\left(\lambda^{(t)}\right)\right) > L\left(\lambda^{(t)}\right). \tag{4.4}$$

In words, the theorem says that each Baum–Welch iteration improves the model strictly in terms of likelihood unless $\lambda^{(t)} = \tau\left(\lambda^{(t)}\right)$, i.e., a fixed point is met. But by the optimization calculus and computations in Section 16.2 we have found that at a fixed point λ^* of the Baum–Welch iteration the likelihood has a stationary point. For an arbitrary initial model $\lambda^{(0)}$ We may express (4.1) as

$$\lambda^{(t)} = \tau^t\left(\lambda^{(0)}\right).$$

Consider now a subsequence t_i such that

$$\tau^{t_i}\left(\lambda^{(0)}\right) \rightarrow \lambda^{\infty},$$

as $i \rightarrow \infty$. Then $\tau^{t_i+1}\left(\lambda^{(0)}\right)$ converges to $\tau(\lambda^{\infty})$ by continuity of τ. By Theorem 16.3.1 above

$$L\left(\tau^{t_i}\left(\lambda^{(0)}\right)\right) \leq L\left(\tau^{t_i+1}\left(\lambda^{(0)}\right)\right) \leq L\left(\tau^{t_{i+1}}\left(\lambda^{(0)}\right)\right),$$

and by the continuity of L this gives

$$L(\lambda^{\infty}) \leq L(\tau(\lambda^{\infty})) \leq L(\lambda^{\infty}),$$

or

$$L(\lambda^{\infty}) = L(\tau(\lambda^{\infty})).$$

Since this equality can by the theorem above hold if and only if $\lambda^{\infty} = \tau(\lambda^{\infty})$ we have shown that all limit points of the sequence $\tau^t\left(\lambda^{(0)}\right)$ are fixed points of τ and a fortiori stationary points of $L(\lambda)$. Thus we have more or less convinced ourselves of theorem 16.1.1 as stated above. ∎

16.5 The Global Convergence Theorem

16.5.1 Zangwill's Theory of Algorithms

Next we summarize a section of (Luenberger 1984) explaining a theory owed to W.I. Zangwill. Let Λ be an open subset of a p-dimensional Euclidean space R^p. An **algorithm τ on Λ** (in Zangwill's sense) is a mapping from points in Λ to subsets of Λ, so that we can write $\lambda' \in \tau(\lambda)$. When mapping a point to a point τ is simply a function.

We say that an algorithm τ on Λ is **closed** if $\lambda \in \Lambda$, $\zeta \in \Lambda$, $\lambda^{(n)} \to \lambda^*$, $\zeta^{(n)} \to \zeta$ and $\zeta^{(n)} \in \tau(\lambda^{(n)})$ imply that $\zeta \in \tau(\lambda^*)$. For a point to point mapping, continuity of τ implies closedness.

Let C^* be the set of fixed points of τ. A function g from Λ to a subset of the real axis is called an **ascent function for the algorithm** τ if

1) g is continuous.

2) $g(\zeta) > g(\lambda)$ for $\zeta \in \tau(\lambda)$ and $\lambda \notin C^*$.

3) $g(\zeta) \geq g(\lambda)$ for $\zeta \in \tau(\lambda)$ and $\lambda \in C^*$.

Strict inequality in 2) is the crucial detail in part **2)**. Then the following theorem can be proved.

Theorem 16.5.1 (Zangwill) *Let the sequence* $\left\{\lambda^{(n)}\right\}_{n=0}^{\infty}$ *be generated by an algorithm* τ *such that*

$$\lambda^{(n+1)} \in \tau\left(\lambda^{(n)}\right)$$

for some $\lambda^{(0)} \in \Lambda$. *Let* τ *be closed and* $C^* \subset \Lambda$ *be the set of fixed points of* τ. *Then*

i) C^* *is closed;*

ii) *all the accumulation points of* $\left\{\lambda^{(n)}\right\}_{n=0}^{\infty}$ *are in* C^**;*

iii) *if* g *is an ascent function, then* $g\left(\lambda^{(n)}\right)$ *converges monotonically to* $g\left(\lambda^*\right)$ *for some* $\lambda^* \in C^*$.

Proof: One short proof is found in (Luenberger 1984). ∎

The result was used by (Wu 1983) to prove the convergence of the EM-algorithm.

A hidden Markov model is a point $\lambda \in \Lambda$, the manifold for the model parameters in (1.6), and to each λ we have a smooth assignment

$$\lambda \mapsto (\pi(\lambda), A(\lambda), B(\lambda)).$$

As Λ is compact and as

$$P(\mathbf{x}, \mathbf{o} \mid \lambda) = \prod_{j=1}^{J} \pi_j(0)^{r_j(s)} \cdot \prod_{i=1}^{J}\prod_{j=1}^{J} a_{i|j}^{n_{i|j}(s)} \cdot \prod_{j=1}^{J}\prod_{k=1}^{K} b_j(k)^{m_{j|k}(s)}.$$

is continuous in Λ and differentiable in its interior, $P(\mathbf{x}, \mathbf{o} \mid \lambda)$ is bounded above. Here $s \mapsto \mathbf{x}$. The properties of Viterbi Learning in (Juang and L.R. Rabiner 1990) are found by proving that the maximum of $P(\mathbf{x}, \mathbf{o} \mid \lambda)$ is an ascent function. This is treated in exercises that follow below.

16.6 Exercises

1. Find the conditions for convergence in terms of Zangwill's theory for the EM algorithms in Chapter 4 (Wu 1983).

2. Is the likelihood function an ascent function (in Zangwill's sense) for the Baum–Welch algorithm?

3. **Viterbi Learning. Part 2: Algorithm and Convergence**
   This exercise is a continuation of Exercise 5 in Chapter 15. Let $\lambda^{(t)}$ be a current model and let

$$P(\mathbf{x}^*(\lambda^{(t)}), \mathbf{o}; \lambda^{(t)}) = \max_{j_0 \cdots j_n} P\left(X_0 = j_0, \ldots, X_n = j_n, Y_0 = o_0, \ldots, Y_n = o_n; \lambda^{(t)}\right)$$

   or in a compact notation

$$P(\mathbf{x}^*(\lambda^{(t)}), \mathbf{o}; \lambda^{(t)}) = \max_{\mathbf{x}} P(\mathbf{x}, \mathbf{o}; \lambda^{(t)})$$

   be the joint probability of the best decoded path $\mathbf{x}^*(\lambda^{(t)}) = j_0^* \ldots j_n^*$ and the observed emitted sequence $\mathbf{o}$ conditioned by the current model $\lambda^{(t)}$. Next we find $\lambda^{(t+1)}$ so that

$$\lambda^{(t+1)} = \arg\max_{\lambda} P(\mathbf{x}^*(\lambda^{(t)}), \mathbf{o}; \lambda). \tag{6.1}$$

The probability in the right hand side was given explicitly in Exercise 5 in Chapter 15. If, next, $\lambda^{(t+1)}$ becomes the current model and the steps above are repeated, we obtain a sequence of models. If this sequence converges we say that we have found the model by *Viterbi learning.* Let us write the overall procedure as

$$\lambda^{(t+1)} = \tau\left(\lambda^{(t)}\right), \qquad t = 0, 1, \ldots.$$

Fixed points of the algorithm are models such that

$$\lambda^{*} = \tau\left(\lambda^{*}\right)$$

or in other words such that

$$\lambda^{*} = \arg\max_{\lambda} P(\mathbf{x}^{*}\left(\lambda^{*}\right), \mathbf{o}; \lambda). \tag{6.2}$$

(a) Show that τ is closed.

(b) Show that

$$P\left(\mathbf{x}^{*}\left(\lambda^{(t+1)}\right), \mathbf{o}; \lambda^{(t+1)}\right) \geq P\left(\mathbf{x}^{*}\left(\lambda^{(t)}\right), \mathbf{o}; \lambda^{(t)}\right).$$

(c) Show that

$$g\left(\lambda\right) = \max_{\mathbf{x}} P(\mathbf{x}, \mathbf{o}; \lambda)$$

is an ascent function for τ in Zangwill's sense. *Hint*: Interpret (*b*) and the cases for strict inequality in it.

(d) Is Viterbi learning globally convergent?

4. **Smooth Learning for HMM**
A 'smooth' alternative to Baum–Welch learning algorithm based on gradient ascent has been proposed and tested. A gradient ascent algorithm is an iterative algorithm, where the parameter of interest, say θ, is adjusted according to the rule

$$\theta_{\text{new}} = \theta_{\text{current}} - \eta\frac{\partial}{\partial\theta}\psi\left(\theta_{\text{current}}\right).$$

where η is the step size or the learning rate, which can be fixed or adjusted during the learning process.

For HMM this can applied as follows. We assume that $\pi(0)$ is a known distribution and ignore it in this exercise, so that $\lambda = (A, B)$.

(a) *Reparametrization:* Set

$$a_{i|j} = \frac{e^{\theta_{i|j}}}{\sum_{l=1}^{J} e^{\theta_{i|l}}}, b_j\left(o_k\right) = \frac{e^{\nu_{i|k}}}{\sum_{l=1}^{K} e^{\nu_{i|l}}}.$$

Then show that

$$\frac{\partial}{\partial \theta_{i|j}} a_{i|j} = a_{i|j}\left(1 - a_{i|j}\right),$$

$$\frac{\partial}{\partial \theta_{i|l}} a_{i|j} = -a_{i|j} \cdot a_{i|l}, \qquad j \neq l.$$

and similarly for $b_j\left(o_k\right)$.

(b) *Gradient descent for log likelihood:* We are going to derive an algorithm for gradient descent for $L\left(\lambda\right)$ with $a_{i|j}$ as parameters to be estimated. Show that if

$$\left(a_{i|j}\right)_{\text{new}} - \left(a_{i|j}\right)_{\text{current}} = -\eta \frac{\partial}{\partial \theta_{i|j}} L\left(\left(a_{i|j}\right)_{\text{current}}\right).$$

then

$$\begin{aligned}&\left(a_{i|j}\right)_{\text{new}} - \left(a_{i|j}\right)_{\text{current}} \\ &= C_1 \cdot \frac{a_{i|j} \sum_{l=0}^{N} \left(\alpha_l(i) \cdot b_j(o_{l+1}) \cdot \beta_{l+1}(j) - \alpha_l(i) \cdot \beta_l(j)\right)}{L\left(\lambda\right)},\end{aligned}$$

where C_1 is a constant. The right hand side is evaluated at the current estimates of A and B.
Hint: Lemma 16.2.1 can be useful here.

(c) *Gradient descent for log likelihood:* Similarly

$$\begin{aligned}&b_j\left(o_k\right)_{\text{new}} - b_j\left(o_k\right)_{\text{current}} \\ &= C_2 \cdot \frac{\sum_{l=0}^{N} \left(I_{\{Y_l^{(k)}=o_s\}} \alpha_l(i) \cdot \beta_l(j) - b_j\left(o_k\right) \cdot \alpha_l(i) \cdot \beta_l(j)\right)}{L\left(\lambda\right)}.\end{aligned}$$

C_2 is a constant. This can require a substantial amount of additional work since a derivation of a suitable counterpart of Lemma 16.2.1 is needed. (Baldi and Chauvin 1994).

16.7 Appendix: Results in Mathematical Analysis

For convenience of reference we recall a few known results from analysis, see, e.g., (Khuri 1993).

16.7.1 Geometric Mean–Arithmetic Mean Inequality

Let $a_1, a_2, \ldots, a_n$ be non-negative real numbers and let $\mu_i \geq 0$ and $\sum_{i=1}^{n} \mu_i = 1$. Then

$$\prod_{i=1}^{n} a_i^{\mu_i} \leq \sum_{i=1}^{n} \mu_i a_i. \tag{7.1}$$

and equality obtains if and only if $a_1 = a_2 = \ldots = a_n$.
Proof: Taking logarithm of the left hand side in (7.1) we obtain

$$\log \prod_{i=1}^{n} a_i^{\mu_i} = \sum_{i=1}^{n} \mu_i \log a_i.$$

But $-\log x$ is a convex function for $x > 0$, since its second derivative is strictly positive. Thus Jensen's inequality gives

$$-\sum_{i=1}^{n} \mu_i \log a_i \geq -\log \left(\sum_{i=1}^{n} \mu_i a_i \right).$$

In other words we have shown that

$$\log \prod_{i=1}^{n} a_i^{\mu_i} \leq \log \left(\sum_{i=1}^{n} \mu_i a_i \right),$$

which is (7.1). ∎
A special case of (7.1) is

$$(a_1 \cdot a_2 \cdot \ldots \cdot a_n)^{1/n} \leq \frac{1}{n} (a_1 + a_2 + \ldots + a_n). \tag{7.2}$$

The left hand side of (7.2) is called the *geometric mean* and the right hand side is called the *arithmetic mean*. This explains the name for (7.1).

16.7.2 Hölder's inequality

Let $p > 1$ and $q > 0$ be integers such that $\frac{1}{p} + \frac{1}{q} = 1$ (conjugate integers). Then for any non-negative real numbers $a_1 \ldots \cdot a_n$ and $b_1 \ldots \cdot b_n$,

$$\sum_{i=1}^{n} a_i b_i \leq \left(\sum_{i=1}^{n} a_i^p\right)^{1/p} \cdot \left(\sum_{i=1}^{n} b_i^q\right)^{1/q}. \tag{7.3}$$

Proof: This is proved with (7.1) as starting point, (Khuri 1993, pp. 231−232). ■

The special case encountered in Chapter 2 above is the *Cauchy−Schwartz inequality* obtained with $p = q = 2$.

16.7.3 Euler's Theorem on Homogeneous Functions

A function of n real variables $f(x_1, x_2, \ldots, x_n)$ is said to be *homogeneous of degree d*, if for every real ρ

$$f(\rho \cdot x_1, \rho \cdot x_2, \ldots, \rho \cdot x_n) = \rho^d \cdot f(x_1, x_2, \ldots, x_n). \tag{7.4}$$

Euler's theorem on homogeneous functions of degree d says that

$$\sum_{i=1}^{n} x_i \frac{\partial}{\partial x_i} f = d \cdot f. \tag{7.5}$$

Proof of (7.5): Differentiate $f(\rho \cdot x_1, \rho \cdot x_2, \ldots, \rho \cdot x_n)$ with respect to ρ. The identity (7.4) gives

$$f_\rho'(\rho \cdot x_1, \rho \cdot x_2, \ldots, \rho \cdot x_n) = \sum_{i=1}^{n} x_i \frac{\partial}{\partial x_i} f = d \cdot \rho^{d-1} \cdot f(x_1, x_2, \ldots, x_n).$$

Then set $\rho = 1$. ■

16.8 References and Further Reading:

- P. Baldi and Y. Chauvin (1994): Smooth On-Line Learning Algorithms for Hidden Markov Models. *Neural Computation*, 6, pp. 307−318.

- L.E. Baum, T. Petrie, G. Soules and N. Weiss (1970): A Maximization Technique Occurring in the Statistical Analysis of Probabilistic Functions of Markov Chains. *The Annals of Mathematical Statistics*, 41, pp. 164–171.
- L.E. Baum, and J.A. Eagon (1968): An Inequality with Applications to Statistical Estimation for Probabilistic Functions of Markov Processes and to a Model for Ecology. *Bulletin of the American Mathematical Society*, 73, pp. 360–363.
- L.E. Baum, and G.R. Sell (1968): Growth Transformations for Functions on Manifolds. *Pacific Journal of Mathematics*, 27, pp. 211–227.
- B.H. Juang and L.R. Rabiner (1990): The segmental K-means Algorithm for estimating Parameters of Hidden Markov Models. *IEEE Transactions on Acoustics, Speech and Signal Processing*, 38, pp. 1639–1641.
- A.I. Khuri (1993): *Advanced Calculus with Applications in Statistics*. John Wiley and Sons, New York.
- S.E. Levinson, L.R. Rabiner and M.M. Sondhi (1983): An Introduction to the Applications of the Theory of Probabilistic Functions of a Markov Chain to Automatic Speech Recognition. *The Bell System Technical Journal*, 62, pp. 1053–1074.
- D.G. Luenberger (1984): *Linear and Nonlinear Programming 2. ed.* Addison-Wesley. Reading, Mass..
- L.R. Rabiner (1989): A Tutorial on Hidden Markov Models and Selected Applications in Speech Recognition. *Proceedings of the IEEE*, 77, pp. 257–286.
- C.F.J Wu (1983): On the Convergence Properties of the EM Algorithm. *The Annals of Statistics*, 11, pp. 95–103.

Chapter 17

Asymptotics of Learning

17.1 Introduction

In the preceding it has been shown how to effectively compute for $\widehat{\lambda}_{\mathrm{ML}}^{(n)}$, the maximum likelihood hidden Markov model defined by

$$\widehat{\lambda}_{\mathrm{ML}}^{(n)} = \arg\max_{\lambda} P\left(Y_0 = o_0 \ldots, Y_n = o_n; \lambda\right) \tag{1.1}$$

based on the sequence $o_0 \ldots o_n$. As in the learning of probabilities of multinomial processes in Chapter 3 and the learning of transition matrices of Markov chains in Chapter 8 the asymptotic properties of the maximum likelihood estimate can be of interest. By asymptotics of the maximum likelihood estimate we understand the properties of $\widehat{\lambda}_{\mathrm{ML}}^{(n)}$ as n increases without bound.

By *consistency* of the maximum likelihood estimate one usually means that

$$\widehat{\lambda}_{\mathrm{ML}}^{(n)} \to \lambda_o \qquad \text{as } n \to \infty \tag{1.2}$$

almost surely, where the limiting model λ_o is the *true generator* of the sequence $o_0 \ldots o_n \cdots$, the length of which is increasing.

It is, of course, difficult to insist that there is some standard or non-standard HMM or a hidden semi-Markov model which has, in reality, generated the sequences analyzed in genetics and molecular biology. Yet there is a legitimate interest in checking the property in (1.2). Because if it were not possible to find with an infinite amount of data their true generator, when the model family actually contains one such, the method of learning would seem to have a serious defect. Another asymptotic property which can be established in the same setting is the asymptotic normal distribution of $\widehat{\lambda}^{(n)}_{\mathrm{ML}}$, but this will not be elaborated here, cf. (Baum and Petrie 1966, Bickel et al. 1998). Algorithms for estimating variances of the estimated parameters and computing the Fisher information are given in (Hughes 1997).

In this chapter we shall give an outline of the argument establishing (1.2) following (Baum and Petrie 1966, Petrie 1969), see also (Leroux 1992, Muri 1997). The presentation is not self-contained, since the ergodic properties (stated in full generality, e.g., in (Gray 1988)) required in the proofs are beyond the scope of this text. The proofs can also be made more streamlined by invoking martingale convergence, but the same comment applies here.

In addition, some finer details like, e.g., the domain of definition of

$$P\left(Y_0 \mid Y_{-1}, \ldots, ; \lambda\right),$$

which involves infinite product spaces and the sigma fields generated by the cylinder sets in infinite product spaces, have been omitted. Hence a large degree of mathematical rigour is absent, but the main line of reasoning becomes perhaps more transparent.

One requirement for (1.2) to hold is that the transition probability matrix A has to be *ergodic* in the sense defined in Chapter 7. This excludes the left to right architectures developed for sequence families. The results obtained in the ergodic case are of interest for models of segmentation for DNA sequences, see, e.g., (Muri 1997).

17.2 Additional Assumptions

We need to impose a few additional assumptions on the standard framework. These assumptions say simply that all the arrays in A and B are positive. We incorporate this assumption by renewing the description of the manifold Λ given in (1.6) in Chapter 16.

(Iδ) The stationary Markov Chain $\{X_n\}_{n=0}^{\infty}$ assuming values in a finite state space $S = \{1, 2, \ldots, J\}$ and has the transition probability matrix

$$A = \left(a_{i|j}\right)_{i=1, \quad j=1}^{J,J}. \tag{2.3}$$

For each i there is the constraint

$$a_{i|j} \geq 0, \sum_{j=1}^{J} a_{i|j} = 1. \tag{2.4}$$

In addition we assume that for all i and j

$$a_{i|j} \geq \delta > 0, \tag{2.5}$$

which implies that the chain is *ergodic, irreducible and aperiodic* in the sense outlined in Chapter 7.

(IIδ) For each $j \in S$ we have

$$\pi_j(0) = P(X_0 = j), \qquad \sum_{j=1}^{J} \pi_j(0) = 1. \tag{2.6}$$

For stationarity we must choose $\pi(0) = \pi$, where the invariant distribution is in view of (2.5) both unique and positive.

(IIIδ) The random process $\{Y_n\}_{n=0}^{\infty}$ with a finite state space $\mathcal{O} = \{o_j\}_{j=1}^{K}$ is conditionally independent given $\{X_n\}_{n=0}^{\infty}$, i.e., for any sequence of states $j_0 j_1 \dots j_n$ the probability of the sequence $o_0 o_1 \dots o_n$ is

$$P(Y_0 = o_0, \dots, Y_n = o_n \mid X_0 = j_0, \dots, X_n = j_n, B) = \prod_{l=0}^{n} b_{j_l}(l) \tag{2.7}$$

with

$$b_j(k) = P(Y_n = o_k | X_n = j) \geq \delta > 0. \tag{2.8}$$

The emission probability matrix is $B = \{b_j(k)\}_{j=1,k=1}^{J,K}$ satisfies for each row the constraint

$$b_j(k) \geq 0, \qquad \sum_{k=1}^{K} b_j(k) = 1. \tag{2.9}$$

We denote now by Λ_δ the manifold of all A, B that satisfy (Iδ)–(IIIδ) above or

$$\Lambda_\delta = \{(A, B) \mid \text{satisfying (2.4), (2.5), (2.8), (2.9)}\}. \tag{2.10}$$

Note that we assume the Markov chain to be stationary and hence, for reasons explained in Chapter 8, we exclude the initial distribution from the manifold. We assume that we are observing a path of $\{Y_n\}_{n=0}^{n}$ for increasing n, the path being generated, in the sense of Section 10.2.2 in Chapter 10, by a model (*the true, but unknown, generator*) $\lambda_o \in \Lambda_\delta$.

The effect of the lower bound δ in Λ_δ is to make certain posterior probabilities geometrically ergodic, as will be demonstrated below (see also Kaijser 1975). A more probabilistic proof of this property can be given invoking the well known coupling argument (Douc et al. 2001).

17.3 Steps of the Proof: An Outline

There is a deep question of *identifiability* involved in learning of the parameters of a family of hidden Markov models. By lack of identifiability we mean that there could be $\lambda \neq \lambda'$ such that

$$P(Y_0 = o_0, \ldots, Y_n = o_n; \lambda) = P\left(Y_0 = o_0, \ldots, Y_n = o_n; \lambda'\right)$$

for all $o_0 \ldots o_n$. Here we cannot, in general, hope to find the true generator λ_o uniquely even with an infinitely long sequence. We shall not go further in characterizing conditions for identifiability to hold (Petrie 1969). The lack of identifiability is acknowledged by introducing

$$M_{\widehat{\lambda}_{\mathrm{ML}}^{(n)}} = \{\widehat{\lambda}_{\mathrm{ML}}^{(n)} \in \Lambda_\delta | \lambda_{\mathrm{ML}}^{(n)} \in \arg\max_\lambda \frac{1}{n+1} \log P(Y_0 = o_0, \ldots, Y_n = o_n; \lambda)\}.$$

This is the equivalence class of maximum likelihood estimates in Λ_δ given $o_0 \ldots o_n$. Hence if there is the consistency $\widehat{\lambda}_{\mathrm{ML}}^{(n)} \to \lambda_o$ then it holds for elements in a topological quotient space as introduced in (Redner 1981).

The proof to be presented is based on the method introduced by Abraham Wald in (Wald 1949) and extended to Markov chains in (Billingsley 1961). A generalized version of the method is recapitulated in (Grenander 1981, Chapter 7). The first step will be to prove a Shannon–McMillan–Breiman theorem (**AEP**) (see Chapter 2) for the log likelihood function of the observed process in an ergodic HMM in the sense that

$$\frac{1}{n+1} \log P(Y_0 = o_0, \ldots, Y_n = o_n; \lambda) \to H_{\lambda_o}(\lambda) \tag{3.11}$$

where $H_{\lambda_o}(\lambda)$ is a notation for the limit (a negative crossentropy), and $H_{\lambda_o}(\lambda) \leq H_{\lambda_o}(\lambda_o)$. Clearly $H_{\lambda_o}(\lambda_o) - H_{\lambda_o}(\lambda)$ is a kind of a generalized *Kullback distance.* There is, in fact, a general statement of this theorem in (Cover and Thomas 1991, pp. 474–477) which encompasses all stationary and ergodic processes with a finite alphabet. We shall present the proof in (Baum and Petrie 1966), since the preparatory lemmata are customized to the special properties of HMMs.

Since the underlying Markov chain is ergodic and stationary, we can talk about an *ergodic HMM*, since even the emitted process will turn out to have an ergodic property (see Leroux 1992). One way of starting to get convinced about ergodicity of $\{Y_n\}_{n=0}^{\infty}$ is to regard $\{Y_n\}_{n=0}^{\infty}$ as a function of a Markov chain which under the assumptions (Iδ)−(IIIδ) is ergodic in the special sense of Chapter 7, see Exercise 3 in Chapter 10 and also Proposition 8.6.1 in Chapter 8.

In order to obtain (3.11) several auxiliary steps are needed. The first of these is to obtain a limit of the form

$$\log P\left(Y_0|Y_{-1},\ldots;\lambda\right) = \lim_{n\to\infty} \log P\left(Y_0|Y_{-1},\ldots,Y_{-n};\lambda\right),$$

where the emitted process has been extended to the negative integers as a stationary process. Both $P\left(Y_0|Y_{-1},\ldots;\lambda\right)$ and $P\left(Y_0|Y_{-1},\ldots,Y_{-n};\lambda\right)$ are ergodic and have expectations that can be identified as entropy rates. The last mentioned result yields also that the limit

$$\lim_{n\to\infty} \sup_{\lambda'\in S(\lambda,\epsilon)} P\left(Y_0 = o_0|Y_{-1},\ldots Y_{-n};\lambda'\right)$$

exists, where $S(\lambda,\epsilon)$ is an open sphere with radius ϵ centered at λ using the Euclidean distance on Λ_δ. The Euclidean distance is appropriate, since Λ_δ is a product of a space of $J\times J$ matrices A and a space of $J\times K$ matrices B. This space has a compactness property that makes it possible to cover closed and bounded subsets of it by a finite number of open spheres like $S(\lambda,\epsilon)$. Therefore we can investigate $\widehat{\lambda}_{\mathrm{ML}}^{(n)}$ by maximization of the likelihood function over a finite number of spheres outside M_{λ_o}, which is the set of $\lambda\in\Lambda_\delta$ such that $H_{\lambda_o}(\lambda) = H_{\lambda_o}(\lambda_o)$, and show that the likelihood function assumes higher values inside M_{λ_o}, if n is large enough. The statement about the consistency of the maximum likelihood estimate is an assertion about set convergence of the form

$$M_{\widehat{\lambda}_{\mathrm{ML}}^{(n)}} \to M_{\lambda_o}.$$

17.4 Some Auxiliary Statements

Lemma 17.4.1 *For all $i,j\in S$, all $o_0\ldots o_n\in\mathcal{O}^{n+1}$ and all $\lambda\in\Lambda_\delta$ there exists a number $\mu_\delta\in(0,1)$ such that*

$$P\left(X_{t+1}=j|X_t=i,Y_0=o_0,\ldots,Y_n=o_n;\lambda\right) > \mu_\delta.$$

Proof: In this proof we may omit the model λ and the sequence $o_0\ldots o_n$ from the notation. As before we designate $Y_0=o_0,\ldots,Y_n=o_n$ by $Y_0^{(n)}$. We shall first find an upper bound for the ratio

$$A \equiv \frac{P\left(X_{t+1} = j | X_t = i, Y_0^{(n)}\right)}{P\left(X_{t+1} = j' | X_t = i, Y_0^{(n)}\right)}.$$

The symbol A is introduced in order to simplify the typesetting. Since the same conditioning event is found both in the numerator and denominator we obtain

$$A = \frac{P\left(X_{t+1} = j, X_t = i, Y_0^{(n)}\right)}{P\left(X_{t+1} = j', X_t = i, Y_0^{(n)}\right)}.$$

There it holds

$$\begin{aligned} & P\left(X_{t+1} = j, X_t = i, Y_0^{(n)}\right) \\ & = P\left(Y_0^{(n)} | X_{t+1} = j, X_t = i\right) P\left(X_{t+1} = j, X_t = i\right). \end{aligned} \tag{4.12}$$

Using the factorization (5.1) in Proposition 13.5.1 in Chapter 13 we obtain that this is equal to

$$P\left(Y_{t+1}^{(n)} | X_{t+1} = j\right) P\left(Y_0^{(t)} | X_t = i\right) P\left(X_{t+1} = j, X_t = i\right).$$

Thus we have

$$\begin{aligned} A &= \frac{P\left(Y_{t+1}^{(n)} | X_{t+1} = j\right) P\left(X_{t+1} = j, X_t = i\right)}{P\left(Y_{t+1}^{(n)} | X_{t+1} = j'\right) P\left(X_{t+1} = j', X_t = i\right)} \\ &= \frac{P\left(Y_{t+1}^{(n)}, X_{t+1} = j | X_t = i\right)}{P\left(Y_{t+1}^{(n)}, X_{t+1} = j' | X_t = i\right)}, \end{aligned} \tag{4.13}$$

where we have also used Proposition 13.6.1 in Chapter 13. Next another marginalization becomes useful.

$$P\left(Y_{t+1}^{(n)}, X_{t+1} = j | X_t = i\right) = \sum_{k=1}^{J} P\left(Y_{t+1}^{(n)}, X_{t+2} = k, X_{t+1} = j | X_t = i\right)$$

In the right hand side we have

$$\sum_{k=1}^{J} P\left(Y_{t+1}^{(n)}|X_{t+2}=k, X_{t+1}=j, X_t=i\right) P\left(X_{t+2}=k, X_{t+1}=j|X_t=i\right)$$
$$= \sum_{k=1}^{J} P\left(Y_{t+2}^{(n)}|X_{t+2}=k\right) P\left(Y_{t+1}|X_{t+1}=j\right) P\left(X_{t+2}=k, X_{t+1}=j|X_t=i\right)$$
$$= \sum_{k=1}^{J} P\left(Y_{t+3}^{(n)}|X_{t+2}=k\right) P\left(Y_{t+2}|X_{t+2}=k\right) P\left(Y_{t+1}|X_{t+1}=j\right) \cdot \mathcal{P},$$

where $\mathcal{P} = a_{i|j}a_{j|k}P(X_t=i)$ is introduced for reasons of space, and we have again invoked Proposition 13.6.1 as well as equation (5.5) in Chapter 13 and the Markov property

$$\begin{aligned} &P\left(X_{t+2}=k, X_{t+1}=j|X_t=i\right) \\ &= P\left(X_{t+2}=k|X_{t+1}=j, X_t=i\right) P\left(X_{t+1}=j|X_t=i\right) P\left(X_t=i\right) \\ &= a_{i|j}a_{j|k}P\left(X_t=i\right) = \mathcal{P}. \end{aligned} \tag{4.14}$$

Therefore we have in (4.13), since $P(X_t=i)$ can be moved outside summation in both numerator and denominator

$$A = \frac{\sum_{k=1}^{J} P\left(Y_{t+3}^{(n)}|X_{t+2}=k\right) P\left(Y_{t+2}|X_{t+2}=k\right) P\left(Y_{t+1}|X_{t+1}=j\right) a_{i|j}a_{j|k}}{\sum_{k=1}^{J} P\left(Y_{t+3}^{(n)}|X_{t+2}=k\right) P\left(Y_{t+2}|X_{t+2}=k\right) P\left(Y_{t+1}|X_{t+1}=j'\right) a_{i|j'}a_{j'|k}}.$$

Owing to (2.5) and (2.8) all the terms in the sums are positive and the ratio of the sums is therefore easily seen to be bounded above by the largest fraction

$$A \leq \max \frac{P\left(Y_{t+2}|X_{t+2}=k\right) P\left(Y_{t+1}|X_{t+1}=j\right) a_{i|j}a_{j|k}}{P\left(Y_{t+2}|X_{t+2}=k\right) P\left(Y_{t+1}|X_{t+1}=j'\right) a_{i|j'}a_{j'|k}} \leq \frac{1}{\delta^4},$$

where we have used (2.5) and (2.8) and the fact that probabilities are smaller than 1. Then we obtain by manipulation of the identity

$$1 = \sum_{j=1}^{J} P\left(X_{t+1}=j|X_t=i, Y_0^{(n)}\right)$$

that

$$\frac{1}{P\left(X_{t+1}=j'|X_t=i,Y_0^{(n)}\right)}$$
$$=1+\sum_{j=1,j\neq j'}^{J}\frac{P\left(X_{t+1}=j|X_t=i,Y_0^{(n)}\right)}{P\left(X_{t+1}=j'|X_t=i,Y_0^{(n)}\right)}$$
$$\leq 1+\frac{J-1}{\delta^4}. \tag{4.15}$$

This gives

$$P\left(X_{t+1}=j'|X_t=i,Y_0^{(n)}\right)\geq\frac{1}{1+\frac{J-1}{\delta^4}}\equiv\mu_\delta,$$

where $0<\mu_\delta<1$ unless $J=1$, a case without interest. ∎

Let now C_n denote any event (a cylinder set) that is uniquely determined by $(X_t)_{t\geq n}$ or $(Y_t)_{t\geq n}$. We introduce

$$M_d^+(\lambda)=\max_u P\left(C_n|X_{n-d}=u,Y_0^n;\lambda\right) \tag{4.16}$$

and

$$M_d^-(\lambda)=\min_u P\left(C_n|X_{n-d}=u,Y_0^n;\lambda\right), \tag{4.17}$$

where Y_0^n is a simplified notation for the generic event $Y_0=o_0,\ldots,Y_n=o_n$.

Lemma 17.4.2 *For all $o_0\ldots o_n\in\mathcal{O}^{n+1}$ and all $\lambda\in\Lambda_\delta$*

$$M_d^+(\lambda)-M_d^-(\lambda)\leq\rho^{d-1} \tag{4.18}$$

with $\rho<1$, which is independent of n, λ and C_n.

Proof: Marginalization and the chain rule (2.16) in Chapter 1 entail, omitting for the moment the model λ,

$$P\left(C_n|X_{n-d-1}=u,Y_0^n\right)=\sum_{j=1}^{J}P\left(C_n,X_{n-d}=j|X_{n-d-1}=u,Y_0^n\right)$$

$$=\sum_{j=1}^{J}P\left(C_n|X_{n-d}=j,Y_0^n\right)P\left(X_{n-d}=j|X_{n-d-1}=u,Y_0^n\right).$$

Therefore

$$
\begin{aligned}
&M_{d+1}^{+}(\lambda) - M_{d+1}^{-}(\lambda) = \\
&\max_{u,v}\left[P\left(C_n|X_{n-d-1}=u,Y_0^n\right) - P\left(C_n|X_{n-d-1}=v,Y_0^n;\lambda\right)\right] \\
&= \max_{u,v}\left[\sum_{j=1}^{J} P\left(C_n|X_{n-d}=j,Y_0^n\right)\left(P\left(X_{n-d}=j|X_{n-d-1}=u,Y_0^n\right)\right.\right. \\
&\left.\left.-P\left(X_{n-d}=j|X_{n-d-1}=v,Y_0^n\right)\right)\right].
\end{aligned}
$$

Next we consider

$$
\begin{aligned}
S_+ &= \{j \in S | P\left(X_{n-d}=j|X_{n-d-1}=u,Y_0^n\right) - P\left(X_{n-d}=j|X_{n-d-1}=v,Y_0^n\right) \geq 0\}, \\
S_- &= \{j \in S | P\left(X_{n-d}=j|X_{n-d-1}=u,Y_0^n\right) - P\left(X_{n-d}=j|X_{n-d-1}=v,Y_0^n\right) < 0\}.
\end{aligned}
$$

By these we obtain

$$
\begin{aligned}
&\sum_{j\in S_+}\left(P\left(X_{n-d}=j|X_{n-d-1}=u,Y_0^n\right) - P\left(X_{n-d}=j|X_{n-d-1}=v,Y_0^n\right)\right) \\
&= 1 - \sum_{j\in S_-} P\left(X_{n-d}=j|X_{n-d-1}=v,Y_0^n\right) - \sum_{j\in S_+} P\left(X_{n-d}=j|X_{n-d-1}=v,Y_0^n\right),
\end{aligned}
$$

since

$$
\begin{aligned}
&-\sum_{j\in S_+}\left(P\left(X_{n-d}=j|X_{n-d-1}=u,Y_0^n\right) - P\left(X_{n-d}=j|X_{n-d-1}=v,Y_0^n\right)\right) \\
&= \sum_{j\in S_-}\left(P\left(X_{n-d}=j|X_{n-d-1}=u,Y_0^n\right) - P\left(X_{n-d}=j|X_{n-d-1}=v,Y_0^n\right)\right).
\end{aligned}
\tag{4.19}
$$

But lemma 17.4.1 above gives

$$
\sum_{j\in S_-} P\left(X_{n-d}=j|X_{n-d-1}=v,Y_0^n\right) \geq q^- \cdot \mu_\delta
$$

and

$$
\sum_{j\in S_+} P\left(X_{n-d}=j|X_{n-d-1}=v,Y_0^n\right) \geq q^+ \cdot \mu_\delta,
$$

where q^+ and q^- are the number of elements in S_+ and S_-, respectively. Therefore we obtain

$$\sum_{j \in S_+} \left(P\left(X_{n-d} = j | X_{n-d-1} = u, Y_0^n\right) - P\left(X_{n-d} = j | X_{n-d-1} = v, Y_0^n\right)\right)$$
$$\leq 1 - 2\mu_\delta, \tag{4.20}$$

where we have used the fact that $q^+ + q^- = J \geq 2$. Consequently we can bound from (4.19)

$$M_{d+1}^+(\lambda) - M_{d+1}^-(\lambda) \leq U,$$

where the auxiliary quantity U is

$$U \equiv \max_{u,v} \left[\sum_{j \in S_+} M_d^+(\lambda) \left(P\left(X_{n-d} = j | X_{n-d-1} = u, Y_0^n\right) \right.\right.$$
$$\left. - P\left(X_{n-d} = j | X_{n-d-1} = v, Y_0^n\right)\right) +$$
$$\left. \sum_{j \in S_-} M_d^-(\lambda) \left(P\left(X_{n-d} = j | X_{n-d-1} = u, Y_0^n\right) - P\left(X_{n-d} = j | X_{n-d-1} = v, Y_0^n\right)\right) \right]$$
$$= \max_{u,v} \left[\sum_{j \in S_+} \left(P\left(X_{n-d} = j | X_{n-d-1} = u, Y_0^n\right) \right.\right.$$
$$\left.\left. - P\left(X_{n-d} = j | X_{n-d-1} = v, Y_0^n\right)\right) \left(M_d^+(\lambda) - M_d^-(\lambda)\right)\right],$$

where we have used (4.19). But the quantity in the rightmost side of the latest inequality is in view of the inequality in (4.20) bounded above by

$$\max_{u,v}(1 - 2\mu_\delta) \left(M_d^+(\lambda) - M_d^-(\lambda)\right) = (1 - 2\mu_\delta) \left(M_d^+(\lambda) - M_d^-(\lambda)\right).$$

In other words we have obtained

$$M_{d+1}^+(\lambda) - M_{d+1}^-(\lambda) \leq (1 - 2\mu_\delta) \left(M_d^+(\lambda) - M_d^-(\lambda)\right),$$

which by iteration entails that

$$M_d^+(\lambda) - M_d^-(\lambda) \leq (1 - 2\mu_\delta)^{d-1},$$

which is the assertion of the lemma as claimed once we take $\rho = (1 - 2\mu_\delta)$ and $\rho < 1$. ∎

Corollary 17.4.3 *For all $\lambda \in \Lambda_\delta$, for all k, and for $l < 0$ we have*

$$| P(C_n|Y_k, Y_{k-1}, \ldots, Y_l; \lambda) - P(C_n|Y_k, Y_{k-1}, \ldots, Y_{l+1}; \lambda) | < \rho^{n-l-1}$$

and

$$| P(C_n|Y_k, Y_{k-1}, \ldots, Y_l; \lambda) - P(C_n|Y_{k-1}, \ldots, Y_l; \lambda) | < \rho^{n-k-1}.$$

Proof: We start by the marginalization and the chain rule

$$\begin{aligned} P\left(C_n|Y_k^{l+1}; \lambda\right) &= \sum_{j=1}^{J} P\left(C_n, X_l = j|Y_k^{l+1}; \lambda\right) \\ &= \sum_{j=1}^{J} P\left(C_n|X_l = j, Y_k^{l+1}; \lambda\right) P\left(X_l = j|Y_k^{l+1}; \lambda\right) \end{aligned}$$

and the inequalities

$$\begin{aligned} M_{n-l}^{-} &= \min_j P\left(C_n|X_l = j, Y_k^{l+1}; \lambda\right) \\ &\leq P\left(C_n|Y_k^{l+1}; \lambda\right) \\ &\leq \max_j P\left(C_n|X_l = j, Y_k^{l+1}; \lambda\right) = M_{n-l}^{+}, \end{aligned}$$

and the result follows by the preceding lemma. ■

Corollary 17.4.4 *The limit*

$$\lim_{l \to -\infty} P(C_n|Y_k, Y_{k-1}, \ldots, Y_l; \lambda) = P(C_n|Y_k, Y_{k-1}, \ldots; \lambda) \tag{4.21}$$

exists for all emitted sequences and is continuous in λ. For all $\lambda \in \Lambda_\delta$ we have

$$| P(C_n|Y_k, Y_{k-1}, \ldots; \lambda) - P(C_n|Y_{k-1}, \ldots; \lambda) | < \rho^{n-k-1}.$$

Proof: Both statements follow by corollary 17.4.3. The limiting function is continuous in λ, since convergence is uniform by the estimates in corollary 17.4.3. ■

The lemma and its corollaries have the following properties $(a)-(c)$ as consequences:

(a) For all $\lambda \in \Lambda_\delta$ the limit

$$P(Y_0 = o_0 | Y_{-1}, \ldots ; \lambda) = \lim_{n \to \infty} P(Y_0 = o_0 | Y_{-1}, \ldots Y_{-n+1}; \lambda)$$

exists. As in (Leroux 1992) this can be proved by martingale convergence without the corollaries above. In addition $H_{\lambda_o}(\lambda)$ given by

$$H_{\lambda_o}(\lambda) = E_{\lambda_o}[\log P(Y_0 = o_0 | Y_{-1}, \ldots ; \lambda)]$$

exists. This statement is proved by bounded convergence and the fact that the function $p \log p$ is continuous and bounded in $0 \leq p \leq 1$. Here

$$E_{\lambda_o}[\log P(Y_0 | Y_{-1}, \ldots ; \lambda)] = \sum_{Y_{-1}, \ldots} P(Y_0, Y_{-1}, \ldots ; \lambda_o) h_{\lambda_o}(Y_{-1}, \ldots ; \lambda), \tag{4.22}$$

where

$$h_{\lambda_o}(Y_{-1}, \ldots ; \lambda) = \sum_{Y_0} P(Y_0 | Y_{-1}, \ldots ; \lambda_o) \log P(Y_0 | Y_{-1}, \ldots ; \lambda). \tag{4.23}$$

The quantity $-H_{\lambda_o}(\lambda_o)$ is the *entropy rate* of the process $\{Y_n\}_{n=-\infty}^{\infty}$ generated by the model λ_o, for definition and properties we refer to (Cover and Thomas 1991, ch. 4.) and to section 7.5.3 in Chapter 7.

(b) For all $\lambda \in \Lambda_\delta$

$$| P(Y_0 = o_0 | Y_{-1}, \ldots Y_{-n+1}; \lambda) - P(Y_0 = o_0 | Y_{-1}, \ldots Y_{-n}; \lambda) | < \rho^{n-1}$$

and the limit

$$\lim_{n \to \infty} \sup_{\lambda' \in S(\lambda, \epsilon)} P\left(Y_0 = o_0 | Y_{-1}, \ldots Y_{-n+1}; \lambda'\right)$$

exists for all λ, where $S(\lambda, \epsilon)$ is an open sphere with radius ϵ centered at λ using the Euclidean distance on Λ_δ.

(c) For all $\lambda \in \Lambda_\delta$ and for all k

$$\lim_{k \to -\infty} \log P(Y_n | Y_{n-1}, \ldots, Y_k; \lambda) = \log P(Y_n | Y_{n-1}, \ldots ; \lambda)$$

exists and

$$|\log P(Y_n | Y_{n-1}, \ldots, Y_0; \lambda) - \log P(Y_n | Y_{n-1}, \ldots ; \lambda)| < C \cdot \rho^{n-1}. \tag{4.24}$$

17.5 The Consistency Result

The preceding findings are now used to prove the desired consistency of the maximum likelihood model. We start with the Shannon–McMillan–Breiman theorem. The crucial property taken for granted without a proof is that the ergodic property (Gray 1988) holds for $\log P(Y_i \mid Y_{i-1}, \ldots, ; \lambda)$.

Proposition 17.5.1 *For all* $\lambda \in \Lambda_\delta$

$$\frac{1}{n+1} \log P(Y_0 = o_0, \ldots, Y_n = o_n; \lambda) \to H_{\lambda_o}(\lambda) \tag{5.25}$$

as $n \to \infty$.

Proof: By the elementary decomposition

$$\log P(Y_0, \ldots, Y_n; \lambda) = \sum_{i=1}^{n} \log P(Y_i \mid Y_{i-1}, \ldots, Y_0; \lambda),$$

and the properties (a)–(c) in the preceding section we obtain

$$\left| \frac{1}{n+1} \log P(Y_0, .., Y_n; \lambda) - \frac{1}{n+1} \sum_{i=1}^{n} \log P(Y_i \mid Y_{i-1}, .., ; \lambda) \right|$$

$$= \left| \frac{1}{n+1} \left(\sum_{i=0}^{n} \log P(Y_i \mid Y_{i-1}, .., Y_0; \lambda) - \sum_{i=0}^{n} \log P(Y_i \mid Y_{i-1}, .., ; \lambda) \right) \right|$$

$$\leq \frac{1}{n+1} \sum_{i=0}^{n} \left| \log P(Y_i \mid Y_{i-1}, .., Y_0; \lambda) - \log P(Y_i \mid Y_{i-1}, ..; \lambda) \right| \to 0,$$

where the final limit follows in view of (4.24). Since the ergodic property (theorem) holds for $\log P(Y_i \mid Y_{i-1}, \ldots, ; \lambda)$ we have the convergence

$$\frac{1}{n+1} \sum_{i=0}^{n} \log P(Y_i \mid Y_{i-1}, \ldots; \lambda) \to E_{\lambda_o} P(Y_0 \mid Y_{-1}, \ldots; \lambda) = H_{\lambda_o}(\lambda)$$

almost surely with respect to λ_o as $n \to \infty$. This establishes the assertion. ∎

Proposition 17.5.2 *For all* $\lambda \in \Lambda_\delta$

$$H_{\lambda_o}(\lambda) \leq H_{\lambda_o}(\lambda_o) \tag{5.26}$$

and

$$H_{\lambda_o}(\lambda) = H_{\lambda_o}(\lambda_o) \Leftrightarrow \lambda \in M_{\lambda_o} \tag{5.27}$$

Proof: The inequality (5.26) in view of (4.22) is another expression of Jensen's inequality, or essentially, of the property that Kullback's distance is non-negative, as established in Chapter 2. The condition for equality is seen as follows.

$$H_{\lambda_o}(\lambda) = H_{\lambda_o}(\lambda_o) \Leftrightarrow \frac{P(Y_0 \mid Y_{-1}, \ldots; \lambda)}{P(Y_0 \mid Y_{-1}, \ldots; \lambda_o)} = 1$$

almost surely w.r.t λ_o. By stationarity this is equivalent to

$$\Leftrightarrow \frac{P(Y_0, Y_{-1}, \ldots, Y_{-l} \mid Y_{-l-1}, \ldots; \lambda)}{P(Y_0, Y_{-1}, \ldots, Y_{-l} \mid Y_{-l-1}, \ldots; \lambda_o)} = 1.$$

But by a summation over the values of the coordinates $Y_{-k-1}, \ldots, Y_{-l}$ this gives the equivalence

$$\Leftrightarrow \frac{P(Y_0, Y_{-1}, \ldots, Y_{-k} \mid Y_{-l-1}, \ldots; \lambda)}{P(Y_0, Y_{-1}, \ldots, Y_{-k} \mid Y_{-l-1}, \ldots; \lambda_o)} = 1.$$

Corollary 17.4.4 implies from this that

$$P(Y_0, Y_{-1}, \ldots, Y_{-k}; \lambda) = P(Y_0, Y_{-1}, \ldots, Y_{-k}; \lambda_o)$$

and this implies the assertion. ∎

For the standard properties and definitions of compactness, Euclidean distance and continuity invoked in the proof of the final consistency statement we refer to (Khuri 1993).

Proposition 17.5.3 *For all* $\lambda \in \Lambda_\delta$

$$M_{\hat{\lambda}^{(n)}_{\mathrm{ML}}} \to M_{\lambda_o}, \tag{5.28}$$

as $n \to \infty$.

Proof: Let $S(\lambda, \epsilon)$ be an open sphere with radius ϵ centered at λ. Owing to to the result in (**b**) we have for all $\lambda \in \Lambda_\delta$ that

$$\lim_{n\to\infty} \sup_{\lambda' \in S(\lambda,\epsilon)} \log P\left(Y_n | Y_{n-1}, \ldots Y_0; \lambda'\right)$$

exists. By an adaptation of the argument invoked in the proof of Proposition 17.5.1 we obtain

$$\frac{1}{n+1} \sum_{i=0}^{n} \sup_{\lambda' \in S(\lambda,\epsilon)} \log P\left(Y_i | Y_{i-1}, \ldots Y_0; \lambda'\right) \to H_{\lambda_o}(\lambda, \epsilon) \tag{5.29}$$

as $n \to \infty$ almost surely with respect to λ_o. Here $H_{\lambda_o}(\lambda, \epsilon)$ is a generic notation for the limit.

Let us now suppose that $\lambda \notin M_{\lambda_o}$. Then Proposition 17.5.2 entails that $H_{\lambda_o}(\lambda) < H_{\lambda_o}(\lambda_o)$. Suppose that there exists $\epsilon > 0$ such that $H_{\lambda_o}(\lambda, \epsilon) < H_{\lambda_o}(\lambda_o)$. Let K be a compact subset of Λ_δ. Then we introduce a subset of K

$$M_{\lambda_o}(K) := \{\lambda \in K | H_{\lambda_o}(\lambda) = H_{\lambda_o}(\lambda_o)\},$$

which can be taken as non-empty, and an ϵ-neighborhood of $M_{\lambda_o}(K)$

$$M_{\lambda_o}(K, \epsilon) := \{\lambda \in K | d(\lambda, M_{\lambda_o}(K)) < \epsilon\},$$

where $d(\lambda, M_{\lambda_o}(K))$ denotes the shortest Euclidean distance between λ and the set $M_{\lambda_o}(K)$. The set $K \backslash M_{\lambda_o}(K, \epsilon)$ (this is the set of λ that lie in K but not in $M_{\lambda_o}(K, \epsilon)$) is again a compact set, and can therefore be covered by a finite number, say k, of spheres $S(\lambda_i, \epsilon_i)$ such that $H_{\lambda_o}(\lambda_i, \epsilon_i) < H_{\lambda_o}(\lambda_o)$ for $i = 1, \ldots, k$. Once more we have

$$\begin{aligned}
&\frac{1}{n+1} \sup_{\lambda \in S(\lambda_i, \epsilon_i)} \log P(Y_n, Y_{n-1}, \ldots Y_0; \lambda) \\
&= \frac{1}{n+1} \sup_{\lambda \in S(\lambda_i, \epsilon_i)} \log \sum_{i=0}^{n} P(Y_i | Y_{i-1}, \ldots Y_0; \lambda) \\
&\leq \frac{1}{n+1} \sum_{i=0}^{n} \sup_{\lambda \in S(\lambda_i, \epsilon_i)} \log P(Y_i | Y_{i-1}, \ldots Y_0; \lambda) \\
&\to H_{\lambda_o}(\lambda_i, \epsilon_i),
\end{aligned}$$

by the preceding observations. And therefore, if n is sufficiently large,

$$\frac{1}{n+1} \sup_{\lambda \in K \backslash M_{\lambda_o}(K, \epsilon)} \log P(Y_n, Y_{n-1}, \ldots Y_0; \lambda) < H_{\lambda_o}(\lambda_o).$$

This means that we can find a higher value of the likelihood function inside $M_{\lambda_o}(K, \epsilon)$ than in $K \backslash M_{\lambda_o}(K, \epsilon)$. Let us set

$$A_n = \left\{ \widehat{\lambda}_K | \widehat{\lambda}_K = \operatorname{argmax}_{\lambda \in K} \frac{1}{n+1} \log P(Y_n, Y_{n-1}, \ldots Y_0; \lambda) \right\}.$$

Then we have proved that if n is large enough,

$$A_n \subset M_{\lambda_o}(K, \epsilon),$$

almost surely with regard to λ_o. In other words we have that $\widehat{\lambda}_{\text{ML}}$ is, for sufficiently large n, found inside an arbitrarily small neighborhood of $M_{\lambda_o}(K)$, which establishes the claim. ∎

Algorithms for estimation of λ by minimization of the ergodic limit of the averaged log likelihood ratio have been developed in (Kwong and Man 1996, Juang and Rabiner 1985).

There might be some interest in investigating the asymptotic properties of $\widehat{\lambda}_{\text{ML}}^{(n)}$ constrained by the family of HMMs, in a case where the observed sequence is generated by an ergodic mechanism not described by any ergodic HMM (cf. Kieffer 1989, Grenander 1981, Chapter 7).

17.6 Exercises

1. Prove the following result:

   Proposition 17.6.1 *If $\{X_n\}_{n=0}^{\infty}$ is a stationary and ergodic Markov chain and $Y_n = g(X_n)$, then the entropy rate $H_\infty(Y)$ of $\{X_n\}_{n=0}^{\infty}$ exists,*

   $$H(Y_n|Y_{n-1},\ldots,Y_0,X_0) \leq H_\infty(Y) \leq H(Y_n|Y_{n-1},\ldots,Y_0)$$

   and

   $$H_\infty(Y) = \lim H(Y_n|Y_{n-1},\ldots,Y_0,X_0) = \lim H(Y_n|Y_{n-1},\ldots,Y_0).$$

   Hint: If $\{Y_n = g(X_n)\}_{n=0}^{\infty}$ is an ergodic Markov chain, then its entropy rate can be computed as in Chapter 7. Otherwise one may consult (Cover and Thomas 1991, pp. 69–71). Expressions for the entropy rate of an ergodic HMM are also given in (Blackwell 1957, Kaijser 1975).

2. Prove that for all $\lambda \in \Lambda_\delta$ the limit

   $$P(Y_0 = o_0|Y_{-1},\ldots;\lambda) = \lim_{n\to\infty} P(Y_0 = o_0|Y_{-1},\ldots Y_{-n+1};\lambda)$$

   exists using corollary 17.4.3.

2. Prove the bound in (4.24).

17.7 References and Further Reading:

L.E. Baum, T. Petrie (1966): Statistical inference for probabilistic functions of finite state Markov chains. *The Annals of Mathematical Statistics*, 30, pp. 1554–1563.

D. Blackwell (1957): The entropy of functions of finite-state Markov chains. *Transactions of the First Prague Conference on Information Theory, Statistical Decision Functions, Random Processes.* Prague, pp. 13–20.

P. Billingsley (1961): *Statistical Inference for Markov Chains.* University of Chigago Press.

P. Bickel, Y. Rytov and T. Rydén (1998): Asymptotic normality of the maximum likelihood estimator for general hidden Markov models. *Annals of Statistics*, 26, pp. 1614–1653.

T.M. Cover and J.A. Thomas (1991): *Elements of Information Theory.* John Wiley and Sons, New York.

R. Douc, É. Moulines and T. Rydén (2001): Asymptotic properties of the maximum likelihood estimator in autoregressive models with Markov regime. *Preprint* `http://www.maths.lth.se/matstat/staff/tobias/`

R.M. Gray (1988): *Probability, Random Processes and Ergodic Properties.* Springer Verlag, New York, Berlin.

U. Grenander (1981): *Abstract Inference.* John Wiley & Sons, New York, Chichester, Brisbane, Toronto.

B.H. Juang and L.R. Rabiner (1990): The segmental K-means Algorithm for estimating Parameters of Hidden Markov Models. *IEEE Transactions on Acoustics, Speech and Signal Processing*, 38, pp. 1639–1641.

B.H. Juang and L.R. Rabiner (1985): A Probabilistic Distance Measure for Hidden Markov Models. *AT & T Technical Journal*, 64, pp. 391–408.

T. Kaijser (1975): A Limit Theorem for Partially Observed Markov Chains. *The Annals of Probability*, 3, pp. 677–696.

A.I. Khuri (1993): *Advanced Calculus with Applications in Statistics.* John Wiley and Sons, New York.

J.C. Kieffer (1989): An Ergodic Theorem for Constrained Sequences of Functions. *Bulletin (New Series) of the American Mathematical Society*, 21, pp. 249–254.

Q.H. Kwong and K.F. Man (1996): Training approach for hidden Markov models. *Electronics Letters*, 15th August, 32, pp. 1554–1555.

B.G. Leroux (1992): Maximum-likelihood estimation for hidden Markov models. *Stochastic Processes and their Applications*, pp. 127–143.

F. Muri (1997): *Comparison d'algorithmes d'identification de chaines de Markov cachées et application á la détection de régions homogènes dans les sequences d'ADN.* Ph.D. Thesis, University of Paris V.

T. Petrie (1969): Probabilistic functions of finite state Markov chains. *The Annals of Mathematical Statistics*, 40, pp. 97–115.

R. Redner (1981): Note on the Consistency of the Maximum Likelihood Estimate for Non-identifiable Distributions. *The Annals of Mathematical Statistics*, 9, pp. 225–228.

A. Wald (1949): A Note on the consistency of maximum likelihood estimate. *Annals of Mathematical Statistics*, 20, pp. 595–601.

Chapter 18

Full Probabilistic HMM

18.1 Introduction

The Smith–Waterman algorithm presented in Exercise 5.6 of Chapter 5 maximizes an alignment scoring function over all possible local alignments between two sequences. The scoring depends on a similarity function and a linear gap penalty. The optimal alignment scores can be viewed, see Chapter 5, as measures of similarity or distance. Some of the optimal scores have been shown in Chapter 5 to define a metric over the sequence space.

In (Bucher and Hofmann 1996) a different interpretation of scoring with local alignment is given. This interpretation is based on an analogy to the computational characteristics of hidden Markov models. Other similar approaches to alignment scores and algorithms are found in (Bishop and Thompson 1986, Bruno 1996).

18.2 Revision of Local Alignment

By pairwise local alignment we mean finding optimally aligned substrings of two sequences. Let $\mathbf{x}$ and $\mathbf{y}$ be two strings with n and m symbols, respectively.

A local alignment between $\mathbf{x}$ and $\mathbf{y}$ is defined by an *alignment path* represented by an *ordered set of index pairs*

$$\mathbf{u} = \{(u_1, v_1), \ldots, (u_l, v_l)\}.$$

Each pair identifies a pair of matched symbols in the pairwise sequence alignment and satisfies the requirements

$$u_{k+1} > u_k, \qquad v_{k+1} > v_k, \qquad u_l \leq n, \qquad v_l \leq m. \tag{2.1}$$

The scoring function can with this notation be written as

$$H(\mathbf{x}, \mathbf{y}, \mathbf{u}) = \sum_{i=1}^{l} s(x_{u_i}, y_{v_i}) + H_G(\mathbf{u}), \tag{2.2}$$

where

$$H_G(\mathbf{u}) = -\left[\sum_{i=1}^{l-1} \left(\widehat{g}(u_{k+1} - u_k - 1) + \widehat{g}(v_{k+1} - v_k - 1)\right)\right] \tag{2.3}$$

with the gap penalty

$$\widehat{g}(n) = \alpha + \beta \cdot n, \qquad n \geq 1, \qquad \widehat{g}(0) = 0. \tag{2.4}$$

Then the best local alignment for $\mathbf{x}$ and $\mathbf{y}$ has the score

$$H^*(\mathbf{x}, \mathbf{y}) = \max_{\mathbf{u}} H(\mathbf{x}, \mathbf{y}, \mathbf{u}).$$

The idea is to regard $\mathbf{u}$ as a path of a stochastic process such that the probability of observing the pair $(\mathbf{x}, \mathbf{y})$ is computed as

$$\mathbf{P}_{\mathsf{ASS}}(\mathbf{x}, \mathbf{y}) = \sum_{\mathbf{u}} \mathbf{P}_{\mathsf{ASS}}(\mathbf{x}, \mathbf{y}, \mathbf{u}),$$

where the summation is extended over all alignment paths $\mathbf{u}$ satisfying (2.1). Here ASS stands for 'alignment specific scoring system'.

We need to construct a probability distribution $\mathbf{P}_{\mathsf{ASS}}(\mathbf{x}, \mathbf{y}, \mathbf{u})$ such that the sum $\sum_{\mathbf{u}}$ can be evaluated without an exhaustive enumeration and such that a method of logodds scoring with regard to a null model as in section 12.4.2 in Chapter 12 exists.

18.3 Probability Defined by Alignment

The null model is chosen to have the following form

$$\mathbf{P}_0(\mathbf{x}, \mathbf{y}) = P_0(\mathbf{x}, \mathbf{y} | n, m) \cdot P_L(n, m). \tag{3.1}$$

In this expression $P_L(n,m)$ designates the joint distribution for the sequence lengths and $P_0(\mathbf{x},\mathbf{y}|n,m)$ designates the null-model probability of the pair $(\mathbf{x},\mathbf{y})$. The distribution $P_0(\mathbf{x},\mathbf{y}|n,m)$ has the form

$$P_0(\mathbf{x},\mathbf{y}|n,m) = \prod_{i=1}^{n} f(x_i) \prod_{j=1}^{m} g(y_j),$$

where $f(\cdot)$ and $g(\cdot)$ are probability distributions on a finite alphabet $\mathcal{O}$. We require that $f(\cdot)$ and $g(\cdot)$ satisfy the condition that there exists a positive real number $\lambda \neq 1$ such that

$$\sum_{x\in\mathcal{O}} \sum_{y\in\mathcal{O}} f(x) \cdot g(y)\, \lambda^{s(x,y)} = 1. \tag{3.2}$$

This condition emerged for the first time in the context of significance testing for pairwise local alignments, see, e.g., (Karlin and Altschul 1990). It is known that this equation can be solved with regard to λ if $\sum_{x\in\mathcal{O}} \sum_{y\in\mathcal{O}} f(x) \cdot g(y)\, s(x,y) < 0$ and if there is some positive probability for a positive score.

Then we introduce $\mathbf{P}_{\mathrm{ASS}}(\mathbf{x},\mathbf{y},\mathbf{u})$ as

$$\mathbf{P}_{\mathrm{ASS}}(\mathbf{x},\mathbf{y},\mathbf{u}|n,m) = \frac{\lambda^{H(\mathbf{x},\mathbf{y},\mathbf{u})} \cdot \mathbf{P}_0(\mathbf{x},\mathbf{y}|n,m)}{B(n,m)}, \tag{3.3}$$

where $B(n,m)$ is the normalization given by

$$B(n,m) = \sum_{\mathbf{x},\mathbf{y},\mathbf{u}} \lambda^{H(\mathbf{x},\mathbf{y},\mathbf{u})} \cdot \mathbf{P}_0(\mathbf{x},\mathbf{y}|n,m).$$

It can be shown that

$$B(n,m) = \sum_{\mathbf{u}} \lambda^{H_G(\mathbf{u})}, \tag{3.4}$$

where $H_G(\mathbf{u})$ is given in (2.3).

18.4 Scoring

The score P_{SW} for the alignment is defined by

$$P_{\mathrm{SW}} = \frac{\mathbf{P}_{\mathrm{ASS}}(\mathbf{x},\mathbf{y})}{\mathbf{P}_0(\mathbf{x},\mathbf{y})} = \frac{\sum_{\mathbf{u}} \mathbf{P}_{\mathrm{ASS}}(\mathbf{x},\mathbf{y},\mathbf{u}|n,m)}{P_0(\mathbf{x},\mathbf{y}|n,m)},$$

taking the length distribution $P_L(n,m)$ to be the same for both $\mathbf{P}_{\text{ASS}}$ and $\mathbf{P}_0$. If we insert from (3.3) we get

$$P_{\text{SW}} = \frac{\sum_{\mathbf{u}} \lambda^{H(\mathbf{x},\mathbf{y},\mathbf{u})}}{B(n,m)} = \frac{\sum_{\mathbf{u}} \lambda^{H(\mathbf{x},\mathbf{y},\mathbf{u})}}{\sum_{\mathbf{u}} \lambda^{H_G(\mathbf{u})}} \tag{4.1}$$

using (3.4) in the last step. From this we can derive what is known as the full probabilistic HMM, but this will not be completed in this text.

18.5 References and Further Reading:

M.J. Bishop and E.A.Thompson (1986): Maximum Likelihood Alignment of DNA Sequences. *Journal of Molecular Biology*, 190, pp. 159–165.

W.J. Bruno (1996): Modelling Residue Usage in Aligned Protein Sequences via Maximum Likelihood. *Molecular Biology and Evolution*, 13, pp. 1361–1375.

P. Bucher and K. Hoffmann (1996): A Sequence Similarity Search Algorithm Based on a Probabilistic Interpretation of an Alignment Scoring System. *International Conference on Intelligent Systems for Molecular Biology*, pp. 444–51.

S. Karlin and S.F.Altschul (1990): Methods for assessing the statistical significance of molecular sequence features by using general scoring schemes. *Proceedings of the National Academy of Sciences, USA*, 87, pp. 2264–2268.

Index

Printed in the United Kingdom
by Lightning Source UK Ltd.
102588UKS00001B/146

9 781402 001352